AF709125

Phase Lock Loops and Frequency Synthesis

Phase Lock Loops and Frequency Synthesis

Věnceslav F. Kroupa

Academy of Sciences of the Czech Republic, Prague

WILEY

Telephone (+44) 1243 779777

Email (for orders and customer service enquiries): cs-books@wiley.co.uk
Visit our Home Page on www.wileyeurope.com or www.wiley.com

Other Wiley Editorial Offices

John Wiley & Sons Inc., 111 River Street, Hoboken, NJ 07030, USA

Jossey-Bass, 989 Market Street, San Francisco, CA 94103-1741, USA

Wiley-VCH Verlag GmbH, Boschstr. 12, D-69469 Weinheim, Germany

John Wiley & Sons Australia Ltd, 33 Park Road, Milton, Queensland 4064, Australia

John Wiley & Sons (Asia) Pte Ltd, 2 Clementi Loop #02-01, Jin Xing Distripark, Singapore 129809

John Wiley & Sons Canada Ltd, 22 Worcester Road, Etobicoke, Ontario, Canada M9W 1L1

Wiley also publishes its books in a variety of electronic formats. Some content that appears in print may not be available in electronic books.

British Library Cataloguing in Publication Data

A catalogue record for this book is available from the British Library

ISBN 0-470-84866-9

Typeset in 10.5/13pt Times by Laserwords Private Limited, Chennai, India
Printed and bound in Great Britain by TJ International, Padstow, Cornwall
This book is printed on acid-free paper responsibly manufactured from sustainable forestry in which at least two trees are planted for each one used for paper production.

To my wife Magda for her encouragement in starting and finishing this work

Contents

Preface

In the last fifty years, many books and a large number of papers have been written on Phase Locked Loops. Why do we need this new one? Would it not be like carrying coal to Newcastle? I think not; otherwise why should I undertake this task. We are at the start of a new epoch of computers, of computers on every desk, and of computers in everyday life. And this new book on PLL theory and practice is intended to follow that direction.

Having before me volumes on PLL, I realized that today's students, researchers, and consumers are overwhelmed with information. I tried, with this new book, to spare their time.

I started with the early achievements in automatic control theory and practice, utilized the basic knowledge about the natural frequency and damping parameters, introduced the normalization, making it possible to solve the PLL problem with the assistance of computers with simple generalized programmes, leaving out complicated mathematics.

Nevertheless, without the specialist knowledge, students and readers will investigate PLL transfer functions, limits of stability, noises both introduced and of the background origin, including spoiling spurious signals.

Special care is given to the problem of the phase-locked loop frequency synthesizers in the high-megahertz and low-gigahertz ranges where modern mobile communications are placed. In this connection, special attention was paid to the theory of the latest fractional-N systems of the first and higher orders.

Finally, I wish that all my efforts, my studies, and my computer simulation are of profit and assistance to students of PLL and practitioners in the field. If the present volume meets this task, then I have not worked in vain.

Věnceslav F. Kroupa
Prague, November 2002

1
Basic Equations of the PLLs

1.1 INTRODUCTION

Phase lock loops (PLLs) belong to a larger set of regulation systems. As an independent research and design field it started in the 1950s [1] and gained major practical application in cochannel TV. On this occasion, we find one of the first fundamental papers [2].

Some 15 years later, we encounter a surveying book by Gardner [3], still mentioned and used. Since a dozen books were published on the topic of PLL problems proper [4] and in connection with frequency synthesis, we would find chapters on PLLs in all of the relevant books. Here we shall only mention some of them [5–8]. But the importance of the topic is testified by the publication of new books on PLLs (e.g., [9–12]) and a wealth of journal articles, the important ones of which will be cited at the relevant places. A major advantage of modern PLLs is the possibility of a widespread use of off-the-shelf *IC chips*. Their application results in low-volume, low-weight, and often power-saving devices. At the same time we also appreciate short switching times and very high-frequency resolution. We shall find PLLs in communications equipment, particularly, in mobile applications in low-gigahertz ranges, in computers, and so on, where we appreciate short switching times and very high-frequency resolution. However, there are shortcomings too: the limited range for high frequencies (today commercial dividers hardly exceed the 5 GHz bound and only laboratory devices work in higher ranges).

In the following paragraphs we summarize the basic properties of PLLs with some design-leading ideas and repeat all the major features and use terminology introduced years ago by mechanical engineers [13] and also used by Gardner [3] and many others.

1.2 BASIC EQUATIONS OF THE PLLs

The task of the PLLs is to maintain coherence between the input (reference) signal frequency, f_i, and the respective output frequency, f_o, via phase comparison. Another

Phase Lock Loops and Frequency Synthesis V.F. Kroupa
 ISBN: 0-470-84866-9

feature of PLLs is the filtering property, particularly with respect to the noise where its behavior recalls a very narrow low-pass arrangement that is not to be realized by other means. The theory was explained in many textbooks as we have mentioned in the previous section.

Each PLL system is composed of four basic parts:

1. the reference generator (RG)
2. the phase detector (PD)
3. the low-pass filter $F_L(f)$ (in higher-order systems)
4. the voltage-controlled oscillator (VCO)

and works as a feedback system shown in Fig. 1.1.

Without any loss of generality, we may assume that input and output signals are harmonic voltages with additional phase modulation

$$v_i(t) = V_i \sin[\omega_i t + \phi_i(t)] \equiv V_i \sin \Psi_i(t) \tag{1.1}$$

where $\phi_i(t)$ and $\phi_o(t)$ are slowly varying quantities.

$$v_o(t) = V_o \cos[\omega_o t + \phi_o(t)] \equiv V_o \cos \Psi_o(t) \tag{1.2}$$

Later we shall prove that realization of the phase lock requires that input and output voltages must be in quadrature, that is, mutually shifted by $\pi/2$.

Phase detector (PD) is a nonlinear element of a different design and construction (we shall deal with PDs later in Chapter 8, Section 8.4). For the present discussion, we assume that the PD is a simple multiplier. In this case the corresponding output voltage will be

$$v_d(t) = K_m v_i(t)\, v_o(t) \tag{1.3}$$

where K_m is the transfer constant with the dimension [1/V]. After introduction of eqs. (1.1) and (1.2) in the above relation, we get

$$\begin{aligned} v_d(t) &= K_m V_i V_o \sin \Psi_i(t) \cos \Psi_o(t) \\ &= \tfrac{1}{2} K_m V_i V_o [\sin[(\omega_i - \omega_o)t + \phi_i(t) - \phi_o(t)] \\ &\quad + \sin[(\omega_i + \omega_o)t + \phi_i(t) + \phi_o(t)]] \end{aligned} \tag{1.4}$$

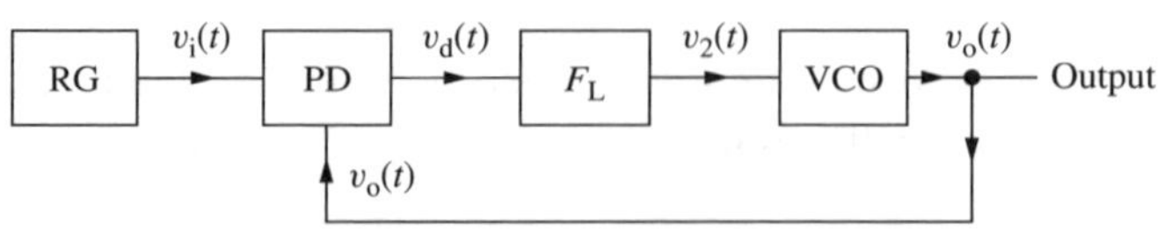

Figure 1.1 Basic feedback network of PLL.

In the simplest case we shall assume that the low-pass filter removes the upper sideband with the frequency $\omega_i + \omega_o$ but leaves the lower sideband $\omega_i - \omega_o$ without change. Evidently the VCO tuning voltage will be

$$v_2(t) = K_d \sin[(\omega_i - \omega_o)t + \phi_i(t) - \phi_o(t)]$$
$$\equiv K_d \sin \Psi_e(t) \tag{1.5}$$

where we have introduced the so-called PD gain $K_d = K_m V_i V_o$ of dimension [V/rad]. Note that the phase difference between the input and the output voltages is

$$\Psi_e(t) = \Psi_i(t) - \Psi_o(t) \tag{1.6}$$

Voltage $v_2(t)$ will change the free running frequency ω_c of the VCO to

$$\dot{\Psi}_o(t) = \omega_c + K_o v_2(t) \tag{1.7}$$

where the proportionality constant K_o is designated as the *oscillator gain* with the dimension [2π Hz/V]. After integration of the above equation and introduction into relation (1.6), we get for the phase difference $\Psi_e(t)$

$$\Psi_e(t) = \Psi_i(t) - \omega_c t - \int K_o v_2(t)\, dt \tag{1.8}$$

which can be rearranged as follows:

$$\Psi_e(t) = \omega_i t - \omega_c t - \int K_o K_d \sin \Psi_e(t)\, dt \tag{1.9}$$

and differentiation reveals

$$\frac{d\Psi_e(t)}{dt} = \Delta\omega - K \sin \Psi_e(t) \tag{1.10}$$

where we have introduced $\omega_i - \omega_o = \Delta\omega$ and $K_d K_o = K$. Note that K is indicated as the gain of the PLL with the dimension [2πHz].

The conclusion that follows from the foregoing discussion is that the phase lock arrangement is described with a nonlinear eq. (1.10), the solution of which for arbitrary values $\Delta\omega$ and K is not known. With certainty we can state that for $\Delta\omega/K \gg 1$ an aperiodic solution does not exist. This conclusion testifies the phase plane arrangement (Fig. 1.2). Without an aperiodic solution, the feedback system in Fig. 1.1 cannot reach the phase stability, that is, the output frequency of the VCO, ω_o, will never be equal to the reference frequency ω_i. However, the DC component in the steering voltage $v_2(t)$ reduces the original difference between frequencies

$$|\omega_i - \omega_c| > |\omega_i - \omega_o| \tag{1.11}$$

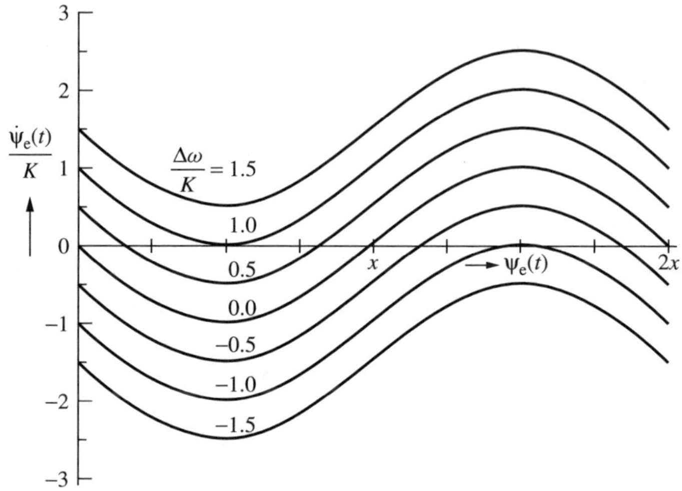

Figure 1.2 Plot of relation (1.10) in the phase plane for different ratios $\Delta\omega/K$.

1.3 SOLUTION OF THE BASIC PLL EQUATION IN THE TIME DOMAIN

To arrive at the solution we have to introduce some simplifications. Nevertheless, we gain more insight into the problem.

1.3.1 Solution in the Closed Form

In the case where $\Delta\omega/K \ll 1$, the differential eq. (1.10) has a solution after application and separation of variables.

$$\frac{\mathrm{d}\Psi_\mathrm{e}(t)}{\Delta\omega - K\sin\Psi_\mathrm{e}(t)} = \mathrm{d}t \tag{1.12}$$

With the assistance of tables [14, p. 804], we get a rather complicated closed-form solution

$$t - t_0 = -\frac{2}{\sqrt{(\Delta\omega)^2 - K^2}}\arctan\left[\sqrt{\frac{\Delta\omega + K}{\Delta\omega - K}}\tan\left(\frac{\pi}{4} - \frac{\Psi_\mathrm{e}}{2}\right)\right] \tag{1.13}$$

where t_0 is a not-yet-defined integration constant. As long as $K > \Delta\omega$, the rhs will be imaginary and with the assistance

$$\tan(-\mathrm{j}x) = -\mathrm{j}\cdot\tanh(x) \tag{1.14}$$

we arrive at

$$
\begin{aligned}
t - t_0 &= -\frac{2}{\sqrt{K^2 - (\Delta\omega)^2}} \operatorname{arctanh}\left[\sqrt{\frac{K + \Delta\omega}{K - \Delta\omega}} \tan\left(\frac{\pi}{4} - \frac{\Psi_e}{2}\right)\right] \\
&= \frac{1}{\sqrt{K^2 - (\Delta\omega)^2}} \ln \frac{1 + \sqrt{(K + \Delta\omega)/(K - \Delta\omega)} \tan(\pi/4 - \Psi_e/2)}{1 - \sqrt{(K + \Delta\omega)/(K - \Delta\omega)} \tan(\pi/4 - \Psi_e/2)}
\end{aligned}
\tag{1.15}
$$

and after computing $\tan(\pi/4 - \Psi_e/2)$ the sought solution is

$$
\Psi_e = 2 \arctan\left[\sqrt{\frac{K - \Delta\omega}{K + \Delta\omega}} \cdot \frac{1 - \exp[-\sqrt{K^2 - (\Delta\omega)^2}(t - t_0)]}{1 + \exp[-\sqrt{K^2 - (\Delta\omega)^2}(t - t_0)]} + \frac{\pi}{2}\right] \tag{1.16}
$$

For the steady state, that is, for $t \to \infty$, the lhs of eq. (1.10) equals zero, with the result

$$
\Psi_{e\infty} = \arcsin \frac{\Delta\omega}{K} \tag{1.17}
$$

1.3.2 Linearized Solution

From the preceding analysis we conclude that the solution of the respective differential equation, in the closed form, is very complicated even for a very simple PLL arrangement. Consequently, we may suppose that for more sophisticated PLL systems it would be practically impossible. However, the situation need not be so gloomy after the introduction of simplifications that are not far from reality. In the first step we find that the time-dependent phase difference $\Psi_e(t)$ at the output of the PD in the closed PLL is small and prone to the simplification

$$
\sin \Psi_e(t) \approx \Psi_e(t) \tag{1.18}
$$

This assumption is supported with the reality that a lot of PDs are linear or nearly linear in the working range (see discussions in Chapter 8). In such a case, the introduction of (1.18) into (1.10) results in the following simplification:

$$
\frac{d\Psi_e(t)}{dt} = \Delta\omega - K\Psi_e(t) \tag{1.19}
$$

Solution of this differential equation is easy,

$$
\Psi_e(t) = e^{-Kt}\left(\Psi_{e0} - \frac{\Delta\omega}{K}\right) + \frac{\Delta\omega}{K} \tag{1.20}
$$

where Ψ_{e0} is the integration constant, that is, the phase at the start for $t = 0$. Further investigation reveals that the phase difference in the steady state compensates the frequency difference (cf. (1.17)).

$$
\Psi_{e\infty} = \frac{\Delta\omega}{K} \tag{1.21}
$$

1.4 SOLUTION OF BASIC PLL EQUATIONS IN THE FREQUENCY DOMAIN

By assuming the phase difference $\Psi_e(t)$, in the locked state, to be always smaller than $\pi/2$, the result is the equality between input and output frequencies

$$\omega_i = \omega_o \tag{1.22}$$

In other words the PLL system is permanently in the phase equilibrium. The situation being such, we can rearrange relation (1.7) to

$$\omega_o + \dot{\phi}_o(t) = \omega_c + K_o v_{2o} + K_d K_o \sin[\phi_i(t) - \phi_o(t)] \tag{1.23}$$

where the term $K_o v_{2o}$ shifts the VCO frequency ω_o to be equal to the input frequency ω_i (of (1.22)). Evidently, in the steady state we get the following relation between the VCO free running frequency and the locked frequency

$$\omega_o = \omega_c + K_o v_{2o} \tag{1.24}$$

Combination with (1.23) reveals

$$\dot{\phi}_o(t) = K \sin[\phi_i(t) - \phi_o(t)] \tag{1.25}$$

where $K = K_d K_o$.

In the steady state the difference ϕ_e

$$\phi_e(t) = \phi_i(t) - \phi_o(t) \tag{1.26}$$

is generally small. Consequently, we may apply the following linearization

$$\dot{\phi}_o(t) = K[\phi_i(t) - \phi_o(t)] \tag{1.27}$$

and employ advantages of the Laplace transform (with a tacit assumption of the zero initial conditions)

$$s\Phi_o(s) = K[\Phi_i(s) - \Phi_o(s)] \tag{1.28}$$

After rearrangement we arrive at the basic PLL transfer function

$$\frac{\Phi_o(s)}{\Phi_i(s)} = H(s) = \frac{K}{s + K} \tag{1.29}$$

or at

$$\frac{\Phi_i(s) - \Phi_o(s)}{\Phi_i(s)} = \frac{\Phi_e(s)}{\Phi_i(s)} = 1 - H(s) = \frac{s}{s + K} \tag{1.30}$$

between input and PD output error.

1.5 ORDER AND TYPE OF PLLs

The PLL system described with relations (1.29) and (1.30) is indicated as PLL of the *first order* since the polynomial in the denominator is of the first order in s (K being a constant).

However, generally PLLs are much more complicated. To get better insight into the PLL properties, we shall simplify, without any loss of generality, the block diagram to that shown in Fig. 1.3 and introduce the Laplace transfer functions of the individual building circuits, suitable for investigation of the small signal properties.

Investigation of the above figure reveals that the input phase $\varphi_i(t)$ is compared with the output phase $\varphi_o(t)$ in the phase detector (ring modulator, sampling circuit, etc.). At its output we get a voltage, $v_d(t)$, proportional to the phase difference of the respective input signals where

$$v_d(t) = [\varphi_i(t) - \varphi_o(t)]K_d \tag{1.31}$$

the proportionality factor, K_d[V/rad], is called the *phase detector gain.*

Next, $v_d(t)$ passes the loop filter, $F(s)$ (a low-pass filter attenuating "carriers" with frequencies $\omega_i = \omega_o$, and ideally all undesired sidebands). Note that the useful signal $v_2(t)$ is a slowly varying "DC" component, the output voltage of which is given by the following convolution:

$$v_2(t) = v_d(t) \otimes h_f(t) \tag{1.32}$$

where $h_f(t)$ is the time response of the loop filter. After applying $v_2(t)$ on the frequency control element of the VCO, we get the output phase

$$\varphi_o(t) = \int \omega_o(t)\,\mathrm{d}t = \omega_c t + \int K_o v_2(t)\,\mathrm{d}t \tag{1.33}$$

with ω_c being the VCO free-running frequency. The proportionality factor, K_o [2π Hz/V], is designated as the *oscillator gain.* Since, in most cases, K_d and K_o are voltage-dependent, the general mathematical model of a PLL is a nonlinear differential equation. Its linearization, justified in small signal cases ("steady state" working modes), provides a good insight into the problem. After reverting to the

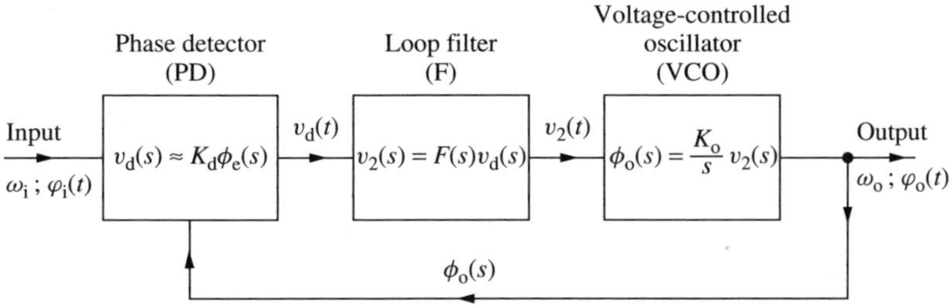

Figure 1.3 Simplified block diagram of the PLL with individual transfer functions.

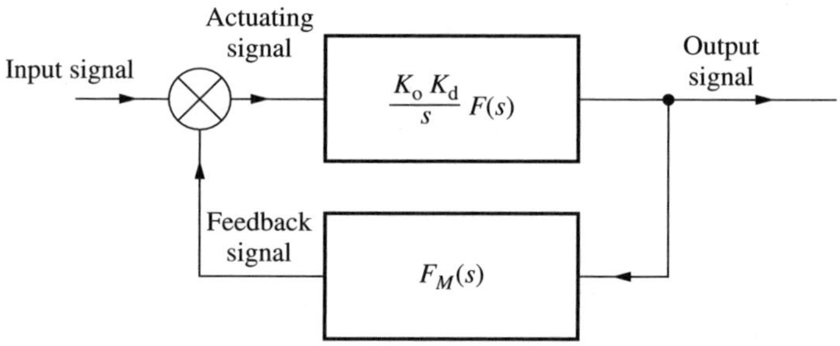

Figure 1.4 Simplified block diagram of the PLL with a transfer function in the feedback path.

whole feedback system (Fig. 1.4), we can write for the relation between the input and the output phases in the Laplace transform notation

$$[\Phi_i(s) - \Phi_o(s)F_M(s)]\frac{K_d K_o F(s)}{s} = \Phi_o(s) \tag{1.34}$$

The ratio, $\Phi_o(s)/\Phi_i(s)$, the PLL transfer function, is given by

$$H(s) = \frac{\dfrac{KF(s)F_M(s)}{s}}{1 + \dfrac{KF(s)F_M(s)}{s}} = \frac{G(s)}{1 + G(s)} \tag{1.35}$$

where we have introduced the forward loop gain $K = K_d K_o$ and the open loop gain $G(s)$

$$G(s) = \frac{KF(s)F_M(s)}{s} \tag{1.36}$$

1.5.1 Order of PLLs

In the simplest case there are no filters in the forward or the feedback paths. The PLL transfer function simplifies to

$$H(s) = \frac{K}{s + K} \tag{1.37}$$

This PLL is designated as the *first-order loop* since the largest power of s in the polynomial of the denominator is of the order *one*. Generally, the transfer functions of the loop filters $F(s)$ are given by a ratio of two polynomials in s. The consequence is that the denominator in $H(s)$ is of a higher order in s and we speak about PLLs of the *second order, third order*, and so on, in accordance with the order of the respective polynomial in the denominator of (1.35).

1.5.2 Type of PLLs

In instances in which the steady state errors are of major interest, the number of poles in the transfer function $G(s)$, that is, the number of integrators in the loop, is of importance. In principle, every PLL has one integrator connected with the VCO (cf. eq. (1.33)). For the phase error at the output of the PD we find

$$\Phi_e(s) = \Phi_i(s) - F_M(s)\Phi_o(s) \tag{1.38}$$

where

$$\Phi_o(s) = \Phi_e(s)\frac{KF(s)}{s} \tag{1.39}$$

After elimination of $\Phi_o(s)$ from the above relations, we get for the phase error $\Phi_e(s)$

$$\Phi_e(s) = \Phi_i(s)\frac{1}{1 + G(s)} \tag{1.40}$$

Introducing the gain, $G(s)$, which is a ratio of two polynomials

$$G(s) = \frac{A(s)}{s^n B(s)} \tag{1.41}$$

we get for the phase error

$$\Phi_e(s) = \Phi_i(s)\frac{s^n B(s)}{A(s) + s^n B(s)} \tag{1.42}$$

and eventually with the assistance of the Laplace limit theorem, we get for the final value of the phase error $\varphi_e(t)$

$$\lim_{t\to\infty}[\phi_e(t)] = \lim_{s\to 0}\left[\Phi_i(s)\frac{s^{n+1}B(s)}{A(s) + s^n B(s)}\right] \tag{1.43}$$

Note that every PLL contains at least one integrator, that is, VCO; consequently, $n \geq 1$ (cf. relation (1.34)).

1.5.3 Steady State Errors

Investigations of the steady state errors in PLLs of different orders and types will proceed after introduction of the Laplace transforms of the respective input phase steps, input frequency steps, and input steady frequency changes into (1.43).

$$\Delta\omega_i = \frac{\Delta\phi_i}{s}; \quad \frac{\Delta\omega_i}{s} = \frac{\Delta\phi_i}{s^2}; \quad \frac{\Delta\dot{\omega}}{s} = \frac{\Delta\omega_i}{s^2} = \frac{\Delta\phi_i}{s^3} \tag{1.44}$$

1.5.3.1 Phase steps

After introducing the Laplace transform of phase steps, $\Delta\phi/s$, into (1.43), we find out that the final value is *zero* in all PLLs.

1.5.3.2 Frequency steps

For the frequency steps, $\Delta\omega/s$, we get

$$\lim_{t\to\infty} \phi_{e2}(t) = \Delta\omega_i \left[\frac{B(0)}{A(0)}\right]_{n=1} = \frac{\Delta\omega_i}{KF(0)F_M(0)} = \frac{\Delta\omega_i}{K_v} \tag{1.45}$$

Evidently in all PLLs of the second order, a frequency step results in a steady state phase error inversely proportional to the so-called *velocity error constant* K_v, in agreement with the terminology used in the feedback control systems (cf. [13]).

In PLLs of type 2, with two integrators in the loop, the DC gain $F(0)$ is very large, so K_v and consequently the steady state error is negligible.

1.5.3.3 Frequency ramps

However, the steady frequency change, $\Delta\omega/s^2$, results in the so-called *acceleration* or *dynamic tracking error* K_a

$$\lim_{t\to\infty} \phi_{e3}(t) = \Delta\dot{\omega}_i \left[\frac{B(0)}{A(0)}\right]_{n=2} = \frac{\Delta\dot{\omega}_i}{K_a} \tag{1.46}$$

PLLs of type 3 can eliminate even the steady state error $\varphi_{e3}(t)$ for $t \to \infty$ to zero. However, PLLs of this type are encountered exceptionally, for example, in time services [15], in space and satellite devices [3], and so on.

Note that the frequency locked loop may be considered as 0 *type PLL.*

1.6 BLOCK DIAGRAM ALGEBRA

Actual PLLs are often much more complicated than block diagrams in Figs. 1.3 or 1.4. For arriving at transfer functions, $H(s)$ and *1−H(s)*, we can apply the rules of *block diagram algebra* [13].

Two or more blocks in series can be combined into one after multiplication of their Laplace transform symbols (see Fig. 1.5a). A typical example is the addition of independent sections to the fundamental low-pass filter.

In the case where two blocks are in parallel, the final combination is provided with a mere addition (see Fig. 1.5b).

Investigation of the relation (1.35) reveals that the feedback block can be put outside of the basic loop [5]

$$H'(s) = \frac{1}{N} H(s) \tag{1.47}$$

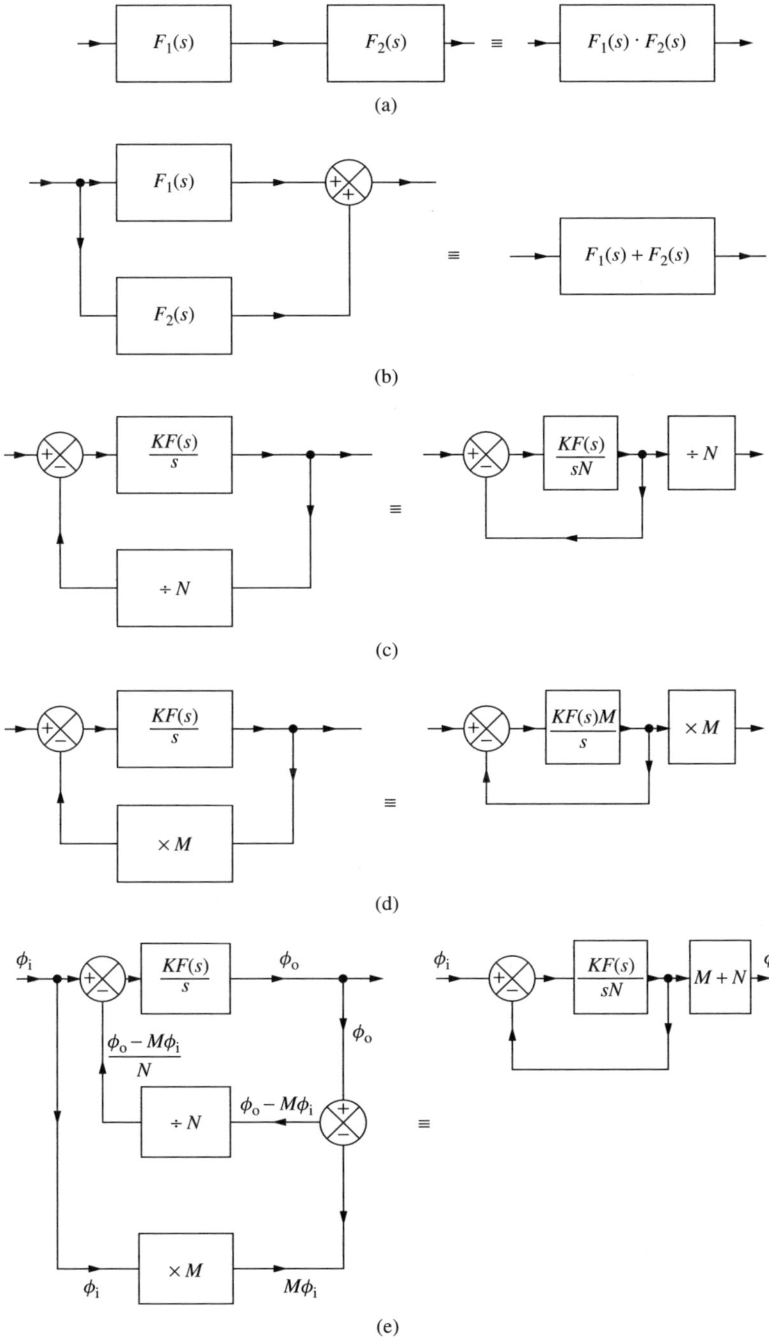

Figure 1.5 Simplification of the block diagrams of PLLs: (a) series connection; (b) parallel connection; (c) and (d) feedback arrangement; (e) more complicated system. (Reproduced from Fig. 1.20 in C.J. Savant Jr., *Basic Feedback Control System Design*. New York, Toronto, London: McGraw-Hill, 1958 by permission of McGraw Hill, 2002).

or

$$H'(s) = MH(s) \tag{1.48}$$

In this way we arrive at the effective transfer functions, $H'(s)$ and $1 - H'(s)$, which contain information about the PLL filtering properties, which will be discussed later. We appreciate this approach in instances in which a simple frequency divider or frequency multiplier is in the feedback path of the PLL. The rearrangement is reproduced in Figs. 1.5(c) and 1.5(d).

Finally, we shall consider the system containing a mixer in the feedback path. Relation between output and input phases is

$$\Phi_o(s) = \left[\Phi_i(s) - \frac{\Phi_o(s) - M\Phi_i(s)}{N}\right]\frac{KF(s)}{s} \tag{1.49}$$

and rearrangement leads to the simplification in accordance with Fig. 1.5(e).

REFERENCES

[1] W.J. Gruen, "Theory of AFC synchronization", *Proc. IRE*, **41**, 1043–1048, 1953.
[2] D. Richman, *APC Color Sync for Television Synchronization*, 1953, IRE Conv. Rec., Part 4.
[3] M. Gardner, *Phase-Lock Techniques*. New York: Wiley, 1966, 1979.
[4] W.C. Lindsey and C.M. Chie, eds *Phase-Locked Loops*. New York: IEEE Press, 1985.
[5] V.F. Kroupa, *Frequency Synthesis: Theory, Design and Applications*. London: CH. Griffin, 1973.
[6] V. Manassewtsch, *Frequency Synthesizers: Theory and Design*. 1st ed. New York: John Wiley & Sons, 1975, last ed. 1990.
[7] W.F. Egan, *Frequency Synthesis by Phase Lock*. 1981, New York: John Wiley, 2000.
[8] U.L. Rohde, *Digital PLL Frequency Synthesizers*. Englewood Cliffs, NJ: Prentice Hall, 1983.
[9] J.A. Crawford, *Frequency Synthesizer Design Handbook*. Boston and London: Artech House, 1994.
[10] B.B. Razavi, ed. *Monolithic Phase-Locked Loops and Clock Recovery Circuits: Theory and Design*. IEEE Press, 1996.
[11] W.F. Egan, *Phase-Lock Basics*. John Wiley & Sons, 1999.
[12] R. Best, *Phase-Locked Loops: Design, Simulation, and Applications*. New York: McGraw-Hill, 1999.
[13] C.J. Savant Jr., *Basic Feedback Control System Design*. New York, Toronto, London: McGraw-Hill, 1958.
[14] G.A. Korn and T.M. Korn, *Mathematical Handbook*. New York: McGraw-Hill, 1958.
[15] J. Tolman, *The Czechoslovak National Standard of Frequency and Time*. Yearbook of the Academy of Sciences 1967, Praha: Academia, 1969, pp. 127–138.

2

PLLs of the First and Second Order

In practice, the most common PLLs are those of the first and second order. Their advantage is absolute stability. We shall discuss the problem of stability in Chapter 4. At the same time, we appreciate the simple theoretical and practical design. For the sake of continuity, we shall base our discussion on some previous publications [1, 2]. In addition, we would like to extend our treatment to be as general as possible and applicable to computer solutions.

2.1 PLLs OF THE FIRST ORDER

First, we shall briefly investigate the simplest first-order loops. They do not have any effective loop filter. Consequently, their open-loop gain is in accordance with (1.36)

$$G(s) = \frac{K_\mathrm{d} K_\mathrm{o} K_A}{s} = \frac{K}{s} \tag{2.1}$$

and transfer functions are in accordance with (cf. (1.29) and (1.30))

$$H(s) = \frac{K}{s+K}; \quad 1 - H(s) = \frac{s}{s+K} \tag{2.2}$$

Note that the DC gain K_A can be used for changing the corner frequency, of this simple PLL, to any desired value (see Fig. 2.1).

Since the open-loop gain K has a dimension of 2π Hz, normalization of the base-band frequency in respect to it provides nearly all the information about the behavior of the PLLs. After introducing

$$\frac{s}{K} = \sigma = \mathrm{j}x = \frac{\mathrm{j}\omega}{K} \tag{2.3}$$

we get the normalized loop gain

$$G(\sigma) = \frac{1}{\sigma} \tag{2.4}$$

Phase Lock Loops and Frequency Synthesis V.F. Kroupa
 ISBN: 0-470-84866-9

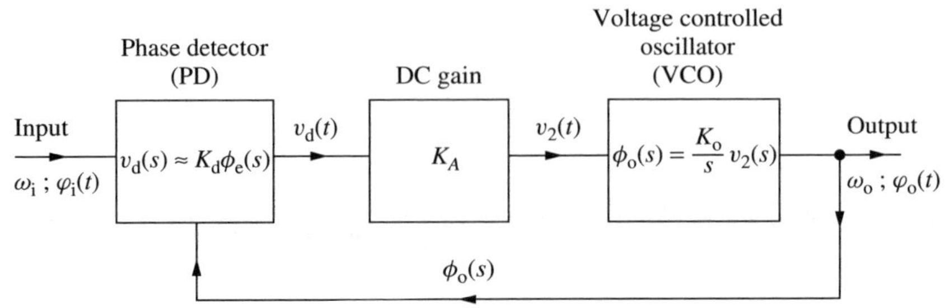

Figure 2.1 The block diagram of the first-order PLL with an additional DC gain K_A.

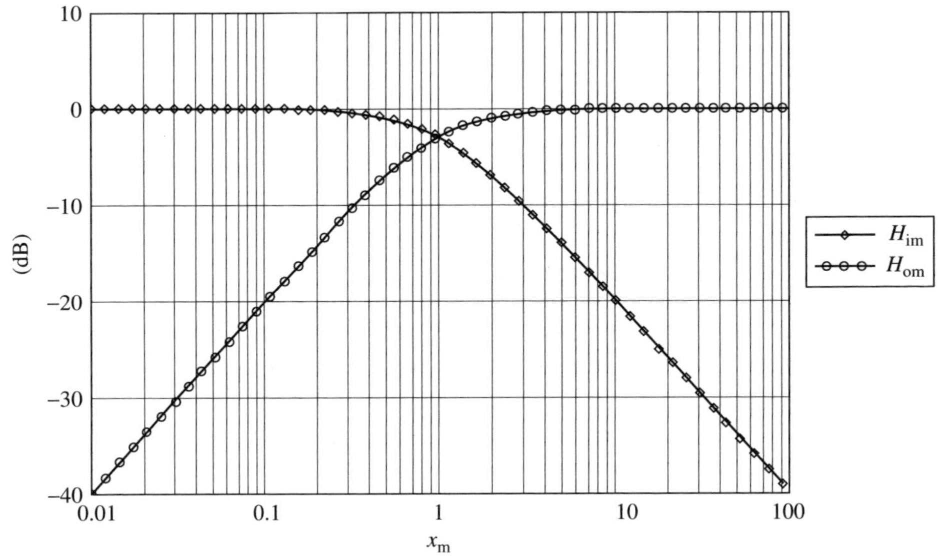

Figure 2.2 Transfer functions $H_i(jx) = 20\log(|H(jx)|)$ (◇), and $H_o(jx) = 20\log(|1 - H(jx)|)$ (○) of the first-order loop.

By introducing eq. (2.3) into (2.2), we arrive at the normalized transfer functions

$$H(\sigma) = \frac{1}{1+\sigma}; \quad 1 - H(\sigma) = \frac{\sigma}{1+\sigma} \tag{2.5}$$

Note that we easily find out that the transfer function $H(jx)$ (defined in (1.29)) behaves as a low-pass filter in respect to the noise and spurious signals accompanying the reference signal, whereas $1 - H(jx)$ behaves as a high-pass filter in respect to the close-to-the-carrier noise of the VCO (see Example 2.1 and Fig. 2.2). The problem of the noise will be discussed later, in Chapter 9, in more detail.

Example 2.1

Investigation of the filtering properties of the first-order PLL reveals

a) $x = \omega/K \ll 1$

$$H(\mathrm{j}x) \cong 1$$

and

$$1 - H(\mathrm{j}x) \cong \mathrm{j}x$$

b) $x = \omega/K \gg 1$

$$H(\mathrm{j}x) \cong 1/\mathrm{j}x$$

and

$$1 - H(\mathrm{j}x) \cong 1$$

2.2 PLLs OF THE SECOND ORDER

Equation (2.1) reveals that the first-order PLL has only one degree of freedom, namely, the DC gain K_A. The other difficulty is the rather modest attenuation in the respective stop bands – only 20 dB/decade in each (see Fig. 2.2). Both problems can be solved by the introduction of a suitable low-pass filter into the forward path. In this manner, we arrive at the second-order PLLs.

2.2.1 A Simple RC Filter

In instances in which we need to increase attenuation of the PLLs for high frequencies (one example might be the attenuation of the leaking carrier), application of the simple RC low-pass filter, shown in Fig. 2.3, provides the desired effect. The respective filter transfer function is

$$F(s) = \frac{1}{1 + sRC} = \frac{1}{1 + sT_1} \tag{2.6}$$

At the same time, the filter time constant T_1 presents an additional degree of freedom for the design of PLL properties. After introduction of $F(s)$ into (1.36), we get the open-loop gain

$$G(s) = K_\mathrm{d} K_A \frac{1}{1 + sT_1} \frac{K_\mathrm{o}}{s} = \frac{K}{s(1 + sT_1)} \tag{2.7}$$

and by putting the above relation for $G(s)$ in (1.35) we get, for the transfer function $H(s)$ of the second-order PLLs,

$$\frac{\Phi_\mathrm{o}(s)}{\Phi_\mathrm{i}(s)} = H(s) = \frac{G(s)}{1 + G(s)} = \frac{K/T_1}{s^2 + s/T_1 + K/T_1} \tag{2.8}$$

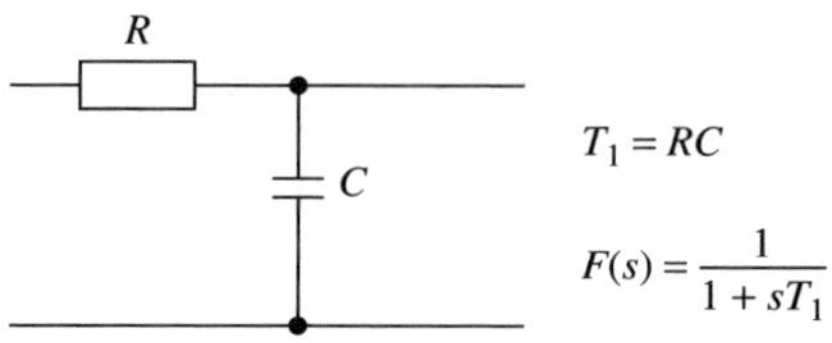

Figure 2.3 Second-order PLL filters: a simple RC filter.

and

$$\frac{\Phi_e(s)}{\Phi_i(s)} = 1 - H(s) = \frac{1}{1 + G(s)} = \frac{s(sT_1 + 1)}{s^2 + s/T_1 + K/T_1} \tag{2.9}$$

After introduction, in agreement with the theory of servomechanism (of dynamic motion equation or damped oscillations), of the natural frequency ω_n and the damping factor ζ

$$\omega_n = \sqrt{\frac{K}{T_1}}; \quad \zeta = \frac{\omega_n}{2K} = \frac{1}{2\sqrt{KT_1}} = \frac{1}{2\omega_n T_1} \tag{2.10}$$

we can rearrange the open-loop gain into

$$G(s) = \frac{\omega_n{}^2}{s^2 + 2\zeta s\omega_n} \tag{2.11}$$

and the PLL transfer function (2.8) into its "characteristic form"

$$H(s) = \frac{\omega_n{}^2}{s^2 + 2\zeta\omega_n s + \omega_n{}^2} \tag{2.12}$$

After normalization of the base-band frequency ω in respect to the natural frequency ω_n, that is,

$$\sigma = \frac{s}{\omega_n} = \frac{\mathrm{j}\omega}{\omega_n} = \mathrm{j}x \tag{2.13}$$

we get for the open-loop transfer function

$$G(\sigma) = \frac{1}{\sigma^2 + 2\zeta\sigma} \quad \text{or} \quad G(\mathrm{j}x) = \frac{1}{-x^2 + 2\mathrm{j}\zeta x} \tag{2.14}$$

and for the PLL transfer function $H(\sigma)$ or $H(\mathrm{j}x)$, we have

$$H(\sigma) = \frac{1}{\sigma^2 + 2\zeta\sigma + 1} \quad \text{or} \quad H(\mathrm{j}x) = \frac{1}{-x^2 + 2\mathrm{j}\zeta x + 1} \tag{2.15}$$

and for $1 - H(\sigma)$ or $1 - H(\mathrm{j}x)$

$$1 - H(\sigma) = \frac{\sigma^2 + 2\zeta\sigma}{\sigma^2 + 2\zeta\sigma + 1} \quad \text{or} \quad 1 - H(\mathrm{j}x) = \frac{-x^2 + 2\mathrm{j}\zeta x}{-x^2 + 2\mathrm{j}\zeta x + 1} \tag{2.16}$$

Example 2.2

Evaluate the asymptotic values for the transfer functions of the second-order PLLs with a simple RC loop filter

a) $x \ll 1$

$$G(\mathrm{j}x) \cong 1/2\mathrm{j}\zeta x$$

$$H(\mathrm{j}x) \cong 1$$

and

$$1 - H(\mathrm{j}x) \cong 2\mathrm{j}\zeta x = \mathrm{j}x(\omega_n/K)$$

b) $x \gg 1$

$$G(\mathrm{j}x) \cong -1/x^2$$

$$H(\mathrm{j}x) \cong -1/x^2$$

and

$$1 - H(\mathrm{j}x) \cong 1$$

Note that the asymptote in the stop band for the transfer functions $H(\mathrm{j}x)$ has the slope −40 dB/decade, independent of the damping factor; however, the steepness of $1 - H(\mathrm{j}x)$ in the stop band is only 20 dB/decade (see Fig. 2.4). The transfer functions for different damping factors ζ are plotted in Fig. 2.5.

2.2.2 Phase Lag-lead RRC or RCC Filter

Application of the proportional–integral or RRC filter, in accordance with Fig. 2.6 having the transfer function

$$F(s) = \frac{1 + sT_2}{1 + sT_1} \tag{2.17}$$

provides a further degree of freedom in designing PLLs. The open-loop gain is

$$G(s) = \frac{K_\mathrm{d} K_A K_\mathrm{o}(1 + sT_2)}{s(1 + sT_1)} \tag{2.18}$$

and with it and the assistance of eq. (1.35) we get the respective transfer function

$$H(s) = \frac{(K/T_1)(1 + sT_2)}{s^2 + s(1 + KT_2)/T_1 + K/T_1} \tag{2.19}$$

We can again introduce the natural frequency and the damping factor

$$\omega_n = \sqrt{K/T_1}; \quad \zeta = \frac{\omega_n}{2}\left(T_2 + \frac{1}{K}\right) \tag{2.20}$$

and arrive at the characteristic form of the open-loop gain

$$G(s) = \frac{s(2\zeta\omega_n - {\omega_n}^2/K) + {\omega_n}^2}{s(s + {\omega_n}^2/K)} \tag{2.21}$$

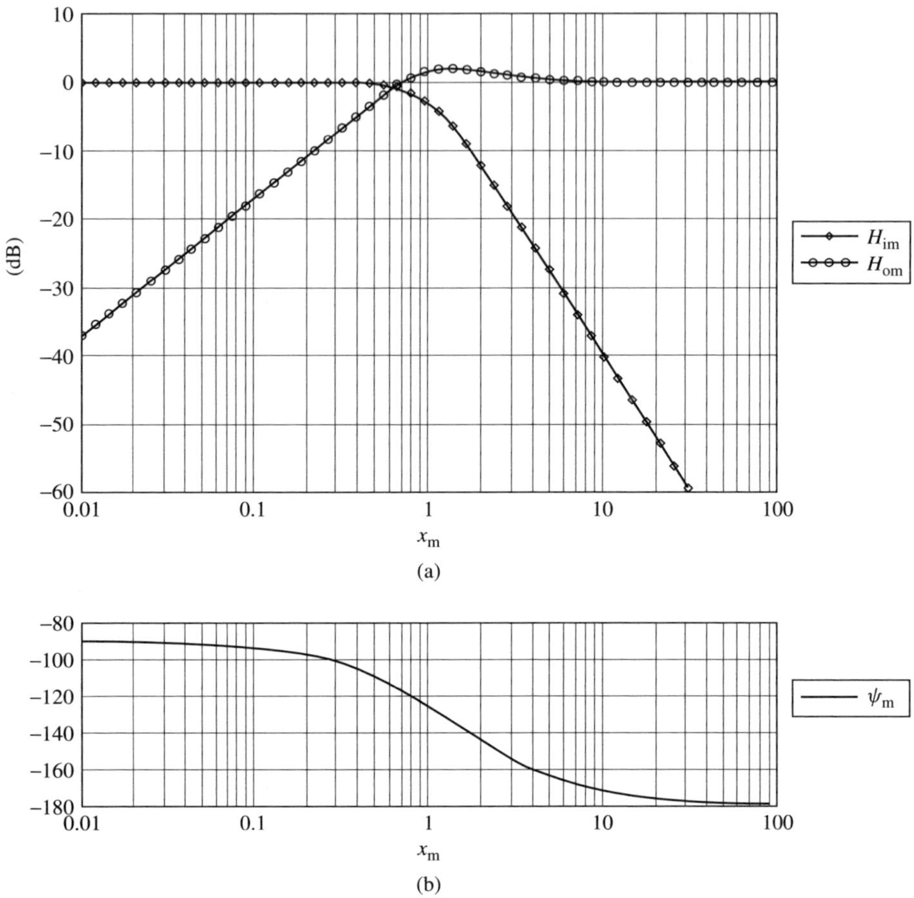

Figure 2.4 (a) Transfer functions $H_i(jx) = 20\log(|H(jx)|)$ (◇) and $H_o(jx) = 20\log(|1 - H(jx)|)$ (○) of the second-order PLL with a simple RC filter and the damping factor $\zeta = 0.7$ and (b) the respective phase characteristic of the open-loop gain $G(jx)$.

and of the transfer functions

$$H(s) = \frac{s\omega_n(2\zeta - \omega_n/K) + {\omega_n}^2}{s^2 + 2\zeta\omega_n s + {\omega_n}^2}; \quad 1 - H(s) = \frac{s^2 + s{\omega_n}^2/K}{s^2 + 2\zeta\omega_n s + {\omega_n}^2} \tag{2.22}$$

In the normalized form, we get with the assistance of (2.13) for the open-loop transfer function

$$G(\sigma) = \frac{1 + 2\sigma\zeta - \sigma\omega_n/K}{\sigma^2 + \sigma\omega_n/K} \quad \text{or} \quad G(jx) = \frac{1 + 2j\zeta x - jx\omega_n/K}{-x^2 + jx\omega_n/K} \tag{2.23}$$

and for the closed-loop transfer functions

$$H(jx) = \frac{jx(2\zeta - \omega_n/K) + 1}{-x^2 + 2j\zeta x + 1}; \quad 1 - H(jx) = \frac{-x^2 + jx\omega_n/K}{-x^2 + 2j\zeta x + 1} \tag{2.24}$$

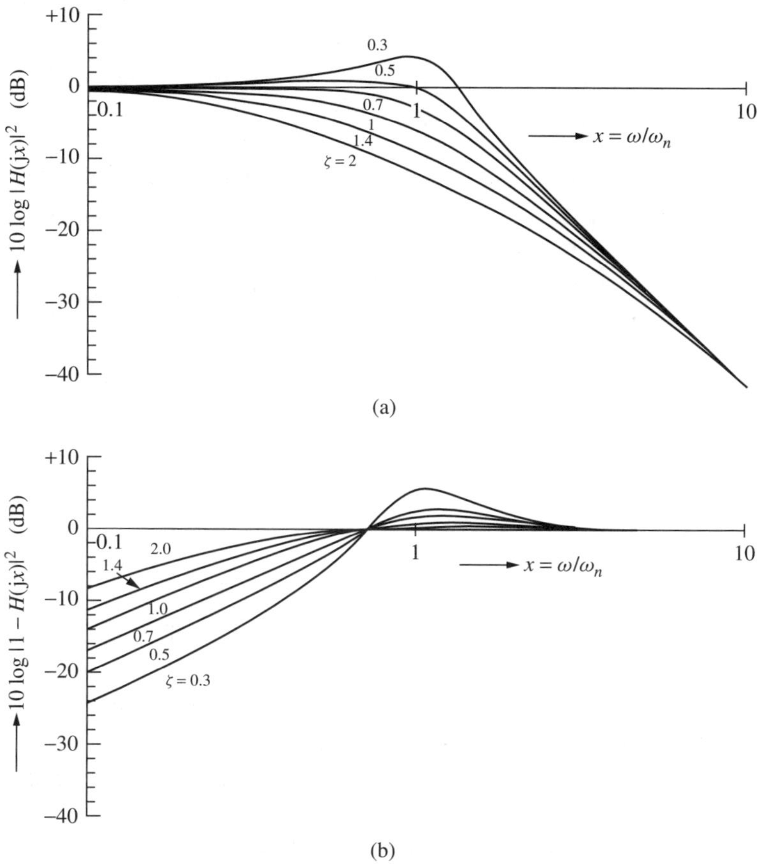

Figure 2.5 Transfer functions of the second-order PLLs with a simple RC filter for different damping factors ζ: (a) $H_i(jx) = 20\log(|H(jx)|)$ and (b) $H_o(jx) = 20\log(|1 - H(jx)|)$.

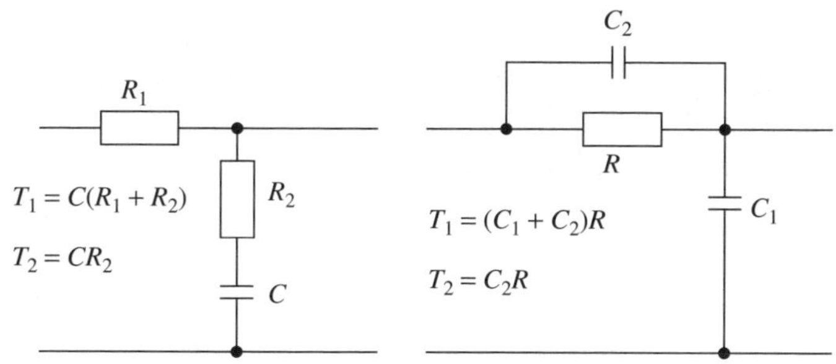

$$F(s) = \frac{1 + sT_2}{1 + sT_1}$$

Figure 2.6 Second-order PLL filters: phase lag-lead or proportional–integral networks (RRC or RCC combination).

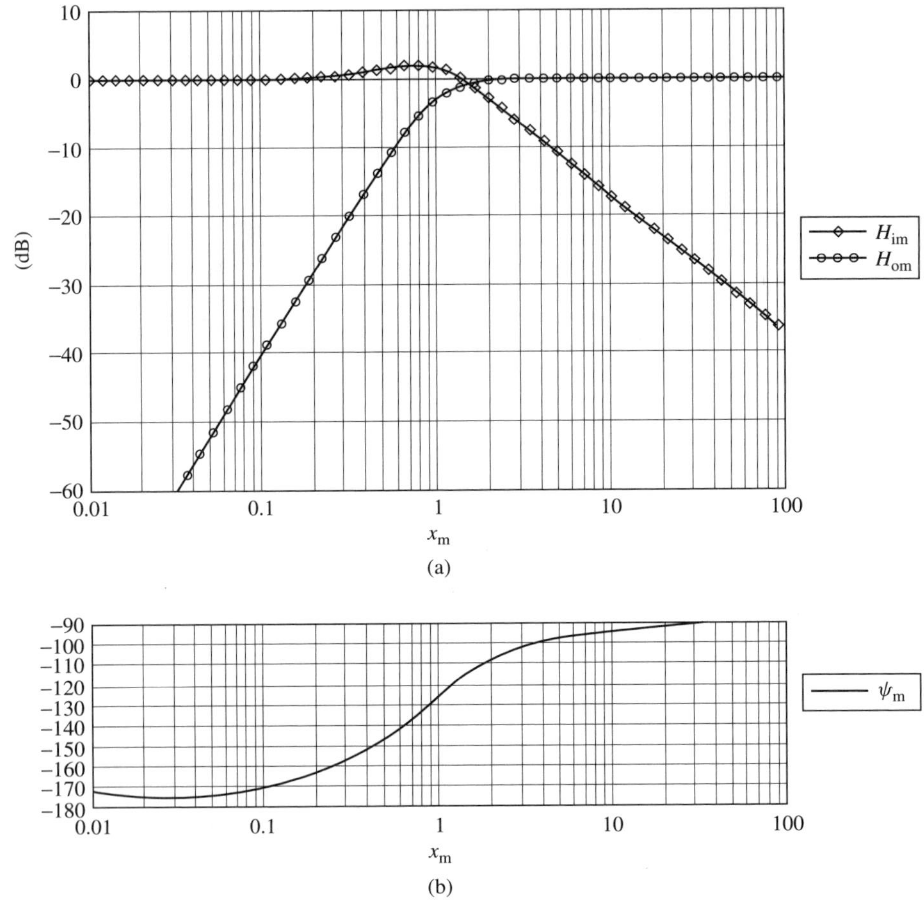

Figure 2.7 (a) Transfer functions $H_i(jx) = 20\log(|H(jx)|)$ (◇) and $H_o(jx) = 20\log(|1 - H(jx)|)$ (○) and (b) phase characteristic of the open-loop gain $G(jx)$ of the second-order PLL with an RRC filter.

These normalized transfer functions are plotted in Fig. 2.7. Note that the freedom for the independent choice of ω_n and ζ resulted in the reduced slope of the stop band of $H(jx)$ on one hand and in a reduced phase margin on the other hand (the problem of the phase margin will be discussed in more detail in Chapter 4). Inspection of all relations from (2.21) to (2.24) reveals a term, ω_n/K. For large gain, K, this term may be neglected and we call these loops *high-gain loops*.

2.3 PLLs OF THE SECOND ORDER OF TYPE 2

Up to now we have discussed PLLs with one integrator formed by the VCO, as explained in Chapter 1. The varactor input voltage $v_2(t)$ changes oscillator frequency; however, the output signal is, eventually, phase-modulated, that is, it exhibits the integral of frequency modulation – the situation will be explained in more detail in

Chapter 9, Section 9.4.7. Nevertheless, we might encounter PLL systems with more integrators in the loop, as discussed in Section 1.5.2.

2.3.1 PLLs of the Second Order of Type 2 with Voltage Output PD

The second integrator is introduced with the assistance of the loop filter, containing an operation amplifier as shown in Fig. 2.8. Its transfer function is

$$F(s) = \frac{1 + sT_2}{sT_1 + 1/A} \tag{2.25}$$

where A is the gain of the operation amplifier ($A \gg 1$) and the time constants are

$$T_1 = \left(R_1 + \frac{R_1 + R_2}{A}\right) C \approx R_1 C; \quad T_2 = R_2 C \tag{2.26}$$

Note that the filter behaves as an integrator equalized with an RC section for all AC signals with the effective loop gain $K = K_d K_A K_o$; however, for the DC working mode, the gain is $K_{DC} = K_d K_A K_o F(0) = K_d K_A K_o A$. With (1.45) we easily find out that the *velocity error constant* K_v is very large, practically infinite. After introduction of $F(s)$ into (1.36), we arrive at the open-loop gain

$$G(s) = \frac{K(1 + sT_2)}{s[sT_1 + s(T_1 + T_2)/A]} \approx \frac{K(1 + sT_2)}{s[sT_1 + 1/A]} \tag{2.27}$$

Comparison of the above relation with (2.1) reveals that for high frequencies the effective gain is reduced, and simulates that of the first-order loop

$$K_r \approx K \frac{T_2}{T_1} \tag{2.28}$$

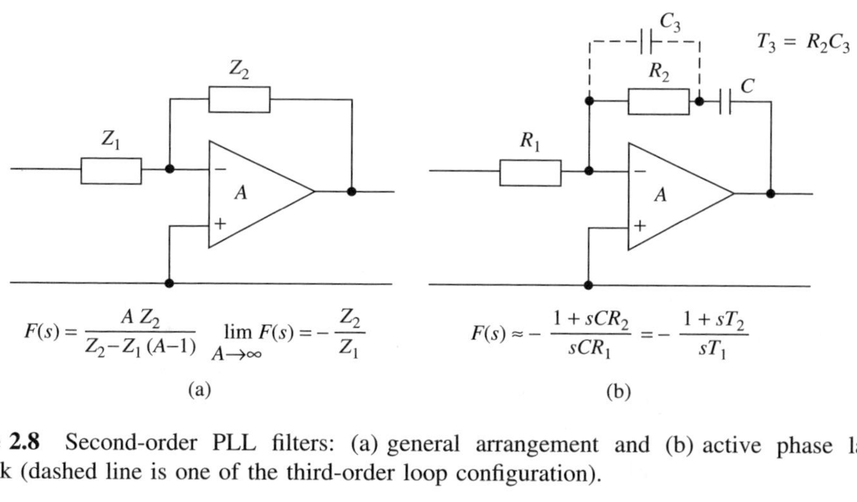

Figure 2.8 Second-order PLL filters: (a) general arrangement and (b) active phase lag-lead network (dashed line is one of the third-order loop configuration).

Examination of the properties of these *second-order loops of type 2* proceeds with the evaluation of the PLL transfer function. Application of the gain $G(s)$ reveals

$$H(s) = \frac{K(sT_2 + 1)}{s^2 T_1 + s(KT_2 + 1/A) + K} \tag{2.29}$$

Introduction of the natural frequency ω_n and damping factor ζ gives

$$\omega_n = \sqrt{K/T_1}; \quad 2\zeta\omega_n = \frac{KT_2 + 1/A}{T_1} \tag{2.30}$$

from which

$$\zeta = \omega_n \frac{T_2}{2} \tag{2.31}$$

Application of the above relations reveals for the open-loop gain

$$G(s) = \frac{2\zeta s\omega_n + {\omega_n}^2}{s^2} \tag{2.32}$$

and for the closed-loop PLL transfer functions

$$H(s) = \frac{2\zeta s\omega_n + {\omega_n}^2}{s^2 + 2\zeta s\omega_n + {\omega_n}^2}; \quad 1 - H(s) = \frac{s^2}{s^2 + 2\zeta s\omega_n + {\omega_n}^2} \tag{2.33}$$

In the normalized form, for the open-loop transfer function we get

$$G(\sigma) = \frac{1 + 2\sigma\zeta}{\sigma^2} \quad \text{or} \quad G(\mathrm{j}x) = \frac{1 + 2\mathrm{j}\zeta x}{-x^2} \tag{2.34}$$

and for the closed-loop transfer functions

$$H(\mathrm{j}x) = \frac{1 + 2\mathrm{j}\zeta x}{-x^2 + 2\mathrm{j}\zeta x + 1}; \quad 1 - H(\mathrm{j}x) = \frac{-x^2}{-x^2 + 2\mathrm{j}\zeta x + 1} \tag{2.35}$$

After plotting the transfer functions $H_\mathrm{i}(x)$ and $H_\mathrm{o}(x)$, we find that they coincide with those plotted in Fig. 2.7 for the PLLs with the high gain, K (high-gain loops). However, we find a substantial difference with the phase characteristic that starts, owing to the two integrators in $G(s)$, at nearly -180° (cf. Fig. 2.9). This is very important in instances with unintentionally introduced poles or delays, for example, with the use of the sampled PDs, since the stability of the PLL system deteriorates.

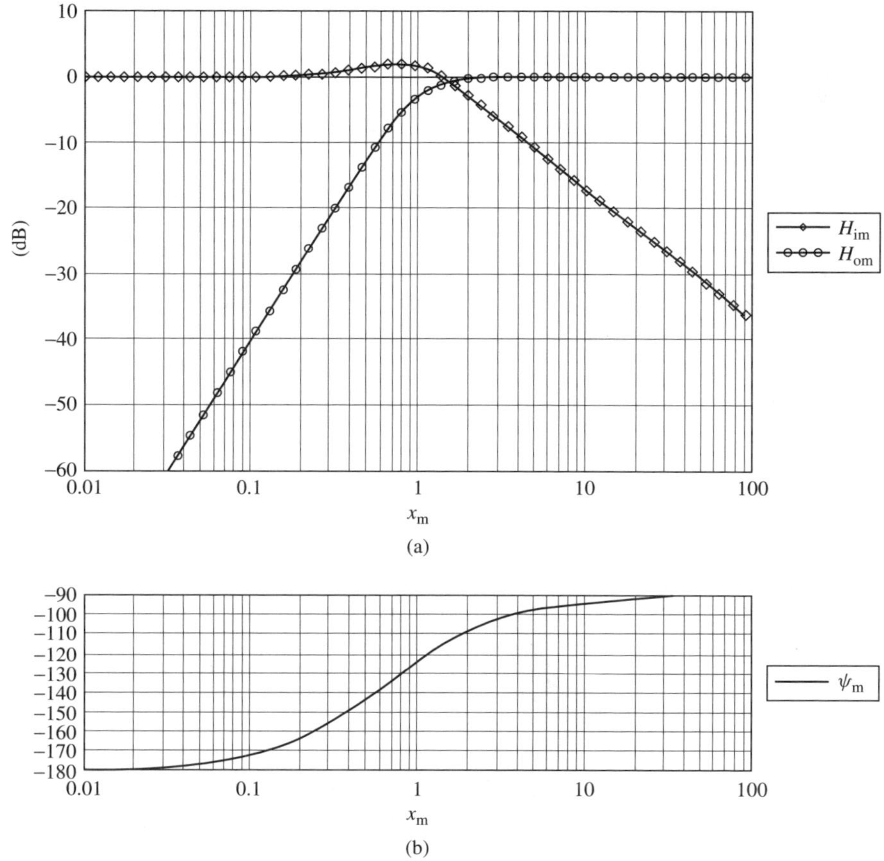

Figure 2.9 (a) Transfer functions $H_i(jx) = 20\log(|H(jx)|)$ (◇) and $H_o(jx) = 20\log(|1 - H(jx)|)$ (○) of the second-order PLL of type 2 and (b) phase characteristic of the open-loop gain $G(jx)$.

This problem will be discussed in the following chapters. PLL transfer functions for high-gain second-order loops are plotted in Fig. 2.10.

Furthermore, note that we have investigated the PLL system with a second ideal integrator; however, this is not the case with actual hardware. Some difficulties might occur with the stability of the zero level, leakage of the charge from the integrating capacitor, and so on. In addition, we must take into account that we have considered the linearized system, which might fail in all cases in which this condition is not met, for example, the limiting values of the input or output voltages of the operational amplifier and so on.

2.3.2 PLLs of the Second Order of Type 2 with Current Output Phase Detector

Here the integration is realized with a simple RC filter fed by current from a generator with infinite input resistance. An idealized arrangement is shown in Fig. 2.11. Note

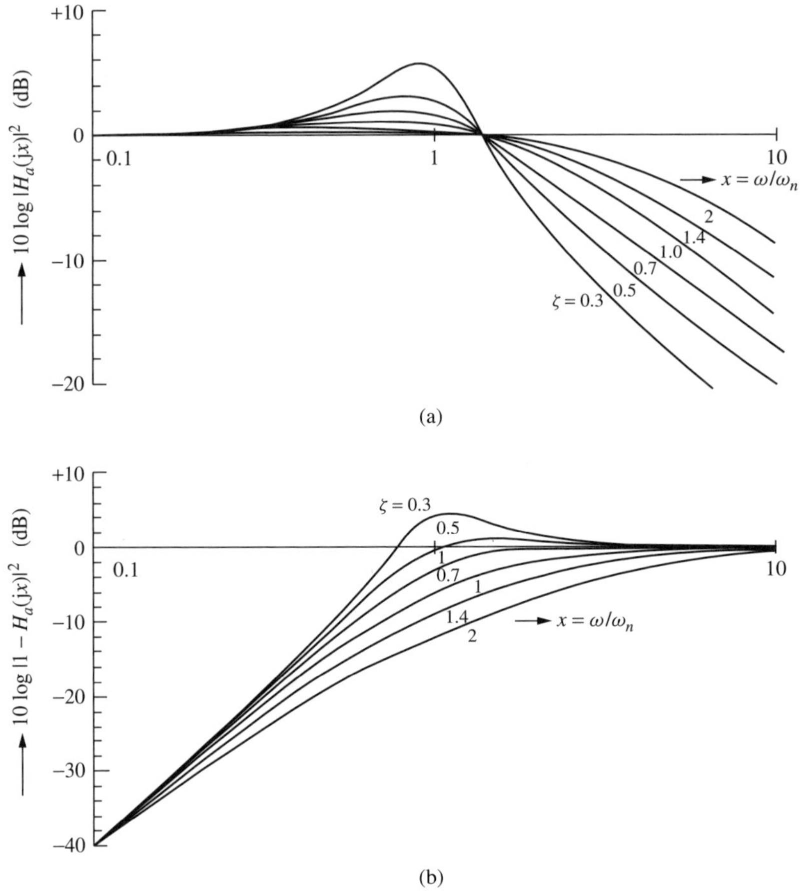

Figure 2.10 Transfer functions of the second-order PLL type 2 for different damping factors ζ: (a) $H_{\rm i}({\rm j}x) = 20\log(|H({\rm j}x)|)$ and (b) $H_{\rm o}({\rm j}x) = 20\log(|1 - H({\rm j}x)|)$.

that the input current is not a continuous one but is supplied via a three-stage switch from two current generators (e.g., phase-frequency detectors). The PD current is proportional to the phase difference $\phi_{\rm e}$

$$i_{\rm d} = \frac{I_{\rm p}\phi_{\rm e}}{2\pi} \tag{2.36}$$

and the Laplace transform of the steering voltage $v_2(t)$ is

$$V_2(s) = I_{\rm d}(s)Z(s) \tag{2.37}$$

However, this is only the case when the working conditions of the PLLs are linear. In Fig. 2.11 the actual *RC integrator* is shunted with the leaking resistor $R_{s,\rm leak}$, which retains all undesired conductances, for example, of the current sources, due to the

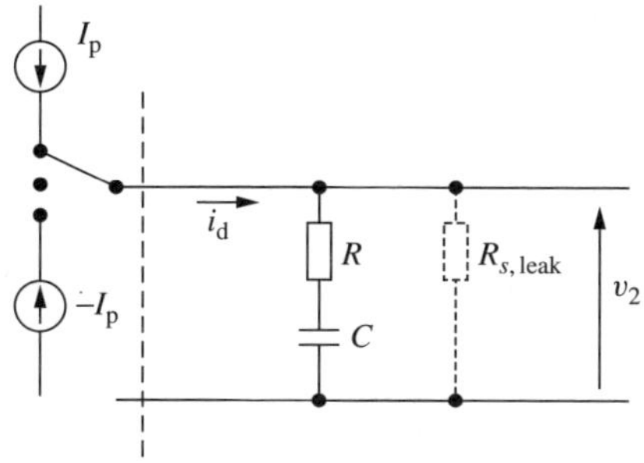

Figure 2.11 Second-order PLL: the passive integrating RC filter.

capacity C, due to the VCO input, and so on. In these circumstances we evaluate the impedance $Z(s)$ as

$$Z(s) = R_{s,\text{leak}} \frac{(1 + sRC)/sC}{R_{s,\text{leak}} + (1 + sRC)/sC} = \frac{(1 + sRC)/sC}{sC + (1 + sRC)/R_{s,\text{leak}}} \tag{2.38}$$

which is easily approximated for a large $R_{s,\text{leak}}$ to

$$Z(s) \approx \frac{1 + sRC}{sC} \tag{2.39}$$

The open-loop gain is found after replacing the phase detector gain with the current gain in (1.36)

$$K_{\text{di}} = \frac{I_{\text{p}}}{2\pi} \ [\text{A}/2\pi] \tag{2.40}$$

that is,

$$G(s) \doteq K_{\text{di}} K_{\text{o}} \frac{1 + sT_2}{s^2 C} \tag{2.41}$$

where we have inserted, in agreement with (2.18), the time constant

$$T_2 = RC \tag{2.42}$$

To arrive at a complete analogy for computer solutions of PLLs, we introduce

$$\omega_n^2 = K_{\text{di}} K_{\text{o}}/C \text{ and } \zeta = \omega_n \frac{T_2}{2} \tag{2.43}$$

In that case, all relations (2.32) to (2.35) are also valid for these types of loops. Investigation of the constant K_v with the assistance of the final value theorem leads to

$$K_v = \lim_{s \to 0} sG(s) = K_{\text{di}} K_{\text{o}} R_{s,\text{leak}} \tag{2.44}$$

Evidently, the leaking resistance has a similar function as the gain of the operation amplifier, that is, for very low frequencies, it changes the PLLs of type 2 into that of type 1.

The difficulty with these *current output PDs* is caused by the current pulses I_p (cf. eq. (2.40)) passing the resistor R and generating voltage pulses

$$V_p = I_p R \tag{2.45}$$

which are fed nonfiltered to the tuning element (varactor) of the VCO and cause frequency modulation

$$\Delta\omega = I_p R K_o \tag{2.46}$$

Another complication is the switching operation of the PD and the time delay connected with it – this problem will be discussed in the next chapter.

2.4 SECOND-ORDER PLLs WITH FREQUENCY DIVIDERS IN THE FEEDBACK PATH

By placing a digital frequency divider (DFD) in the feedback path of the PLLs, as schematically indicated in Fig. 2.12, the open-loop gain is

$$G(s) = K_d \cdot F_L(s) \cdot \frac{K_o}{s} \cdot \frac{1}{N} \tag{2.47}$$

and the respective transfer function is

$$H(s) = \frac{G(s)}{1 + G(s)} = \frac{KF_L/sN}{1 + KF_L/sN} = \frac{KF_L/N}{s + KF_L/N} \tag{2.48}$$

We see that the gain $K_d K_o$ is reduced in proportion to the division factor N to a new value

$$K' = K_d K_o / N \tag{2.49}$$

The consequence is that the *natural frequency* ω_n, the corner frequencies of both transfer functions $H(s)$ and $1 - H(s)$ are all reduced N *times* [3]. In the noise theory of PLLs, we introduce the primed transfer functions $H'(s)$ and $1 - H'(s)$ – this is important for the investigation of noise processes, as will be explained in Chapter 9. However, the actual output frequency behaves in accordance with Fig. 1.5(c) that is,

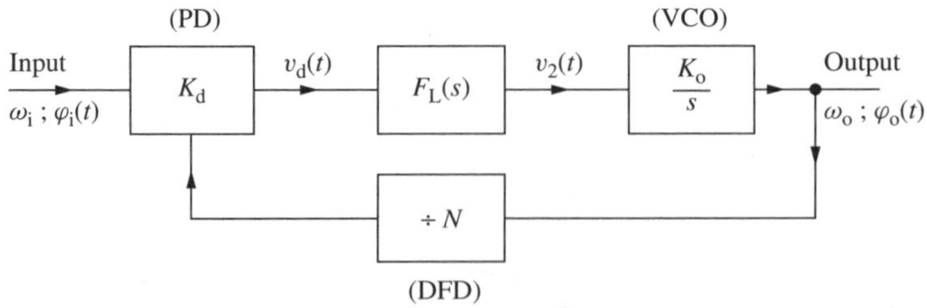

Figure 2.12 Second-order PLL with a digital frequency divider (DFD) in the feedback path.

the output frequency is multiplied by the division factor N. Note that the character of the second-order PLL is maintained. In this way very large multiplication factors, with good suppression of the neighboring harmonics, can be realized [4]. But this technique finds much more important application in *Digital Frequency Synthesis*, which will be discussed in Chapter 12. For the design of the second-order PLL, see Section 3.6.

REFERENCES

[1] F.M. Gardner, *Phaselock Techniques*. 2nd ed. New York: John Wiley, 1979.

[2] W.F. Egan, *Frequency Synthesis by Phase Lock*. New York: Wiley, 1981 and 2000.

[3] V.F. Kroupa, "Low-noise microwave-frequency synthesizers: design principles", *IEE Proceedings-H*, **130**, 483–488, 1983 (reprinted in [9.14]).

[4] V.F. Kroupa, *Frequency Synthesis: Theory, Design et Applications*. London: Ch. Griffin, 1973; New York: John Wiley, 1973.

3

PLLs of the Third and Higher Orders

Practical PLLs are preferably of the high-gain second-order arrangement. Nevertheless, we often encounter higher-order systems. They come into being unintentionally because of the existence of spurious impedances around the VCO or in connection with the transfer function in the feedback path, and because of the time delay, particularly, in modern digital systems. However, in most cases the undesired roots of the transfer function $1 + G(s)$ (see Chapter 4) are in the left-hand half plane, s, or, σ, so far from the imaginary axis that the PLL almost behaves as the original second-order loop, however, with a slightly reduced damping factor and the phase margin. In addition, the pull-in range may be changed, often reduced, as we shall see later in Chapter 6.

However, in many instances we change the PLL into a higher-order loop intentionally. The reason is to increase attenuation of spurious signals in the range of higher Fourier frequencies, $\omega \gg \omega_n$. This is necessary in instances in which the reference frequency is low and requirements on the output spectral purity high. These conditions are encountered, for example, in digital frequency synthesis, PLL frequency synthesizers, and so on.

The problem of the reduced pull-in range can be solved either with the assistance of a frequency discriminator, with a coarse pretuned VCO, by adding the searching voltage to the steering voltage, $v_2(t)$, of the VCO, or by changing the high-gain second-order loop type 1 into type 2. These questions will be discussed in more detail in the following chapters.

Another difficulty encountered with higher-order PLLs is the stability of the feedback system. Often the reduced pull-in range may be troublesome but the information about stability provides the basic guidelines for practical realization. A very simple criterion is the estimated phase margin.

In the following sections we shall investigate the most important higher-order PLLs encountered in practice. Finally, we shall provide the effective open-loop gain $G(jx)$

Phase Lock Loops and Frequency Synthesis V.F. Kroupa
 ISBN: 0-470-84866-9

and present a general solution of the transfer functions $H(\mathrm{j}x)$ and $1 - H(\mathrm{j}x)$ of the higher-order loops based on the basic second-order high-gain PLLs.

It is odd that the higher-order loops are not discussed in depth in the literature. To cite only a few, Egan [1] deals only with the third-order loops, Gardner [2] mentions the third-order loop of type 3 for some special applications, and only Rohde [3] has published long computer programs for the solution of third- and fifth-order loops. The present discussion is based on papers by the author [4, 5] and on his PLL book in Czech [6].

3.1 GENERAL OPEN-LOOP TRANSFER FUNCTION $G(s)$

In nearly all instances it is possible to separate individual blocks, forming the loop, and in accordance with the rules derived in connection with Fig. 1.5 to write the open-loop transfer function, $G(s)$, and proceed with a computer solution of the transfer functions and PLL stability. In the following paragraphs we shall investigate individual additional loop sections and consider the order and the type of the resulting PLLs. The problem of the stability of the feedback system will be discussed in more detail in the next chapter. Here, we shall deal with the expected phase margins. In addition, we shall introduce and use the normalized solution wherever possible.

3.1.1 Additional RC Section

In instances in which input and output impedances are such that each additional RC filter can be considered as independent, the additive transfer function is (cf. Fig. 2.3)

$$F_{\mathrm{RC}}(\mathrm{j}\omega) = \frac{1}{1 + \mathrm{j}\omega\mathrm{RC}} = \frac{1}{1 + \mathrm{j}\omega T}; \quad T = \mathrm{RC} \tag{3.1}$$

and it is added as a factor to the open-loop gain. We shall investigate the problem later in connection with transfer functions of higher-order loops.

3.1.2 Two RC Sections

Often the additive section cannot be considered as independent and we shall consider the problem more closely. The principle block diagram is shown in Fig. 3.1 and the respective transfer function is

$$F(s) = \frac{1}{s^2 T_1 T_3 + s(T_1 + T_{13} + T_3) + 1} \tag{3.2}$$

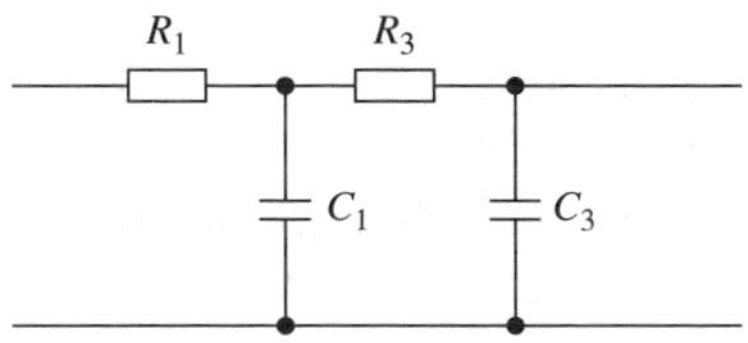

Figure 3.1 PLL filter with two RC sections in series.

where

$$T_1 = R_1C_1,\ T_{13} = R_1C_3 \text{ and } T_3 = R_3C_3 \tag{3.3}$$

In instances in which the time constant, T_1, is predominant, it is much larger than T_3 and by choosing, in addition, the resistance R_3 to be also several times larger than R_1 we have

$$R_1C_1 > R_3C_3 > R_1C_3 \tag{3.4}$$

The result is that the time constant, T_{13}, is small compared to $(T_1 + T_3)$ and the filter is effectively composed of two independent sections

$$F(s) \doteq \frac{1}{(sT_1 + 1)(sT_3 + 1)} \tag{3.5}$$

An important simplification provides the normalization of the additive time constant T_3 in respect to T_1 by introducing

$$T_3/T_1 = \mu < 1 \tag{3.6}$$

Applications will be discussed in Section 3.2.2.

Another solution supplies the situation illustrated in Fig. 3.2 where the two RC sections are separated with the assistance of an operation amplifier, in which case the filter transfer function is

$$F(s) = \frac{R_a + R_b}{R_a} \cdot \frac{1}{1 + sR_1C_1} \cdot \frac{1}{1 + sR_3C_3} \tag{3.7}$$

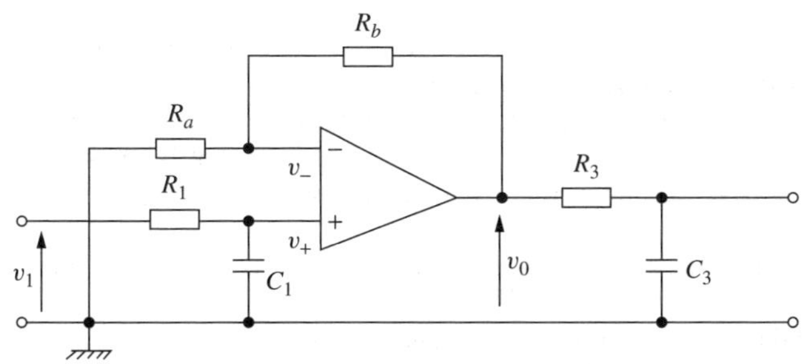

Figure 3.2 PLL filter with two independent RC sections in series.

3.1.2.1 Two RC sections with equal resistors

In this case in accordance with Fig. 3.1 we put

$$R_1 = R_3 \tag{3.8}$$

and consequently we also find the time constants T_{13} and T_3 are equal:

$$T_{13} = T_3 \tag{3.9}$$

By introducing the proportionality factor

$$\lambda = \frac{T_3}{T_1} \tag{3.10}$$

we can rearrange eq. (3.2) into

$$F(s) = \frac{1}{s^2 \lambda T_1{}^2 + sT_1(1+2\lambda) + 1} \tag{3.11}$$

After solution of the quadratic term in the denominator, we get

$$\tau_1 = \frac{2T_1\lambda}{1 + 2\lambda - \sqrt{1+4\lambda^2}} \tag{3.12}$$

$$\tau_2 = \frac{2T_1\lambda}{1 + 2\lambda + \sqrt{1+4\lambda^2}} \tag{3.13}$$

and the transfer function (3.11) can be replaced with (3.5). Evaluation of the effective ratio μ reveals as maximum

$$\mu_{\max} = 0.172 \quad \text{for} \quad \lambda = 0.5 \tag{3.14}$$

For two equal RC sections, the ratio λ is equal, that is, $\lambda = 1$, and the respective μ is only

$$\mu_{\lambda=1} = 0.146 \tag{3.15}$$

We shall see later that in both cases the effective ratio is too small for optimum suppression of spurious signals.

3.1.3 Active Second-order Low-pass Filter

Its basic arrangement is plotted in Fig. 3.3 and its transfer function evaluated in Section 8.1, eq. (8.16), is

$$\begin{aligned} F(\mathrm{j}\omega) &= \frac{1}{1 + \mathrm{j}\omega 2RC_2 - \omega^2 R^2 C_1 C_2} \\ &= \frac{1}{1 + \mathrm{j}\omega 2T_2 - \omega^2 T_1 T_2} \end{aligned} \tag{3.16}$$

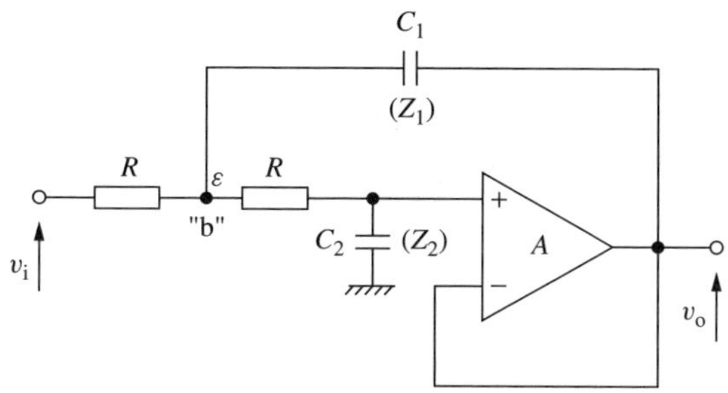

Figure 3.3 Active second-order low-pass filter.

After introduction of the filter natural frequency ω_{nf}

$$\omega_{nf}^2 = 1/T_1T_2 = 1/R^2C_1C_2 \tag{3.17a}$$

and damping d

$$d = \omega_{nf}T_2 = \sqrt{T_2/T_1} = \sqrt{C_2/C_1} \tag{3.17b}$$

We get the transfer function of (3.16) in the normalized form

$$F(jy) = \frac{1}{1 + jy2d - y^2} \quad \text{where } y = \frac{\omega}{\omega_{nf}} = \frac{\omega}{\omega_n} \cdot \frac{\omega_n}{\omega_{nf}} = x\alpha \tag{3.18}$$

which is plotted in Fig. 3.4 for different damping constants together with the respective phase characteristics [7]. Note that for small $y = x\alpha$ the reduction of the phase is quite small and one may expect that application of this low-pass filter in PLLs would not degrade the stability (which will be discussed in the following chapter) since

$$\Psi \doteq 2yd \cdot 180/\pi \doteq -115x\,d\alpha \tag{3.19}$$

and for $x = 1$, $d = 0.6$, and $\alpha = 0.1$ we get the degradation of the phase margin by $\Psi \approx 7°$.

3.1.4 Twin-T RC Filter

In some instances in which we need to reduce or remove one discrete spurious signal, the twin-T RC filter illustrated in Fig. 3.5 might be of assistance. This network exhibits "infinite attenuation" for the following arrangement:

$$\begin{aligned} \omega^2 &= 1/2R_1R_2C_1^2 \\ \omega^2 &= 2/C_1C_2R_2^2 \end{aligned} \tag{3.20}$$

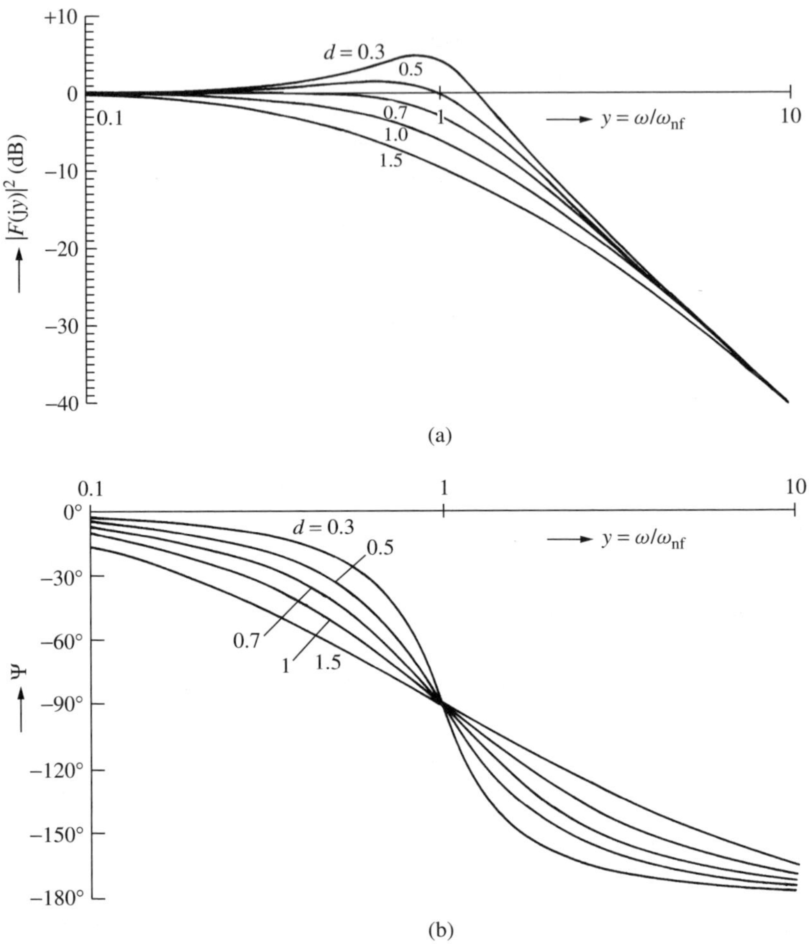

Figure 3.4 (a) Transfer function *a* of the active second-order low-pass filter (cf. (3.16)) as function of the normalized frequency, y, for different damping factors d and (b) its phase characteristics (Reproduced from C.J. Savant Jr., *Basic Feedback Control System Design*. New York, Toronto, London: McGraw-Hill, 1958 by permission of McGraw Hill, 2002).

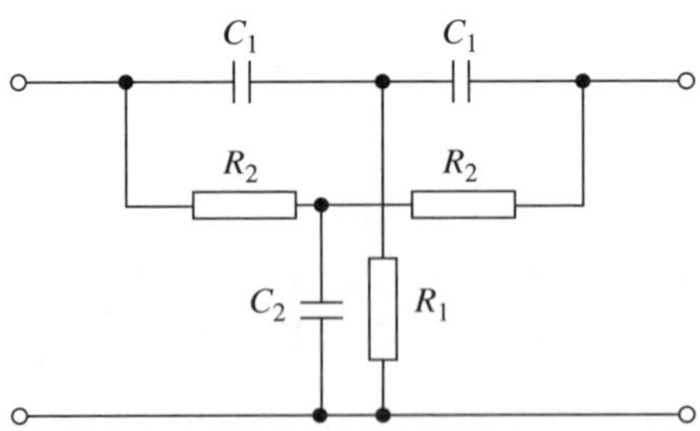

Figure 3.5 Twin-T RC filter.

A further simplification is provided with fixed relations between resistance and capacitance values

$$R = R_1/n = R_2 \text{ and } C = C_1 = C_2/4n \tag{3.21}$$

In this case we get for the "resonant," or more precisely, the "infinite" attenuation frequency, ω_{rf},

$$\omega_{rf} = \frac{1}{RC\sqrt{2n}} \tag{3.22}$$

Further, where the input resistance $R_i \ll R$ and the output resistance $R_{out} \gg R$, we get for the transfer function of the Twin-T

$$F(jy) = \frac{1 - y^2}{1 + 4jy - y^2}$$

$$\text{where} \qquad y = \frac{\omega}{\omega_{rf}} = \frac{\omega}{\omega_n} \cdot \frac{\omega_n}{\omega_{rf}} = x\alpha \tag{3.23}$$

The normalized transfer function and the respective phase characteristics are shown in Fig. 3.6. An active Twin-T filter may increase the "effective Q," that is, narrow the stop band [8, 9].

3.1.5 PLLs with a Selective Filter in the Feedback Path

A simple PLL, with the block diagram shown in Fig. 1.3, is rarely encountered in practice. One complication is a filter in the feedback path. In most cases it would

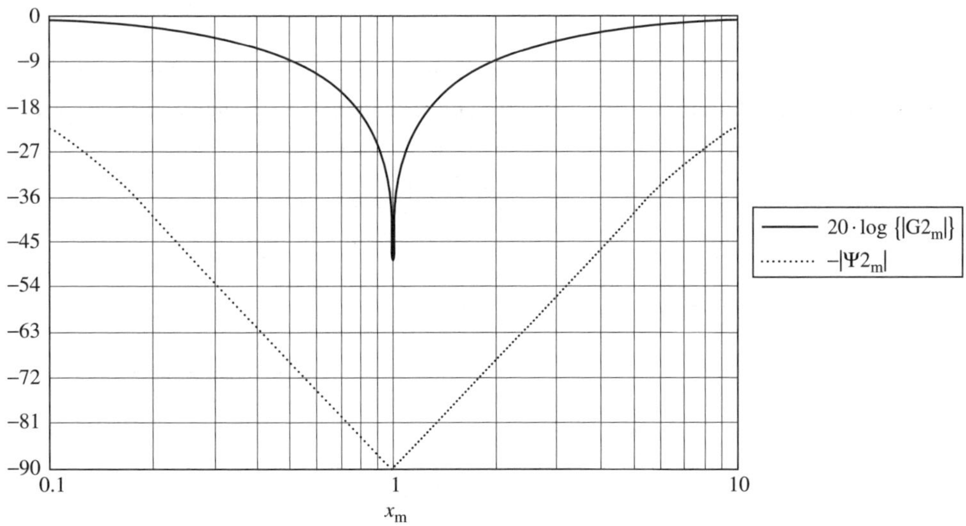

Figure 3.6 Twin-T RC filter: transfer function and phase characteristics with respect to normalized frequency: $x_m = \omega_m/\omega_n$.

be a simple divider; however, mixers, IF filters, and other networks might be met (cf. Fig. 9.20). Here, we shall consider the case with the IF filter shown in Fig. 3.7. Discussion of its influence will be based on the reasoning introduced by [2].

To illustrate the problem in more detail, we shall assume a small disturbance $\phi_i(t)$ at the input and its response $\phi_o(t)$ at the output. For voltages in the important network points of Fig. 3.7 we get

$$\begin{aligned} v_i(t) &= V_i \sin([\omega_i t + \phi_i(t)]; \quad |\phi_i(t)| \ll 1 \\ v_o(t) &= V_o \cos([\omega_o t + \phi_o(t)]; \quad |\phi_o(t)| \ll 1 \\ v_a(t) &= V_a \cos([\omega_a t] \end{aligned} \tag{3.24}$$

Note that for simplicity we neglect the noise introduced by the input signal, $v_a(t)$, to the mixer and in addition we only consider the lower sideband with the frequency $\omega_{oi} \approx \omega_o$. With these assumptions we encounter (at the input to the feedback mixer) the voltage

$$v_m(t) = V_m \cos([\omega_{oi} t + \phi_o(t)] \tag{3.25}$$

For further simplification and for clarity, but without any loss of generality, we may consider that the disturbance, $\phi_o(t)$, is a single harmonic component, that is,

$$\phi_o = \phi_{ov}(t) = -m_p \cos(\nu t) \tag{3.26}$$

where m_p is the corresponding phase modulation index. The situation being such, we can rearrange the voltage, $v_m(t)$, in the following way

$$v_m(t) \doteq V_m[\cos(\omega_{oi} t) + \frac{m_p}{2}[\sin(\omega_{oi} + \nu)t + \sin(\omega_{oi} - \nu)t]] \tag{3.27}$$

In instances in which the IF filter is symmetric around the resonance frequency ω_{oi}, both sidebands encounter the same damping and the same phase shift; however,

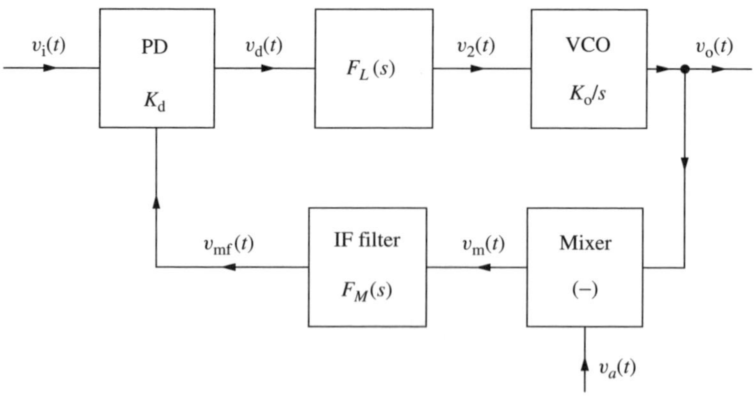

Figure 3.7 The IF filter in the feedback path of the PLL.

the latter with opposite signs is in accordance with the following relation

$$|F_M(\omega_{oi} + \nu)| = |F_M(\omega_{oi} - \nu)| = |F_M(\nu)| \tag{3.28}$$

With the assumption $F_M(\omega_{oi}) = 1$, the voltage at the output of the IF filter will be

$$\begin{aligned} v_{mf}(t) &= V_m(\cos(\omega_{oi}t) + \frac{m_p F_M(\nu)}{2}(\sin[(\omega_{oi} + \nu)t - \Psi_v] + \sin[(\omega_{oi} - \nu)t + \Psi_v])) \\ &= V_m \cos[\omega_{oi} - m_p |F_M(\nu)| \cos(\nu t - \Psi_v)] \end{aligned} \tag{3.29}$$

By considering the original disturbance $\phi_{ov}(t)$, we get from the above relation

$$v_{mf}(t) = V_m \cos[\omega_{oi}t - |F_M(\nu)|\phi_{ov}(t - T_M)] \tag{3.30}$$

When the loop is locked, that is, $\omega_{oi} \approx \omega_o$, the voltage at the output of the phase detector (PD) will be

$$v_d(t) = K_d[\phi_i(t) - |F_M(\nu)|\phi_{ov}(t - T_M)] \tag{3.31}$$

After the application of the Laplace transform, we have

$$V_d(s) = K_d[\Phi_i(s) - F_M(s)\Phi_o(s)] \tag{3.32}$$

and finally with the assistance of the above result and relations (1.31) or (1.34) we arrive at (1.35), that is, at

$$\frac{\Phi_o(s)}{\Phi_i(s)} = \frac{KF(s)/s}{1 + KF(s)F_M(s)/s} \tag{3.33}$$

The final result is that in respect to the small disturbing signals (small phase disturbances), the mixer with the following IF filter behaves like a *black box* with the transfer function $F_M(s)$, which is called *the modulation transfer function*. We find its shape by shifting the transfer function of the real filter from the carrier frequency to the *zero* frequency and retaining the part over the positive frequencies. The principle is illustrated with the assistance of Fig. 3.8.

Example 3.1

We assume the IF filter to be a simple resonant circuit with the transfer function

$$F(j\omega) = \frac{R}{R + j(\omega L - 1/\omega C)} \approx \frac{1}{1 + jQ2v/\omega_o} \tag{3.34}$$

where Q is the quality factor of the resonant circuit, ω_o its resonant frequency, and $v = \omega - \omega_o$ is the detuning. After introducing the *time delay*

$$T_M = 2Q/\omega_o \tag{3.35}$$

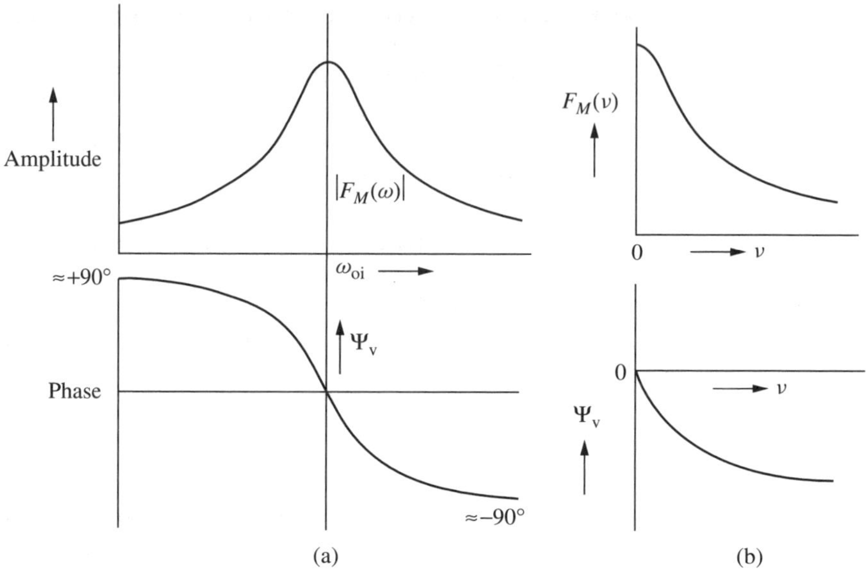

Figure 3.8 (a) The transfer function of an IF filter and (b) the effective modulation transfer function (Reproduced from F.M. Gardner, *Phaselock Techniques*. 2nd ed. New York: John Wiley, 1979 by permission of John Wiley & Sons, Inc, 2002).

we can simplify the modulation transfer function to

$$F_M(s) = \frac{1}{1 + sT_M} \tag{3.36}$$

which is equivalent to the transfer function of the simple RC filter (cf. (3.1)).

3.1.6 Time Delays in PLLs

In practical PLLs we encounter different types of time delays, generally introduced with digital circuits. However, other sources may also be present. The difficulties they cause include reduced phase margins, and consequently PLL stability, constricted pull-in properties, and eventually false locks.

3.1.6.1 *Simple time delay*

Simple time delay, τ, is respected by multiplying the open-loop gain by the factor

$$F_{\mathrm{dl}}(s) = \mathrm{e}^{-s\tau}; \quad F_{\mathrm{dl}}(\mathrm{j}\omega) = \mathrm{e}^{-\mathrm{j}\omega\tau} \tag{3.37}$$

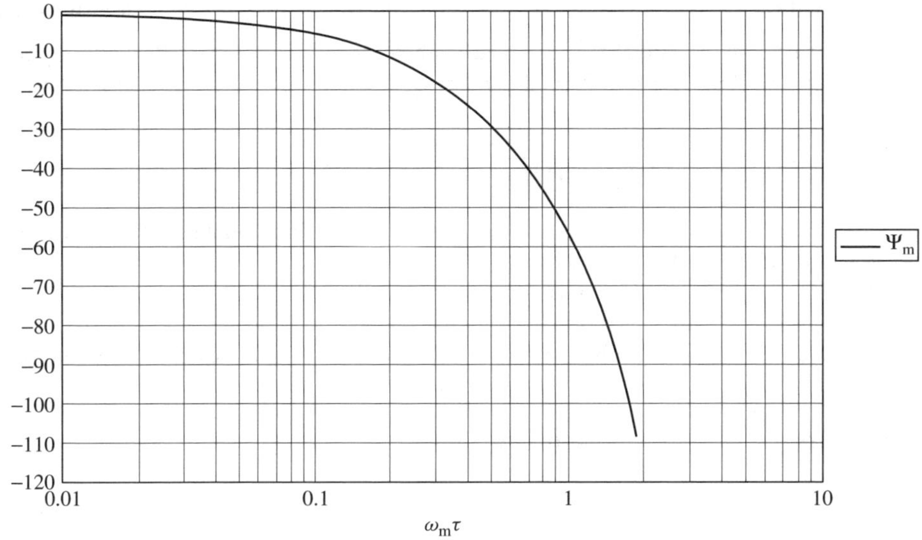

Figure 3.9 Phase shift introduced by a normalized time delay $\omega\tau$.

Evidently this transfer function would only change the phase margin. However, from Fig. 3.9 we see that its influence might be considerable (cf. Chapters 4 and 10).

3.1.6.2 Sampling

In modern electronic technology, many analog arrangements are replaced with digital circuits or processing. This is also true of PLLs. The proper approach would be an investigation with the assistance of the *z-transform*, which will be discussed in Chapter 10. The other possibility is to modify the original Laplace transform of $G(s)$ in the following way (cf. Section 10.1):

$$G_{\text{mod}}(s) = F_{\text{h}}(s)\hat{G}(s) \tag{3.38}$$

where

$$F_{\text{h}}(s) = \frac{1 - e^{-sT}}{s} \tag{3.39}$$

After introduction of the sampling frequency ω_s

$$\hat{G}(s) = \frac{1}{T}\sum_{n=\infty}^{\infty} G(s - \text{j}n\omega_{\text{s}}); \quad \omega_{\text{s}} = \frac{2\pi}{T} \tag{3.40}$$

we arrive at

$$G_{\text{mod}}(s) = \frac{1 - e^{-sT}}{s} \cdot \frac{1}{T}\sum_{n=\infty}^{\infty} G(s - \text{j}n\omega_{\text{s}}) \approx \frac{1 - e^{-sT}}{s} \cdot \frac{1}{T} G(s) \tag{3.41}$$

The simplification is justified since we expect that the higher-order terms of $G_{\text{mod}}(s)$ are attenuated with the loop filter $F(s)$. For small frequencies, s, that is, for $|sT| \ll 1$, we can rearrange $G_{\text{mod}}(s)$ to

$$G_{\text{mod}}(s) \approx \frac{\sinh(sT/2)}{sT/2} G(s)\mathrm{e}^{-sT/2} \approx G(s)\mathrm{e}^{-sT/2} \approx \frac{G(s)}{1 + sT/2} \tag{3.42}$$

In the normalized form we introduce

$$sT/2 = \sigma\omega_n T_{\text{s}}/2 = \sigma\pi\frac{f_n}{f_{\text{s}}} = \sigma\delta \tag{3.43}$$

Consequently, the normalized open-loop gain is

$$G_{\text{mod}}(\sigma) = \frac{\sinh(\sigma\delta)}{\sigma\delta} G(\sigma)\mathrm{e}^{-\sigma\delta} \tag{3.44}$$

The situation with the sampled PLL is illustrated in Fig. 3.10. Finally, we arrive at the often suggested approximation of the sampling process, that is, with the assistance of an additional RC section. In Fig. 3.11 we compare the transfer function $G_{\text{mod}}(s)$ with its simplified version $G(s)/(1 + sT/2)$. Note that it effectively forms the envelope of the $G_{\text{mod}}(s)$. Even the phase characteristics are practically identical, but the degradation of the phase margin for larger normalized frequencies, x_{m}, cannot be neglected.

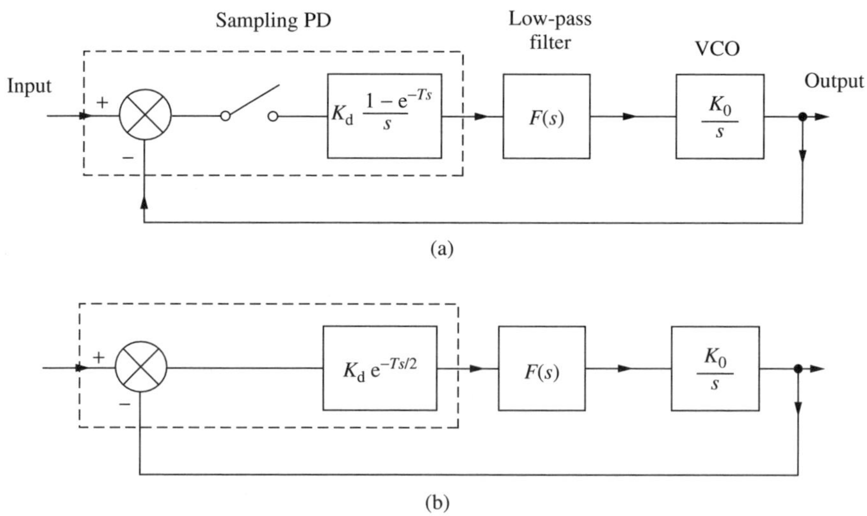

Figure 3.10 PLL with sampling phase detector: (a) block diagram of the loop and (b) the simulated analog system.

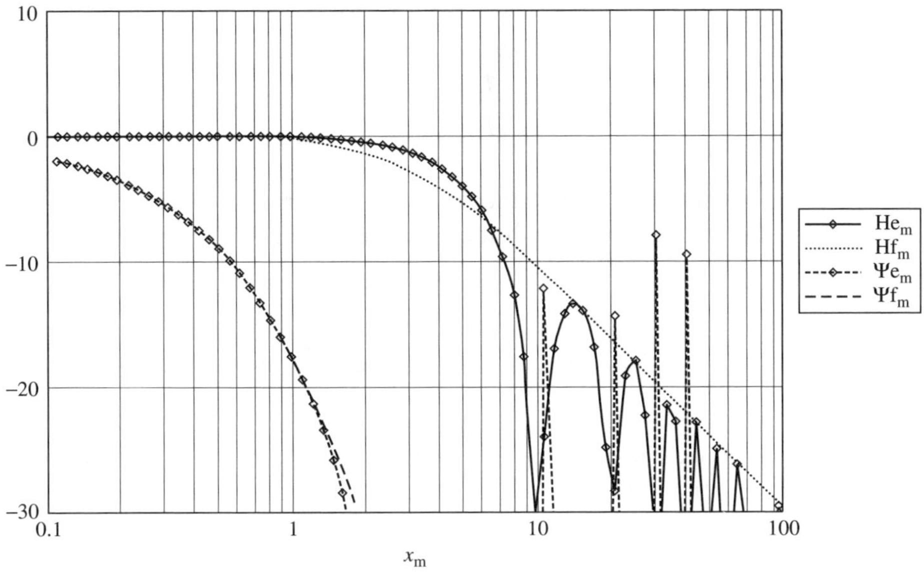

Figure 3.11 Comparison of the transfer function $G_{mod}(s)$ (◇) with that of a simple RC section (dashed) and of the corresponding phase shifts Ψe_m and Ψf_m.

3.2 HIGHER-ORDER TYPE 2 PLLs

We have seen that the first-order loops are used rarely. On the other hand, the set of second-order loops has the advantages of freedom in the choice of the natural loop frequency ω_n, of the damping factor ζ, and of the slopes of the transfer functions. However, in many instances, some undesired spurious loop filtering sections, such as RC section or delays, are present and change the PLL into a higher-order system.

3.2.1 Third-order Loops: Lag-lead Filter with Additional RC Section

The third-order loops realized by the addition of an independent RC section are the ones most often encountered. By starting with the second-order high-gain loop, we have for the open-loop gain

$$G_3(s) = \frac{K_d K_A K_o(1 + sT_2)}{s(1 + sT_1)(1 + sT_3)} = G_2(s)G_{RC}(s) \tag{3.45}$$

In type 2 systems we achieve this goal by changing the feedback path in the integrating OP amplifier system (see the dashed part in Fig. 2.8(b)). For the loop gain we get

$$G_3(s) \approx \frac{K}{s} \cdot \frac{1 + sR_2(C + C_3)}{sR_1C(1 + sR_2C_3)} = \frac{K}{s} \cdot \frac{1 + sT_2}{sT_1(1 + \kappa sT_2)} \tag{3.46}$$

where we have introduced an important design factor

$$\kappa = \frac{T_3}{T_2} \qquad \kappa < 1 \tag{3.47}$$

Since the time constant T_3, in accordance with the design procedure, is smaller than T_2, even this third-order loop is unconditionally stable because $G(s)$ exhibits a positive phase margin

$$\Psi_{\mathrm{pm}} = -\frac{180^\circ}{\pi}[-\pi + \arctan(\omega T_2) - \arctan(\omega\kappa T_2)] > -180^\circ \tag{3.48}$$

After the introduction of the natural frequency ω_n and the damping factor ζ, in accordance with (2.20) and (2.30), we get for the transfer functions

$$H_3(\mathrm{j}x) = \frac{\mathrm{j}x2\zeta + 1}{-\mathrm{j}x^3 2\zeta\kappa - x^2 + \mathrm{j}x2\zeta + 1} \tag{3.49}$$

$$1 - H_3(\mathrm{j}x) = \frac{-\mathrm{j}x^3 2\zeta\kappa - x^2}{-\mathrm{j}x^3 2\zeta\kappa - x^2 + \mathrm{j}x2\zeta + 1} \tag{3.50}$$

and for the phase margin

$$\Psi_{\mathrm{pm}} = -\frac{180^\circ}{\pi}[-\pi + \arctan(2\zeta x) - \arctan(2\zeta\kappa x)] > -180^\circ \tag{3.51}$$

Example 3.2

Compute and plot the open-loop gain and the transfer functions of the third-order loop for $\kappa = 0.3$ and $\zeta = 1.5$. The result is shown in Fig. 3.12. Note that the phase margin is about 25° and the normalized frequency x_{o}, for which the logarithm of the open-loop gain is zero, is about 1.7. After introduction of all these values into relation (3.51), we get for the phase margin, 22°.

Further investigations of relations (3.49) and (3.50) and Fig. 3.12 reveal that both asymptotic slopes of the transfer functions are 40 dB/dec. This property, together with the unconditional stability, is highly appreciated by designers. Therefore, we have provided a deeper study in respect to different damping factors and rations κ. The results are plotted in Figs. 3.13 and 3.14 and may be summarized as follows. The smallest overshoots and the largest safe phase margins are for damping factors ζ in the range 0.5 to 1 of the original second-order loop, whereas the value of the ratio κ is in the neighborhood of 0.3.

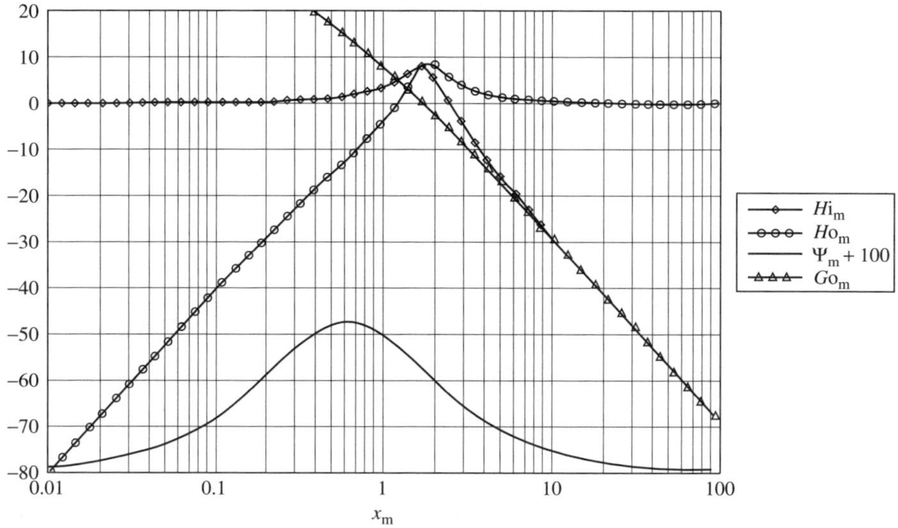

Figure 3.12 Transfer functions $H_i(jx) = 20\log(|H_3(jx)|)$ (◇) and $H_o(jx) = 20\log(|1 - H_3(jx)|)$ (○), and open-loop gain $G_o(jx) = 20\log(G_3(jx))(\Delta)$ and the phase Ψ of the $G_3(jx)$ of the third-order PLL of type 2; $\kappa = 0.3$ and $\zeta = 1.5$.

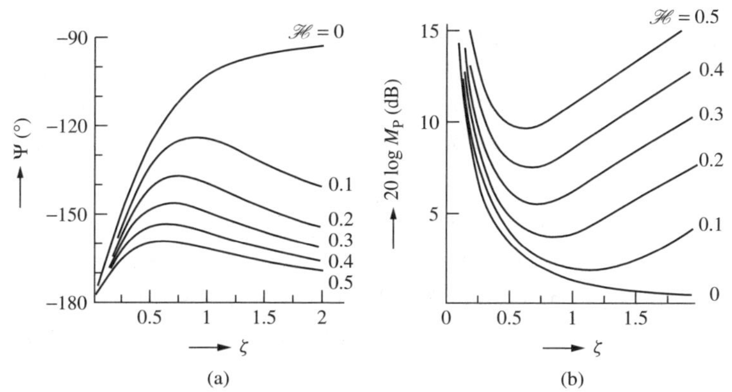

Figure 3.13 Properties of the third-order PLL for different damping constants of the original second-order loop for different κ of the additional RC section: (a) phase of the open-loop gain and (b) magnitude of the overshoot M_p of the transfer function $10\log(|H_3(jx)|^2)$.

3.2.2 Third-order Loop: Second-order Lag Filter Plus RC Section

With the assistance of the filter transfer function (3.5), we have for the open-loop gain of this third-order PLL

$$G_{3,RC}(s) = \frac{K}{s(1 + sT_1)(1 + \mu sT_1)} \tag{3.52}$$

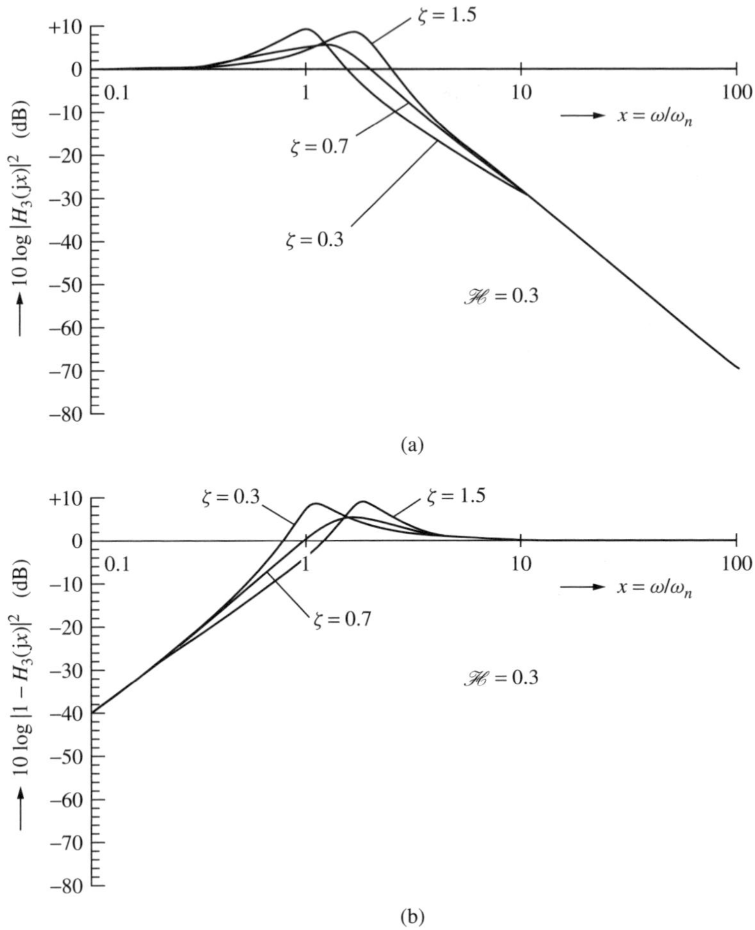

Figure 3.14 Transfer factions of the third-order PLL for three different damping factors ζ of the original second-order PLL for the constant $\kappa = 0.3$: (a) for $10\log(|H_3(\mathrm{j}x)|^2)$ and (b) for $10\log(|1 - H_3(\mathrm{j}x)|^2)$.

After introduction of ω_n and ζ from (2.10) and application of (2.13), we arrive at the normalized loop gain

$$G_{3,\mathrm{RC}}(\mathrm{j}x) = \frac{1}{\mathrm{j}x(\mathrm{j}x + 2\zeta)} \cdot \frac{2\zeta}{(\mathrm{j}x\mu + 2\zeta)} \tag{3.53}$$

The PLL transfer functions are

$$H_{3,\mathrm{RC}}(\sigma) = \frac{1}{\sigma^3\mu/2\zeta + \sigma^2(1+\mu) + 2\zeta\sigma + 1} \tag{3.54}$$

and

$$1 - H_{3,\mathrm{RC}}(\sigma) = \frac{\sigma^3\mu/2\zeta + \sigma^2(1+\mu) + 2\zeta\sigma}{\sigma^3\mu/2\zeta + \sigma^2(1+\mu) + 2\zeta\sigma + 1} \tag{3.55}$$

A typical plot of all transfer functions is shown in Fig. 3.15. Investigation of the asymptotic slopes reveals for $H_{3,\mathrm{RC}}(\sigma)$, -60 dB/dec, and for $1 - H_{3,\mathrm{RC}}(\sigma)$, only $+20$ dB/dec.

On comparison with the previous case of the third-order PLL with the lag-lead filter and one independent RC section, this arrangement reveals, for practically the same phase margin, a better attenuation of the noise and spurious signals accompanying the input frequency, that is, the reference-leaking frequency itself, its harmonics, and eventual mixing products generated in the PD. On the other hand, we face a lower attenuation of the VCO noise and its harmonics. The other difficulty is the same as that pointed out in Chapter 2, namely, a lower degree of freedom.

Example 3.3

Preliminary estimation of the phase margin of the above-discussed PLL, for $\omega_0/\omega_n = x_\mathrm{o} \approx 1$, would reveal

$$\Psi_{\mathrm{pm}} \approx \frac{180^\circ}{\pi}\left[-\frac{\pi}{2} - \arctan\left(\frac{1}{2\zeta}\right) - \arctan\left(\frac{\mu}{2\zeta}\right)\right] \tag{3.56}$$

and for $\zeta = 0.5$ and $\mu = (0.7\zeta) = 0.35$, we have

$$\Psi_{\mathrm{pm}} \approx 180 - 90 - 45 - 20 = +25^\circ \tag{3.57}$$

However, from Fig. 3.15, we get better values, namely, $x_\mathrm{o} = 0.7$ and $\Psi_{\mathrm{pm}} \approx 38^\circ$ and from the polar diagram, discussed in the next chapter, we find $\Psi_{\mathrm{pm}} \approx 36^\circ$ (x_o is the normalized frequency for which the open-loop gain crosses the 0-dB level).

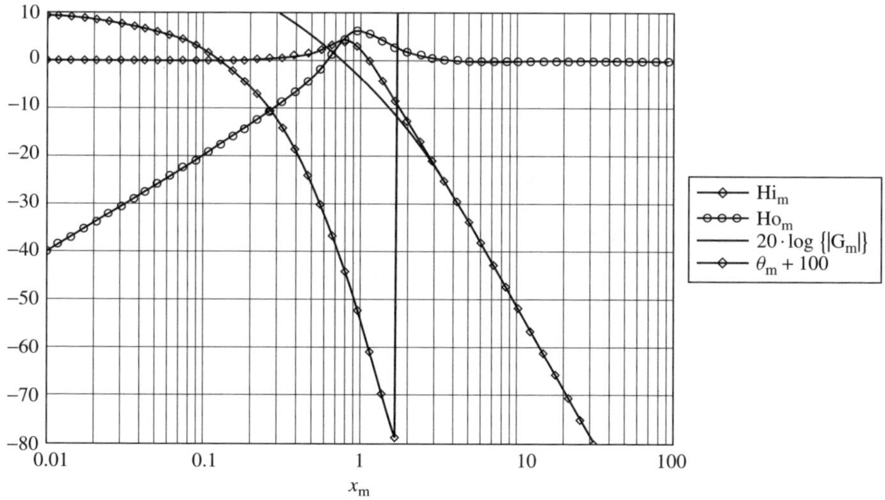

Figure 3.15 Transfer functions $H_\mathrm{i}(\mathrm{j}x) = 20\log(|H_{3,\mathrm{RC}}(\mathrm{j}x)|)$ (◇) and $H_\mathrm{o}(\mathrm{j}x) = 20\log(|1 - H_{3,\mathrm{RC}}(\mathrm{j}x)|)$ (○), open-loop gain $G_\mathrm{m} = 20\log(|G_{3,\mathrm{RC}}(\mathrm{j}x)|)$ (○), and θ_m is the phase of the open-loop gain $G_{3,\mathrm{RC}}(\mathrm{j}x)$ (○) of the third-order PLL loop based on the original second-order PLL with a simple RC; $\zeta = 0.5$ and $\mu = 0.7\zeta = 0.35$.

3.2.3 Fourth-order Loops

There are three types of fourth-order loop. In instances in which we need an increased attenuation at higher frequencies, we may intentionally add additional filter. The simplest one is the application of a mere low-pass RC section. In instances in which we need suppression of a certain spurious signal, a twin-T RC filter may provide a solution to the problem. Finally, we shall discuss the application of an active low-pass filter of the second order introduced in Section 3.1.3.

3.2.3.1 Fourth-order loop: lag-lead filter with two additional RC sections

The addition of another RC section to the third-order loop will increase the slope of $G(s)$ and $H(s)$ in the stop band to -60 dB/dec. We shall limit our discussion to the high-gain second-order loop with two independent RC sections. The open-loop gain will be

$$G_4(\mathrm{j}\omega) = \frac{K/T_1}{-\omega^2} \cdot \frac{1+\mathrm{j}\omega T_2}{1+\mathrm{j}\omega T_3} \cdot \frac{1}{1+\mathrm{j}\omega T_4} \tag{3.58}$$

After introduction of the normalized time constants

$$T_3/T_2 = \kappa; \quad T_4/T_3 = \eta; \quad T_4/T_2 = \kappa\eta \tag{3.59}$$

and original second-order loop damping factor and normalized frequency, $\mathrm{j}\omega/\omega_\mathrm{o} \approx \sigma$, we get the following transfer functions for the fourth-order PLLs:

$$H_4(\sigma) = \frac{1+2\zeta}{\sigma^4(2\zeta\kappa)^2\eta + \sigma^3 2\zeta\kappa(1+\eta)\sigma^2 + \sigma 2\zeta + 1} \tag{3.60}$$

$$1 - H_4(\sigma) = \frac{\sigma^4(2\zeta\kappa)^2\eta + \sigma^3 2\zeta\kappa(1+\eta)\sigma^2}{\sigma^4(2\zeta\kappa)^2\eta + \sigma^3 2\zeta\kappa(1+\eta)\sigma^2 + \sigma 2\zeta + 1} \tag{3.61}$$

Since η is small compared with κ, the phase margin is not considerably deteriorated. After introduction of T_2 with the assistance of (2.20) and for ω/ω_n rearrange as computation Ψ_pm for $\omega/\omega_n \approx x_\mathrm{o} \approx 1$ reveals (cf. eq. (2.20)):

$$\Phi_\mathrm{pm} \approx -\frac{180}{\pi}[-\pi + \arctan(2\zeta) - \arctan(2\zeta\kappa) - \arctan(2\zeta\kappa\eta)] \tag{3.62}$$

Example 3.4

Estimate reduction of the phase margin of the fourth-order loop for $\zeta = 0.7$, $\kappa = 0.3$, and $\eta = 0.2$. With the assistance of the last term in relation (3.62), we have to reduce the 22° from Example 3.2 by approximately $180 \arctan(0.084)/\pi = 4.8°$, that is, $\Psi_\mathrm{pm} \approx 17°$. With the assistance of the polar diagram, we find a smaller value, namely, $\Psi_\mathrm{pm} \approx 14°$.

3.2.3.2 *Fourth-order loop: lag-lead filter with the twin-T RC filter*

In instances in which we need large attenuation at a specific frequency, the addition of the twin-T RC filter, shown in Fig. 3.5, may solve the problem. Investigation of the properties of this PLL will start with the normalized open-loop gain of the second-order PLL type 2, $G_2(jx)$,

$$G_2(jx) = \frac{1 + j2\zeta x}{(jx)^2} \tag{3.63}$$

and thereafter by adding additional gain of the twin-T, $G_T(jx)$ (cf. relation (3.23))

$$G_T(jx) = \frac{1 + (jx\alpha)^2}{1 + 4jx\alpha + (jx\alpha)^2} \tag{3.64}$$

where we have introduced ratios between natural frequency and the "resonant" frequency f_{rf}, defined in (3.22), that is,

$$x = \frac{\omega}{\omega_n} \quad \text{and} \quad \alpha = \frac{\omega_n}{\omega_{rf}} \tag{3.65}$$

With the overall open-loop gain

$$G_{2,T}(jx) = G_2(jx)G_T(jx) \tag{3.66}$$

we can compute the transfer functions $H_i(jx)$ and $H_o(jx)$, which are plotted in Fig. 3.16. In the normalized form, after replacing $jx = j\omega/\omega_n$ with σ, we have

$$H_{4,T}(\sigma) = \frac{(1 + 2\zeta\sigma)(1 + \alpha^2\sigma^2)}{\sigma^4\alpha^2 + \sigma^3 2\alpha(2 + \alpha\zeta) + \sigma^2(1 + \alpha^2) + \sigma 2\zeta + 1} \tag{3.67}$$

Example 3.5

Evaluate the second-order loop type 2 with an additional twin-T filter, the resonant frequency being ten times higher than the loop frequency, that is, $\alpha = 0.1$, and the original damping factor being $\zeta = 0.5$. The plot in Fig. 3.16 reveals (for the phase margin) about 26° only. In addition, both transfer functions have peaks of about 7.8 dB, which indicates a serious underdamping. Computation of the roots of the function $1 + G_{2,T}(\sigma)$ provides

$$\text{polyroots (k)} = \begin{bmatrix} -38.438 \\ -1.947 \\ -0.308 + 1.114i \\ -0.308 - 1.114i \end{bmatrix}$$

Evaluation of the cosine from the complex roots will reveal for the effective damping factor $\zeta_{eff} \approx 0.27$. From Figs. 4.18 and 4.19, we read about the same values.

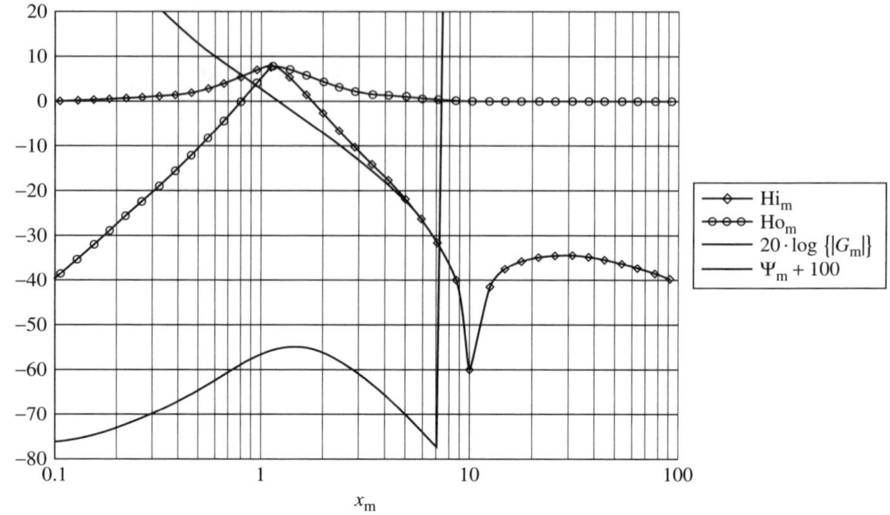

Figure 3.16 Transfer functions $H_i(jx) = 20\log(|H_{4,T}(jx)|)$ (◇) and $H_o(jx) = 20\log(|1 - H_{4,T}(jx)|)$ (○), and open-loop gain $20\log\{|G_m|\}$ (−) with its phase characteristic Ψ of the fourth-order PLL of type 2 ($\zeta = 0.5$) with the twin-T RC filter: $\omega_m/\omega_{ref} = \alpha = 0.1$.

3.2.3.3 Fourth-order loop: lag-lead filter with the second-order active low-pass filter

The problem with PLLs containing an additional twin-T filter is that they behave, outside of the notch deep, as second-order loops. This proves inspection of the relation (3.67) and of Fig. 3.16 for Fourier frequencies approximately five times higher than the zero-transmission frequency ω_{rf}.

This drawback can be eliminated by replacing the twin-T filter with the second-order active low-pass filter, introduced in Section 3.1.3, with the transfer function (3.16) rearranged with the assistance of (3.17) through (3.19) into

$$G_{AF}(\sigma) = \frac{1}{\sigma^2 + 2\sigma\alpha d + 1} \tag{3.68}$$

Its application reveals for the PLL open-loop gain

$$G_{4,AF}(\sigma) = G_2(\sigma)G_{AF}(\sigma) = \frac{1 + 2\zeta\sigma}{\sigma^2} \cdot \frac{1}{\sigma^2 + 2\alpha d\sigma + 1} \tag{3.69}$$

where α is the normalized filter "natural" frequency, in respect to ω_{nf}, and d its damping factor.

The PLL transfer functions are

$$H_{4,AF}(\sigma) = \frac{1 + 2\zeta\sigma}{\sigma^4\alpha^2 + \sigma^3 2d\alpha + \sigma^2 + \sigma 2\zeta + 1} \tag{3.70}$$

and

$$1 - H_{4,\mathrm{AF}}(\sigma) = \frac{\sigma^4\alpha^2 + \sigma^3 2d\alpha + \sigma^2}{\sigma^4\alpha^2 + \sigma^3 2d\alpha + \sigma^2 + \sigma 2\zeta + 1} \tag{3.71}$$

Inspection of the above transfer functions reveals that they are dependent on three or four different variables, that is, original ω_n and ζ of the basic second-order type 2 loop, the damping factor d, and the natural filter frequency ω_{nf}, or $\alpha = \omega_n/\omega_{\mathrm{nf}}$. In such a situation, it is not an easy task to find the optimum for the transfer functions. One of the guidelines is that for very large Fourier frequencies the slope of $|H(\sigma)|^2$ is $-60\,\mathrm{dB/dec}$. Evidently, the larger the α, the sooner the $-60\,\mathrm{dB/dec}$ slope starts. However, the stability deteriorates.

Example 3.6

Application of the Hurwitz criterion from the next chapter reveals for the second subdeterminant, $\Delta 2$, of the $P(\sigma)$ (cf. eq. (4.5))

$$\Delta 2 = (a_3 a_2 - a_4 a_1) 2\,\mathrm{d}\alpha \cdot (1 - \alpha^2) \cdot 2\zeta > 0 \tag{3.72}$$

and consequently for the stability condition

$$d > \alpha\zeta \tag{3.73}$$

From the third subdeterminant, $\Delta 3$, we get a slightly harder condition, namely,

$$d(1 - \alpha) > \alpha\zeta \tag{3.74}$$

Evidently the value of the damping factor, d, of the active low-pass filter is not critical. We arrive at the same conclusion after inspection of Fig. 3.4 that as long as α is small (in the neighborhood of 0.1), the phase margin does not deteriorate substantially even for large d. In accordance with relation (3.20), we get a deterioration of about 7° for $d = 0.6$. This is not the case with transfer functions, since even for small d, the attenuation of $H_{4,\mathrm{AF}}(\sigma)$ is reduced in the range of the Fourier frequencies, $\omega_n < \omega < 10\,\omega_n$ – see a typical computer simulation in Fig. 3.17.

3.2.4 Fifth-order Loops

In the previous section we have discussed second-order lag-lead loops transformed into the fourth-order systems. The same procedure may be applied on the third-order PLLs discussed in Section 3.2.1. Here, we shall investigate only two cases, namely, those instances in which a twin-T RC filter and second-order active low-pass filters are added.

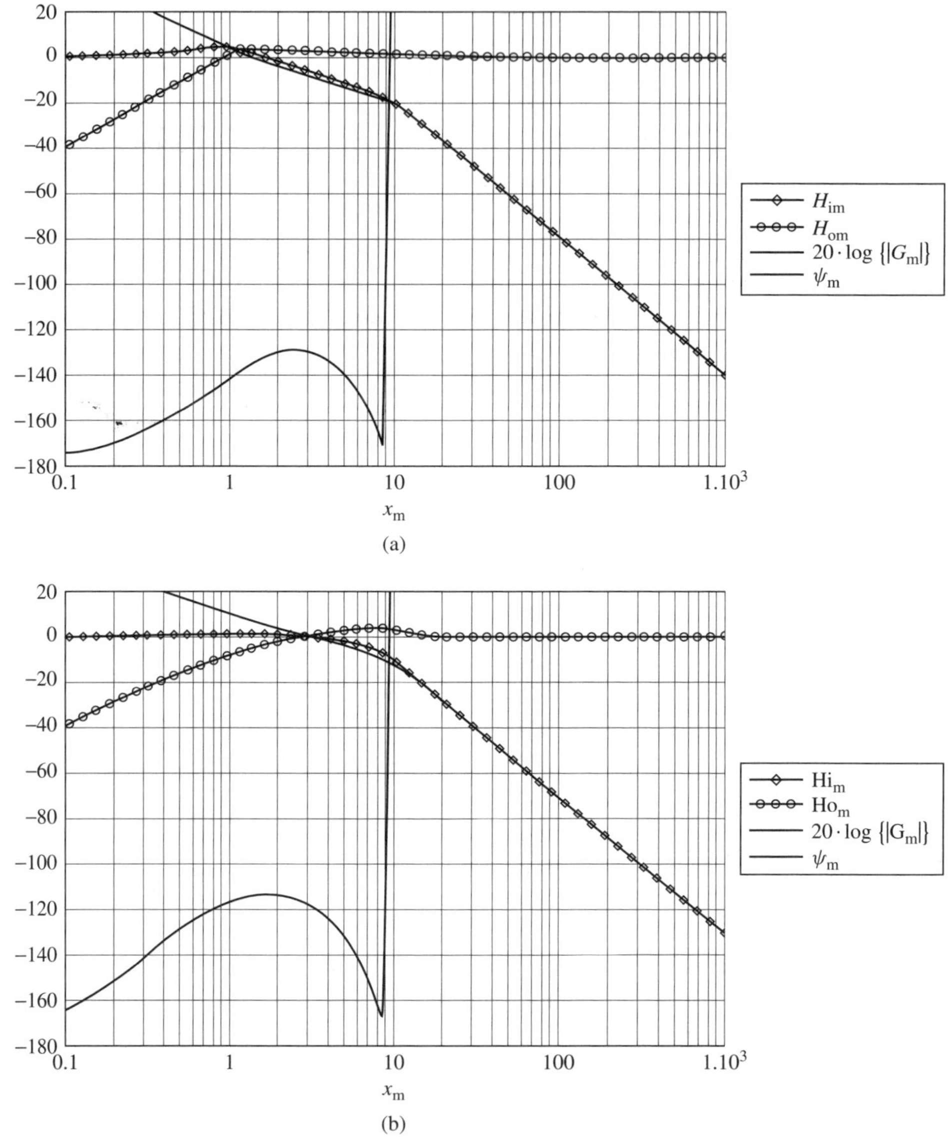

Figure 3.17 Transfer functions $H_i(jx) = 20\ \log(|H_{4,AF}(jx)|)$ (◇) and $H_o(jx) = 20\ \log(|1 - H_{4,AF}(jx)|)$ (○), and open-loop gain $20 \log\{|G_m|\}$ (−) with its phase characteristic Ψ of the fourth-order PLL of type 2: (a) $\zeta = 0.5$ and (b) $\zeta = 1.4$) with the second-order active low-pass RC filter with constants: $\alpha = 0.1$, $d = 0.6$.

3.2.4.1 Fifth-order loop with twin-T RC filter

The open-loop gain will be

$$G_{5T}(\sigma) = G_2(\sigma)G_{RC}(\sigma)G_T(\sigma) \tag{3.75}$$

with $G_2(\sigma)$ recalled in (3.63)

$$G_{RC}(\sigma) = \frac{1}{1 + 2\zeta\kappa\sigma} \tag{3.76}$$

and $G_T(\sigma)$ defined in (3.23) and (3.64). Evidently, we have

$$G_{5T}(\sigma) = \frac{1 + 2\zeta\sigma}{\sigma^2} \cdot \frac{1}{1 + 2\zeta\kappa\sigma} \cdot \frac{1 + (\alpha\sigma)^2}{1 + 4\alpha\sigma + \sigma^2} \tag{3.77}$$

and the transfer function

$$H_{5T}(\sigma) = \frac{G_{5T}(\sigma)}{1 + G_{5T}(\sigma)} = \frac{(1 + 2\zeta\sigma)(1 + \alpha^2\sigma^2)}{A} \tag{3.78}$$

where A is

$$A = \sigma^5\alpha^2\kappa 2\zeta + \sigma^4(\alpha^2 + 8\zeta\alpha\kappa) + \sigma^3(2\zeta\kappa + 4\alpha + 2\zeta\alpha) + \sigma^2(1 + \alpha^2) + 2\zeta\sigma + 1 \tag{3.79}$$

For $\kappa = 0$ we have the already discussed fourth-order loop (cf. (3.67)). Several transfer functions are plotted in Fig. 3.18, compared with those in Fig. 3.16, in which we can see advantages of the fifth-order loops. As long as the original second-order loop ζ is in the range

$$0.5 < \zeta < 1, \quad \alpha \approx 0.03, \quad \kappa \approx 0.2 \tag{3.80}$$

the overshoot will not exceed 5 to 7 dB. However, the difficulty lies with a rather small phase margin, particularly in instances in which a time delay might be present.

Example 3.7

Evaluate the phase margin of the above fifth-order PLL when $\zeta_{orig} = 0.5$ and the constant $\alpha = 0.03$. The plot in Fig. 3.19 reveals for the fifth-order loop the phase margin

$$\Psi_{pm} \approx 32^\circ$$

However, the additional time delay, $\omega_n\tau \approx 0.1$, will reduce the phase margin substantially. With the assistance of Fig. 3.6, we expect a reduction of about 11°. Actually, we find (e.g., by applying the polar plot solution)

$$\Psi_{pm} \approx 19^\circ$$

in good agreement with the expectation.

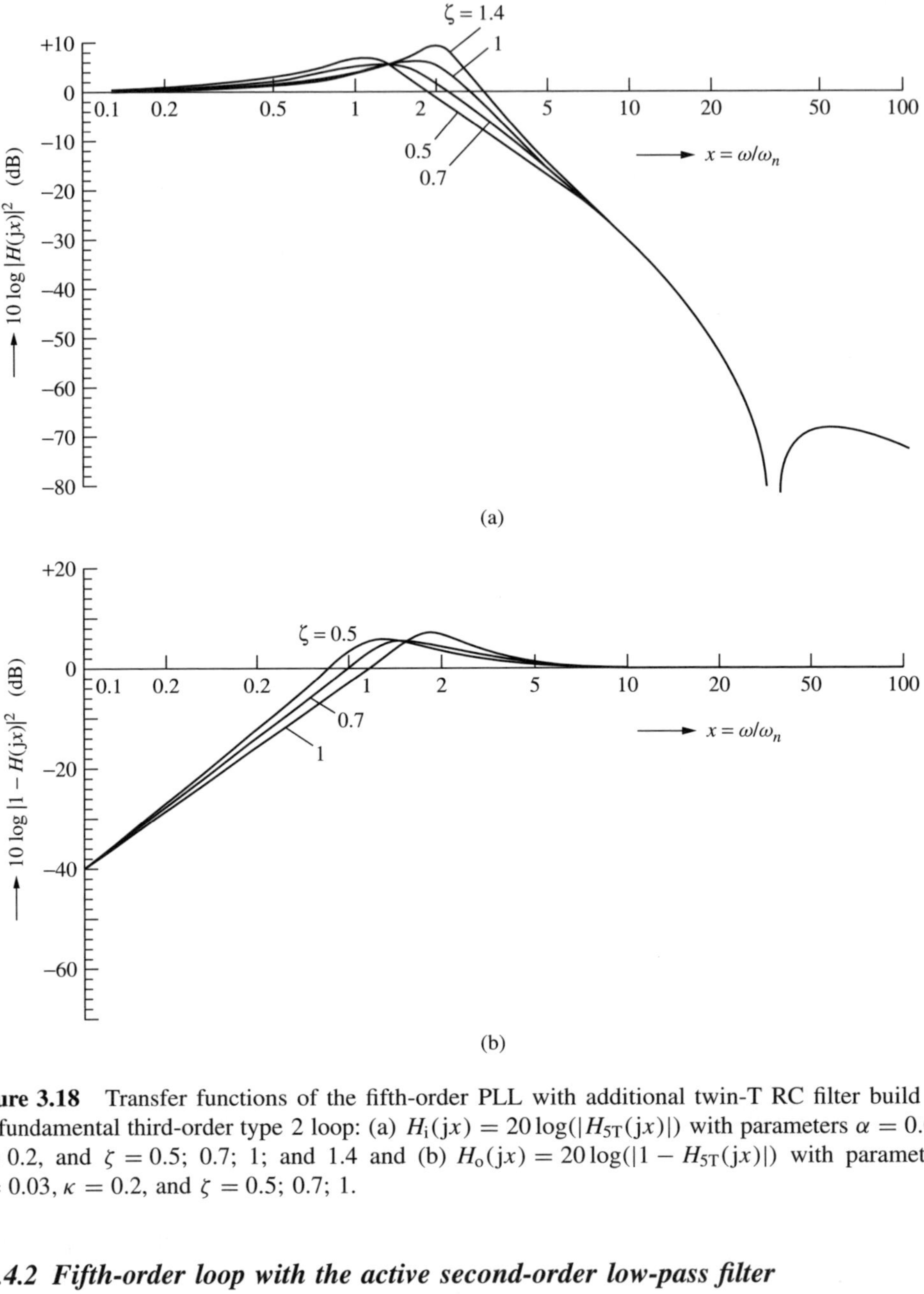

Figure 3.18 Transfer functions of the fifth-order PLL with additional twin-T RC filter build on the fundamental third-order type 2 loop: (a) $H_i(jx) = 20\log(|H_{5T}(jx)|)$ with parameters $\alpha = 0.03$, $\kappa = 0.2$, and $\zeta = 0.5$; 0.7; 1; and 1.4 and (b) $H_o(jx) = 20\log(|1 - H_{5T}(jx)|)$ with parameters $\alpha = 0.03$, $\kappa = 0.2$, and $\zeta = 0.5$; 0.7; 1.

3.2.4.2 Fifth-order loop with the active second-order low-pass filter

As in the previous section, we shall proceed with the case in which the twin-T is replaced with a second-order active low-pass filter, and we get

$$H_{5,\mathrm{AF}}(\sigma) = \frac{1 + \sigma 2\zeta}{\sigma^5\alpha^2 2\zeta\kappa + \sigma^4(\alpha^2 + 4d\zeta\kappa\alpha) - \sigma^3(2d\alpha + 2\zeta\kappa) - \sigma^2 + \sigma 2\zeta + 1} \tag{3.81}$$

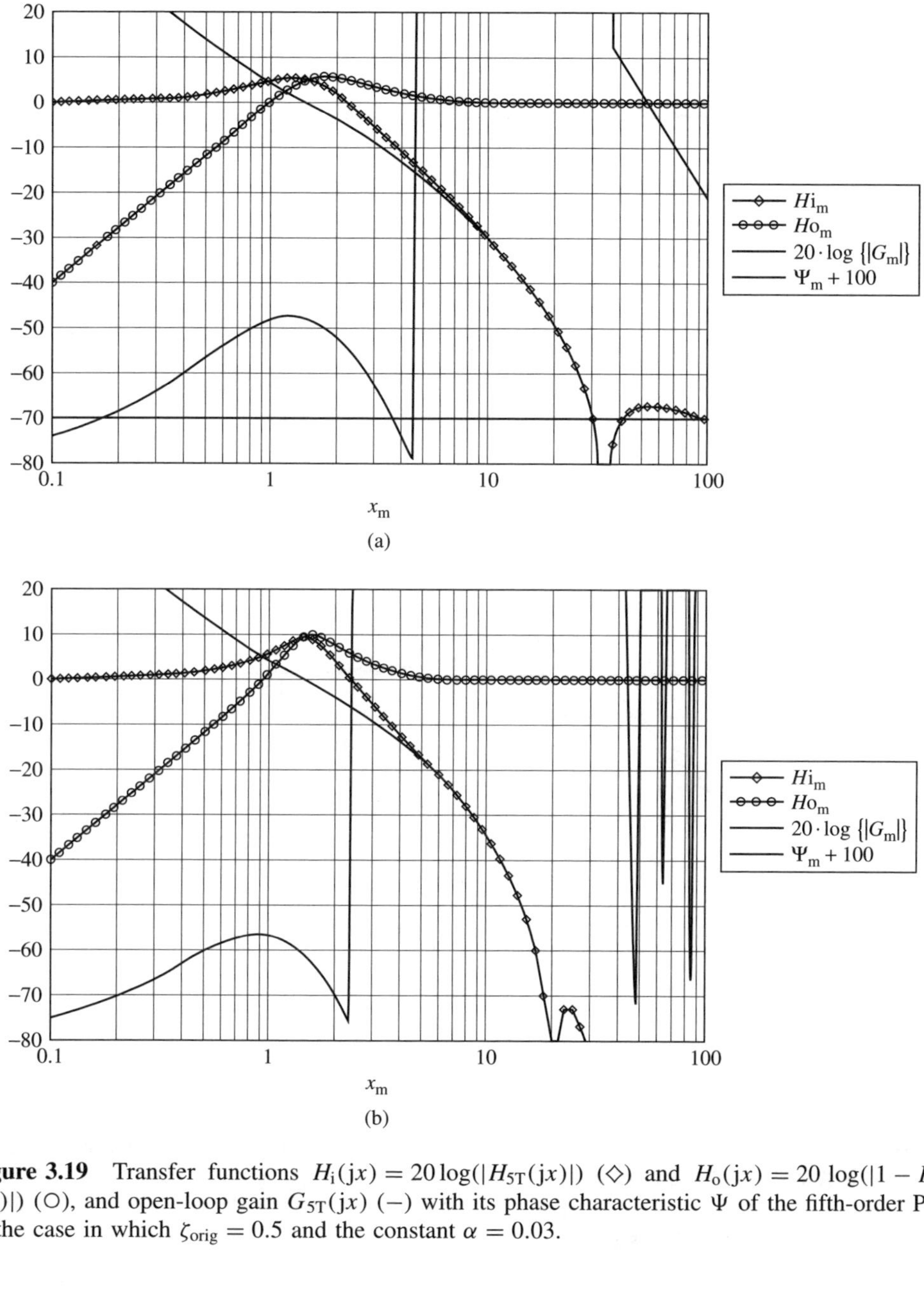

Figure 3.19 Transfer functions $H_i(jx) = 20\log(|H_{5T}(jx)|)$ (◇) and $H_o(jx) = 20\ \log(|1 - H_{5T}(jx)|)$ (○), and open-loop gain $G_{5T}(jx)$ (−) with its phase characteristic Ψ of the fifth-order PLL in the case in which $\zeta_{orig} = 0.5$ and the constant $\alpha = 0.03$.

These PLLs were discussed by Rohde [3] in connection with PLL frequency synthesizers. Here, we shall investigate properties of the transfer functions $|H_{5,AF}(jx)|^2$ and $|1 - H_{5,AF}(jx)|^2$ (see Fig. 3.20). The advantage is a very steep slope in the stop band of $|H(jx)|^2$ – in the range of Fourier frequencies $10 < \omega < 100$ – about −80 dB/dec. The difficulty is a small phase margin as the following example proves.

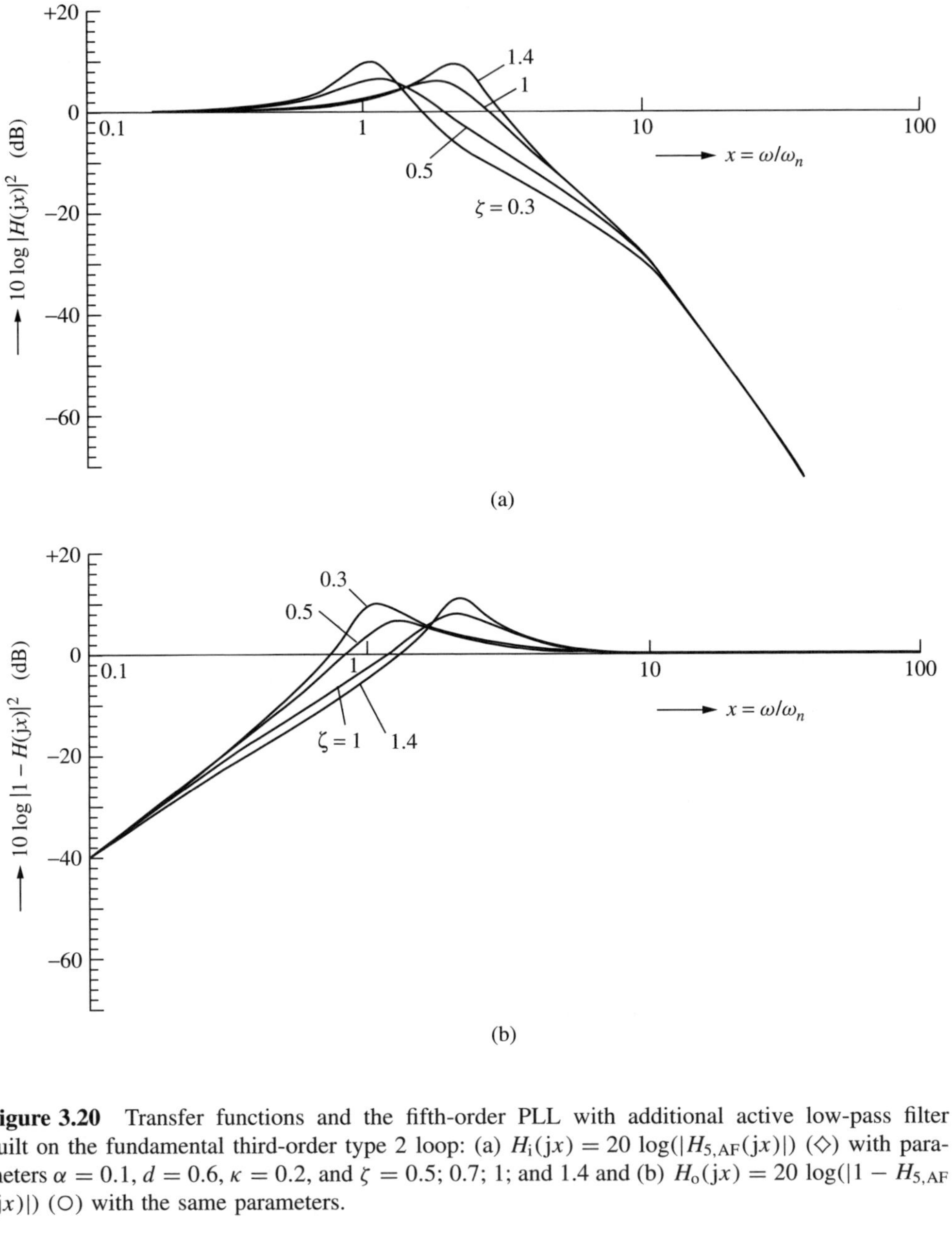

(a)

(b)

Figure 3.20 Transfer functions and the fifth-order PLL with additional active low-pass filter built on the fundamental third-order type 2 loop: (a) $H_i(jx) = 20\log(|H_{5,AF}(jx)|)$ (◇) with parameters $\alpha = 0.1$, $d = 0.6$, $\kappa = 0.2$, and $\zeta = 0.5; 0.7; 1;$ and 1.4 and (b) $H_o(jx) = 20\log(|1 - H_{5,AF}(jx)|)$ (○) with the same parameters.

Example 3.8

Plot the phase margin of the fifth-order PLL with an active second-order low-pass filter (with $\alpha = 0.1$ and $d = 0.6$) built on the fundamental third-order loop with $\kappa = 0.2$ and different damping factors, ζ, for zero time delay and for normalized time delay $\omega_n\tau = 0.5$ (see Fig. 3.21).

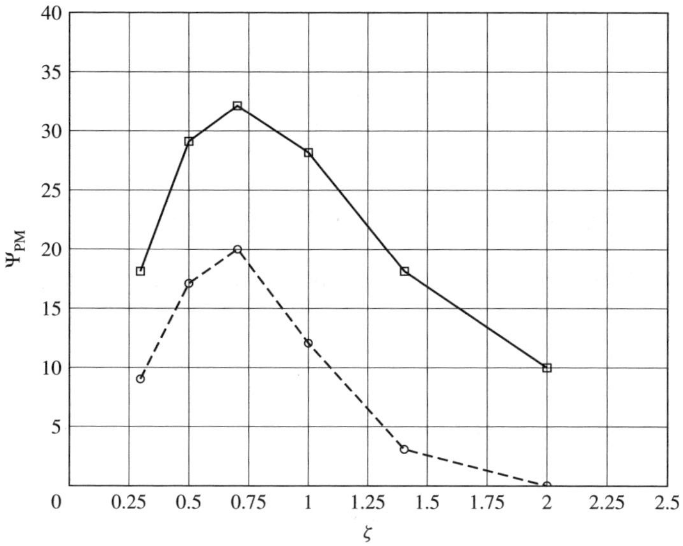

Figure 3.21 Phase margin Ψ_{pm} of the fifth-order PLL loop build on the fundamental third-order loop with $\kappa = 0.2$ and an active second-order low-pass filter as function of the damping factor ζ of the original second-order loop type 2 (□) and of the same loop with a normalized time delay $\omega_n\tau = 0.05$.

3.3 PLLs WITH TRANSMISSION BLOCKS IN THE FEEDBACK PATH

In practice, we have two important cases. Most often we encounter a divider by N in the feedback path, but in some instances it might be an intermediate filter (see Fig. 1.4).

3.3.1 Divider in the Feedback Path

The problem was discussed briefly in Chapter 1 with the assistance of Fig. 1.5(c) and relations (1.35) and (1.36). For the open-loop gain we have

$$G(s) = \frac{KF(s)}{sN} \tag{3.82}$$

We see that the actual loop gain K is reduced to the effective value $K_{ef} = K' = K/N$. Consequently in the second-order loop, in the same proportion, the natural frequency is reduced from ω_n to ω_n' or σ to σ' (or jx to jx') and we can proceed with PLL investigation as explained in Chapter 2 and in the above sections. However, there is one problem; that is, all the dividers used in today's PLLs are of the digital type and, in addition, sampling is imposed on the phase detector. Consequently, we must take into account a time delay or even modify the loop gain as in eq. (3.42). In any case, we increase the order of the PLL system and decrease the phase margin.

3.3.2 IF Filter in the Feedback Path

This problem was discussed in depth in Section 3.1.5. The result is the introduction of a time delay or of an effective RC filter section with the above-mentioned consequences.

3.3.3 IF Filter and Divider in the Feedback Path

This situation is depicted in Fig. 9.22 and the open-loop gain is as follows:

$$G(s) = \frac{KF_{\mathrm{L}}(s)}{s} \cdot \frac{1}{1 + sT_{\mathrm{IF}}} \cdot \frac{1}{N} F_{\mathrm{h}}(s) \tag{3.83}$$

where $F_{\mathrm{h}}(s)$ was defined in relation (3.39).

3.4 SAMPLED HIGHER-ORDER LOOPS

The popular phase frequency PDs are sampled systems with the consequence that the phase margin is reduced. We have shown in Section 3.1.6.2 that the situation may be solved by introducing an auxiliary transfer function F_{h} (cf. (3.39)), which for high sampling frequencies can be simplified into a simple RC filter. Evidently the order of all loops is effectively increased by one degree. Of all the possibilities, we shall investigate, here, only the third-order loop with the current output PD.

3.4.1 Third-order Loops with the Current Output Phase Detector

The difficulty with the fundamental second-order loop discussed in Section 2.3.2 is that current pulses I_{p} generate large voltage pulses on the resistor R

$$V_{2\mathrm{p}} = I_{\mathrm{p}} R \tag{3.84}$$

which are not effectively filtered. Consequently, they introduce spurious frequency modulation with the index

$$\Delta\omega = V_{2\mathrm{p}} K_{\mathrm{o}} \tag{3.85}$$

This problem will be alleviated with the third-order loop by the introduction of one of the filters shown in Fig. 3.22. Application of the arrangement Fig. 3.22(a) reveals

$$Z(s) = \frac{1 + sR(C + C_3)}{sC(1 + sRC_3)} \tag{3.86}$$

from which we find the constant κ

$$\kappa = \frac{C_3}{C + C_3} \tag{3.87}$$

that is,

$$Z(s) = \frac{1 + sT_2}{s(1 + s\kappa T_2)} \tag{3.88}$$

Example 3.9

Evaluate the loading impedance of the filter illustrated in Fig. 3.22(b),

$$Z(s) = \frac{1 + sRC}{sC(1 + C_3/C + sRC_3)} = \frac{1 + sRC}{s(C + C_3)\left(1 + R\dfrac{C_3 C}{C_3 + C}\right)} \tag{3.89}$$

Evidently we have the same relation as in (3.86).

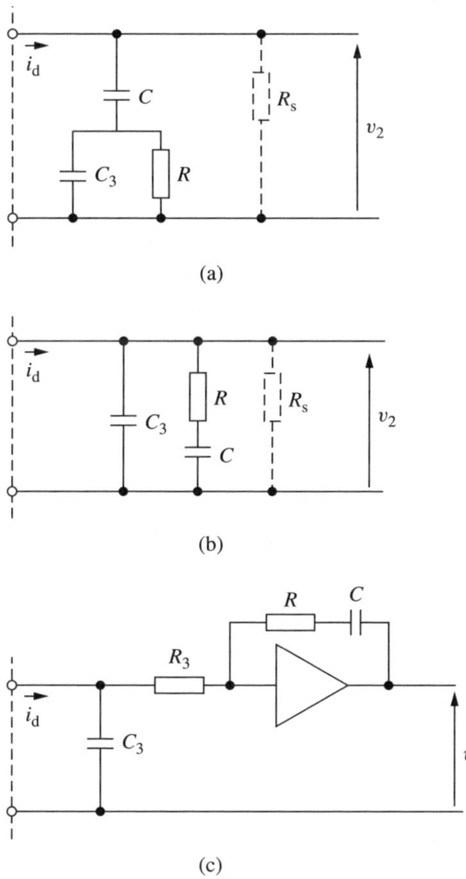

Figure 3.22 Three different filter arrangements of the passive integrator of the sampled PLLs of the third-order type 2.

For application of the normalized PLL solutions, we introduce the natural frequency ω_n and the damping factor ζ

$$\omega_n = \sqrt{\frac{K_\mathrm{d} K_\mathrm{o}}{C}} \qquad 2\zeta = \omega_n T_2 \tag{3.90}$$

and the open-loop gain

$$G_{2,\mathrm{i}}(s) = K_\mathrm{di} K_\mathrm{o} \frac{1 + sT_2}{s^2 C} \cdot \frac{1}{1 + s\kappa T_2} \tag{3.91}$$

After introduction of relation (3.91), and normalization, $s/\omega_n = \sigma$, with the modification due to the sampling (3.44), we arrive at the earlier third-order loop

$$G_{3,\mathrm{i}}(\sigma) = \frac{1}{\sigma^2} \cdot \frac{1 + 2\zeta\sigma}{1 + 2\zeta\kappa\sigma} \cdot \frac{\sinh(\sigma\delta)}{\sigma\delta} \cdot \mathrm{e}^{-\sigma\delta} \tag{3.92}$$

Example 3.10

Let us plot the transfer function characteristics of the third-order sampled PLLs for the original second-order loop with $\zeta = 0.7$ and the third-order loop $\kappa = 0.3$. From Fig. 3.23 we read the phase margin approximately 25°, whereas for the analog third-order loop we read 33° from Fig. 3.13.

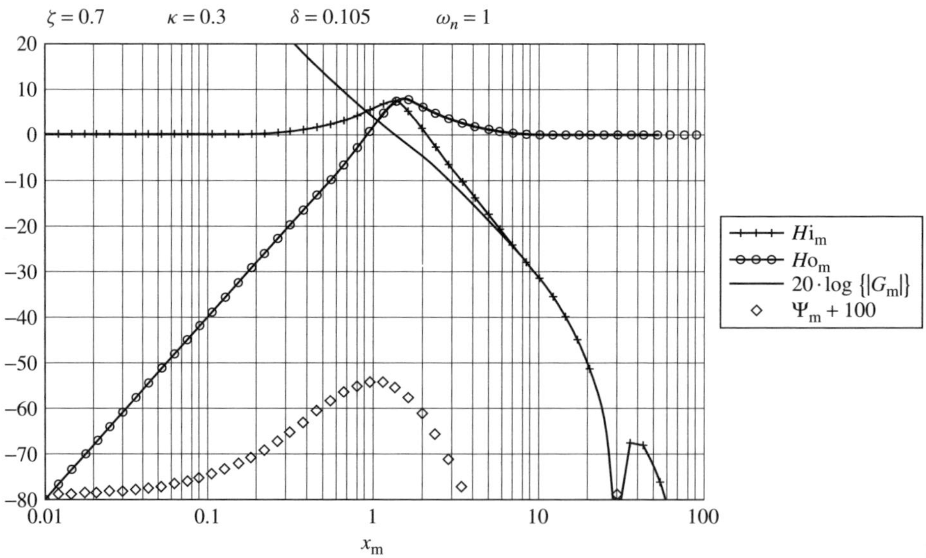

Figure 3.23 Transfer factions and phase noise characteristics of the sampled third-order PLL (for the damping factors $\zeta = 0.7$ of the original second-order PLL) having the constant $\kappa = 0.3$.

On the other hand, there is an important advantage of the third-order arrangement (3.89), that is, much smaller spurious modulation of the VCO [10]

$$\frac{|\Delta\omega_{\mathrm{o},3}|}{|\Delta\omega_{\mathrm{o},2}|} = \frac{(1-\kappa)|\phi_{\mathrm{e}}|}{\kappa\omega_{\mathrm{i}}T_2} \tag{3.93}$$

3.5 HIGHER-ORDER LOOPS OF TYPE 3

PLLs of type 3 are rarely encountered in practice [2, 11] – in standard frequency and time services, in satellite monitoring, and in a few other applications. The difficulty is with the three integrators around the loop that are responsible for the phase characteristic to start from the margin of -3π. Evidently, these loops are only conditionally stable. Gardner [2] investigated the situation with the loop filter having two equal RC sections in series. In that case the open-loop gain is (cf. relation (3.45))

$$G_{3,3}(s) = \frac{K(sT_2+1)^2}{s^3{T_1}^2} \tag{3.94}$$

With the assistance of normalization

$$sT_2 = \sigma = \mathrm{j}x \tag{3.95}$$

we arrive at

$$G_{3,3}(\mathrm{j}x) = -\gamma\frac{(1+\mathrm{j}x)^2}{\mathrm{j}x^3} \tag{3.96}$$

where

$$\gamma = KT_2\left(\frac{T_2}{T_1}\right)^2 \tag{3.97}$$

The respective PLL transfer functions are

$$H_{3,3}(\sigma) = \frac{\gamma(\sigma^2+2\sigma+1)}{\sigma^3+\gamma(\sigma^2+2\sigma+1)} \tag{3.98}$$

and

$$1 - H_{3,3}(\sigma) = \frac{\sigma^3}{\sigma^3+\gamma(\sigma^2+2\sigma+1)} \tag{3.99}$$

For the ideal integrators (i.e., for $A \to \infty$), both velocity and acceleration constant errors are very small, particularly for the mechanical systems [11]. Another advantage is in the case of the swift changes of the input frequency, since the natural frequency, ω_n, can be smaller compared with type 2 systems.

The difficulty is that PLLs of this kind are only conditionally stable. However, in instances in which the gain K, and consequently the constant γ, is small, the phase margin might even be negative and the PLL unstable. This problem will be discussed in more detail in Chapter 5. Here, we illustrate the stability situation with

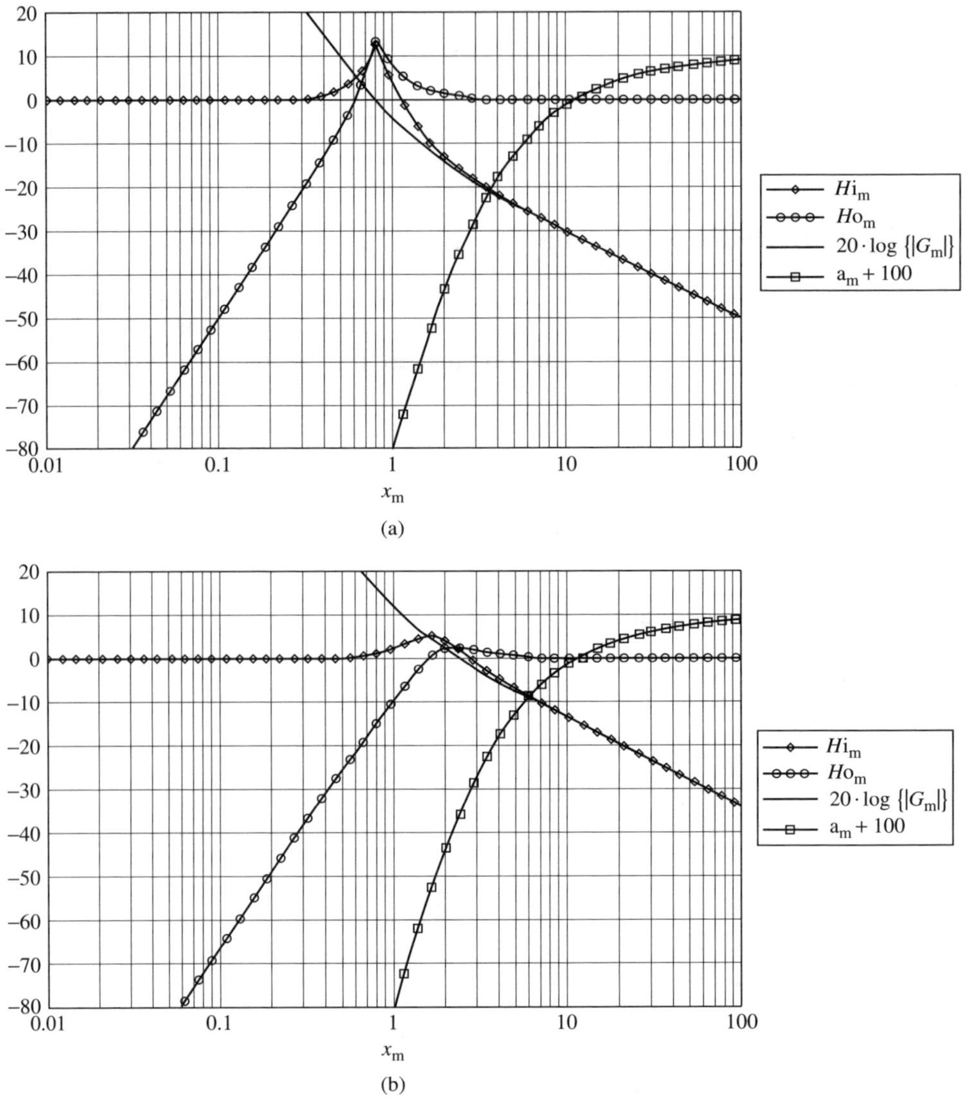

Figure 3.24 PLLs of the third-order type 3: (a) the unstable loop and (b) the conditionally stable loop with phase margin $\Psi_{pm} = 40^\circ$.

the assistance of Fig. 3.24, in which we have depicted in (a) the unstable loop with $\gamma = 0.3$ and in (b) a stable loop with $\gamma = 2$; note the phase is stable with the phase margin, $\Psi_{pm} \approx 40^\circ$.

3.6 COMPUTER DESIGN OF A HIGHER-ORDER PLL

Here, we summarize all the important formulae for solution of the higher-order PLL based on the second-order high-gain loops of type 1 or type 2 with natural frequency

ω_n and the damping factor ζ

$$\omega_{n,\text{eff}} = \sqrt{\frac{K_0 K_\text{d}}{N T_1}} \quad \text{and} \quad \zeta = \frac{\omega_{n,\text{eff}}}{2} T_2 \tag{3.100}$$

and normalization

$$\sigma = s/\omega_{n,\text{eff}} = \text{j}\omega/\omega_{n,\text{eff}} \tag{3.101}$$

The open-loop gain of the normalized second-order loop is

$$G_2(\sigma) = \frac{1 + 2\sigma\zeta}{\sigma^2} \tag{3.102}$$

The normalized open-loop gain of the third-order loop with one RC section is

$$G_3(\sigma) = G_2(\sigma) G_\text{RC}(\sigma) \tag{3.103}$$

where

$$G_\text{RC}(\sigma) = \frac{1}{1 + 2\zeta\kappa\sigma} \tag{3.104}$$

The normalized open-loop gain of the fourth-order PLL due to two additional RC sections is

$$G_4(\sigma) = G_3(\sigma) G_\text{RC}(\sigma) \tag{3.105}$$

With

$$G_\text{RC}(\sigma) = \frac{1}{1 + 2\zeta\eta\sigma} \qquad \eta < \kappa \tag{3.106}$$

we arrive at

$$G_4(\sigma) = \frac{1}{1 + 2\zeta\kappa\sigma} \cdot \frac{1}{1 + 2\zeta\kappa\sigma} \cdot \frac{1}{1 + 2\zeta\eta} \tag{3.107}$$

Extension gain to the fourth-order loop due to the twin-T is

$$G_\text{T}(\sigma) = \frac{1 + (\sigma\alpha)^2}{1 + 4\sigma\alpha + (\sigma\alpha)^2} \qquad \alpha = \frac{\omega_n}{\omega_{\text{rf},\text{T}}} \tag{3.108}$$

Extension gain to the fourth-order loop due to the active low-pass filter is

$$G_\text{AF}(\sigma) = \frac{1}{\sigma^2 + 2\sigma\alpha d + 1} \qquad \alpha = \frac{\omega_n}{\omega_\text{nf}} \tag{3.109}$$

Open-loop gain of the fifth-order loop is

$$G_{5,\text{AF}}(\sigma) = G_2(\sigma) G_\text{RC}(\sigma) G_\text{AF}(\sigma) \tag{3.110}$$

First-order loop

$$K := 1 \qquad \mathrm{m} := 0 \ldots 70 \quad \mathrm{x_m} := \frac{1.2^{\mathrm{m}}}{100} \qquad \sigma_\mathrm{m} := \mathrm{j} \cdot x_\mathrm{m}$$

$$G1_\mathrm{m} := \frac{1}{\sigma_\mathrm{m}} \qquad G_\mathrm{m} := G1_\mathrm{m} \qquad H_\mathrm{m} := \frac{G_\mathrm{m}}{1+G_\mathrm{m}}$$

$$Hi_\mathrm{m} := 20 \cdot \log(|H_\mathrm{m}|) \qquad Ho_\mathrm{m} := 20 \cdot \log(|1 - H_\mathrm{m}|) \qquad \Psi_\mathrm{m} := \frac{180}{\pi} \cdot (\mathrm{Im}(\ln(G_\mathrm{m})))$$

Second-order RC

$$\zeta := 0.7 \qquad \omega_\mathrm{n} := 1$$

$$G2_\mathrm{m} := \frac{1}{(\sigma_\mathrm{m})^2 + 2 \cdot \zeta \cdot \sigma_\mathrm{m}} \qquad G_\mathrm{m} := G2_\mathrm{m} \qquad H_\mathrm{m} := \frac{G_\mathrm{m}}{1+G_\mathrm{m}}$$

$$Hi_\mathrm{m} := 20 \cdot \log(|H_\mathrm{m}|) \qquad Ho_\mathrm{m} := 20 \cdot \log(|1 - H_\mathrm{m}|) \qquad \Psi_\mathrm{m} := \frac{180}{\pi} \cdot (\mathrm{Im}(\ln(G_\mathrm{m})))$$

Second-order RRC high gain

$$G2_\mathrm{m} := \frac{1 + 2 \cdot \zeta \cdot \sigma_\mathrm{m}}{(\sigma_\mathrm{m})^2} \qquad G_\mathrm{m} := G2_\mathrm{m} \qquad H_\mathrm{m} := \frac{G_\mathrm{m}}{1+G_\mathrm{m}}$$

$$Hi_\mathrm{m} := 20 \cdot \log(|H_\mathrm{m}|) \qquad Ho_\mathrm{m} := 20 \cdot \log(|1 - H_\mathrm{m}|) \qquad \Psi_\mathrm{m} := \frac{180}{\pi} \cdot (\mathrm{Im}(\ln(G_\mathrm{m})))$$

Third-order loop

$$\kappa := 0.3$$

$$\mathrm{Grc_m} := \frac{1}{1 + 2 \cdot \zeta \cdot \sigma_\mathrm{m} \cdot \kappa} \qquad G_\mathrm{m} := G2_\mathrm{m} \cdot \mathrm{Grc_m} \qquad H_\mathrm{m} := \frac{G_\mathrm{m}}{1+G_\mathrm{m}}$$

$$Hi_\mathrm{m} := 20 \cdot \log(|H_\mathrm{m}|) \qquad Ho_\mathrm{m} := 20 \cdot \log(|1 - H_\mathrm{m}|) \qquad \Psi_\mathrm{m} := \frac{180}{n} \cdot (\mathrm{Im}(\ln(G_\mathrm{m})))$$

Third-order loop–sampled

$$\omega_\mathrm{s} := 30 \cdot \omega_\mathrm{n} \qquad \zeta := 0.7 \qquad \kappa := 0.3 \qquad \delta := \pi \cdot \frac{\omega_\mathrm{n}}{\omega_\mathrm{s}}$$

$$\mathrm{Gt_m} := \mathrm{e}^{-\sigma_\mathrm{m} \cdot \delta} \cdot \frac{\sinh(\sigma_\mathrm{m} \cdot \delta)}{\sigma_\mathrm{m} \cdot \delta} \qquad G_\mathrm{m} := G2_\mathrm{m} \cdot (\mathrm{Grc_m}) \cdot (\mathrm{Gt_m}) \qquad H_\mathrm{m} := \frac{G_\mathrm{m}}{1+G_\mathrm{m}}$$

$$Hi_\mathrm{m} := 20 \cdot \log(|H_\mathrm{m}|) \qquad Ho_\mathrm{m} := 20 \cdot \log(|1 - H_\mathrm{m}|) \qquad \Psi_\mathrm{m} := \frac{180}{\pi} \cdot (\mathrm{Im}(\ln(G_\mathrm{m})))$$

Fourth-order loop–Twin-T filter

$$\zeta := 0.7 \qquad \alpha := 0.1$$

$$GT_\mathrm{m} := \frac{1 + (\alpha \cdot \sigma_\mathrm{m})^2}{1 + 4 \cdot \alpha \cdot \sigma_\mathrm{m} + (\alpha \cdot \sigma_\mathrm{m})^2} \qquad G_\mathrm{m} := G2_\mathrm{m} \cdot GT_\mathrm{m} \qquad H_\mathrm{m} := \frac{G_\mathrm{m}}{1+G_\mathrm{m}}$$

$$Hi_\mathrm{m} := 20 \cdot \log(|H_\mathrm{m}|) \qquad Ho_\mathrm{m} := 20 \cdot \log(|1 - H_\mathrm{m}|) \qquad \Psi_\mathrm{m} := \frac{180}{\pi} \cdot (\mathrm{Im}(\ln(G_\mathrm{m})))$$

Fourth-order loop – sampled (with active filter)

$$\alpha := 0.05 \qquad \mathrm{d} := 0.6 \qquad \zeta := 0.7$$

$$\mathrm{GAF_m} := \frac{1}{1 + 2 \cdot \alpha \cdot \mathrm{d} \cdot \sigma_\mathrm{m} + (\sigma_\mathrm{m} \cdot \alpha)^2} \qquad G_\mathrm{m} := G2_\mathrm{m} \cdot \mathrm{GAF_m} \qquad H_\mathrm{m} := \frac{G_\mathrm{m}}{1+G_\mathrm{m}}$$

$$Hi_\mathrm{m} := 20 \cdot \log(|H_\mathrm{m}|) \qquad Ho_\mathrm{m} := 20 \cdot \log(|1 - H_\mathrm{m}|) \qquad \Psi_\mathrm{m} := \frac{180}{\pi} \cdot (\mathrm{Im}(\ln(G_\mathrm{m})))$$

Figure 3.25 Mathcad program for computation of $G_{\mathrm{tot}}(\sigma)$, $H_{\mathrm{tot}}(\sigma)$, $(1 - H_{\mathrm{tot}}(\sigma))$, and Ψ_{tot}.

or

$$G_{5T}(\sigma) = G_2(\sigma)G_{RC}(\sigma)G_T(\sigma) \tag{3.111}$$

Extension gain in the presence of the time delay on the sampling process

$$G\tau(\sigma) = \frac{\sinh(\sigma\delta)}{\sigma\delta}e^{-\sigma\delta} \quad \sigma\delta = \sigma\pi\frac{\omega_n}{\omega_{samp}} \tag{3.112}$$

The total open-loop gain is

$$G_{tot}(\sigma) = \prod G_{individual}(\sigma) \tag{3.113}$$

The corresponding transfer functions are easily computed

$$H_{tot}(\sigma) = \frac{G_{tot}(\sigma)}{1 + G_{tot}(\sigma)} = \frac{\Phi_i(\sigma)}{\Phi_o(\sigma)} \tag{3.114}$$

$$1 - H_{tot}(\sigma) = \frac{1}{1 + G_{tot}(\sigma)} = \frac{\Phi_e(\sigma)}{\Phi_o(\sigma)} \tag{3.115}$$

as is the phase

$$\Psi_{tot} = \frac{180}{\pi}\arg(G_{tot}(\sigma)) \tag{3.116}$$

Example 3.11

Computation of $G_{tot}(\sigma)$, $H_{tot}(\sigma)$, $(1 - H_{tot}(\sigma))$, and Ψ_{tot} with the assistance of the Mathcad program is reproduced in Fig. 3.25.

REFERENCES

[1] W.F. Egan, *Frequency Synthesis by Phase Lock*. New York: Wiley, 1981 and 2000.

[2] F.M. Gardner, *Phaselock Techniques*. 2nd ed. New York: John Wiley, 1979.

[3] U.L. Rohde, *Digital PLL Frequency Synthesizers, Theory and Design*. Englewood Cliffs: Prentice Hall, 1983.

[4] V.F. Kroupa, "Spectral properties of third order phaselocked loops", *Proceedings of the Third Summer Symposium on Circuit Theory*, Kladna Czechoslovakia, September 1977.

[5] V.F. Kroupa, "Low-noise microwave-frequency synthesizers: Design principles", *IEE Proceedings-H*, **130**, 483–488, 1983.

[6] V.F. Kroupa, *Theory of Phase-Locked Loops and their Applications in Electronics*. Praha: Academia, 1995 (in Czech).

[7] C.J. Savant Jr., *Basic Feedback Control System Design*. New York, Toronto, London: McGraw-Hill, 1958.

[8] E. Kushnick, "Feedback Improves Notch Filter Q", *Electron. Design*, **February**, 131, 1996.

[9] B. Olsson, "New null filter replaces Twin-T", *Electron. World+Wireless World*, **February**, 119–120, 1995.

[10] F.M. Gardner, "Charge-Pump Phase-Lock Loops", *IEEE Trans. Commun.*, **COM-28**, 1849–1858, 1980 (Reprinted by W.C. Lindsey and C.M. Chie, *Phase-Locked Loops*. New York: IEEE Press, 321–330, 1986).

[11] J. Tolman, "The Czechoslovak National Standard of Frequency and Time", *Yearbook of the Czechoslovak Academy of Sciences 1967*, Praha: Academia, 127–138, 1969.

4
Stability of the PLL Systems

PLLs belong to the set of the control systems used for years. First applications were in mechanical engineering, for example, for maintaining the correct speed of steam engine shafts and so on. In this connection, the book by Savant [1] deserves to be mentioned.

However, applications in electronics, particularly in frequency synthesis systems, require some special treatment, which will be discussed below [2]. Since PLLs are feedback systems with the feedback transfer function $G(s)$, they will oscillate whenever the gain $G(s)$ is equal to -1, that is,

$$1 + G(s) = 0 \tag{4.1}$$

This condition expressed in a complex form is

$$|G(\mathrm{j}\omega)|\,\mathrm{e}^{\mathrm{j}\Psi} = -1 \tag{4.2}$$

that is,

$$|G(\mathrm{j}\omega)| = 1 \quad \text{and} \quad \Psi = (2k+1)\pi \quad (k = \pm 1, \pm 2, \ldots) \tag{4.3}$$

There are many criteria for appreciating the stability of the PLL systems. In contemporary literature it is investigated with the assistance of the simple Bode plots in accordance with the old tradition of servo systems. However, application of modern computers provides a better insight and possible perfections, particularly in cases in which current pump phase detectors are used. In this instance the classic approach might be difficult, which is not the case with computers. Furthermore, we can easily plot transfer functions $|H(s)|^2$ and $|1 - H(s)|^2$ and evaluate gain and phase margins, even for loops of higher orders.

Investigation of the transient effects in the first- and second-order loops revealed that they decayed exponentially with time. The consequence is an unconditional stability of these simple PLL systems. However, this need not be the case with higher-order loops, and the designer should investigate the stability early in the research state. Here,

Phase Lock Loops and Frequency Synthesis V.F. Kroupa
 ISBN: 0-470-84866-9

we shall repeat in short all important criteria and emphasize those suitable for use with the assistance of modern computation techniques. By taking into account that $G(s)$ is usually equal to the ratio of two polynomials in s,

$$G(s) = \frac{A(s)}{s^m B(s)} \tag{4.4}$$

In addition, $B(s)$ is generally of the same or a higher order in s than $A(s)$. Condition (4.1) results in the polynomial $P(s)$

$$P_n(s) = s^m B(s) + A(s) = a_n s^n + a_{n-1} s^{n-1} + \cdots + a_1 s + a_o \tag{4.5}$$

4.1 HURWITZ CRITERION OF STABILITY

The stability condition requires that all real parts of roots in the polynomial, $P_n(s)$, are negative. To this end we shall investigate the determinant Δ_n written in accordance with the following rules:

1. We start the first column with a_{n-1} and proceed with a_{n-3}, and so on, in rows below.
2. We start the second column with a_n and proceed with a_{n-2}, and so on, in rows below.
3. We start the third and fourth columns with zeros but further apply the first and the second columns.
4. We start the fifth and sixth columns with two zeros but further apply the first and the second columns, and so on.
5. We finish as soon as the determinant has n columns and n rows.

Then we evaluate all principal subdeterminants (minors) Δ_1; if they are all larger than zero, then the feedback system is a stable one, that is,

$$\Delta_1, \Delta_2, \ldots, \Delta_n > 0 \tag{4.6}$$

where

$$\Delta_1 = a_{n-1} \tag{4.7}$$

and

$$\Delta_2 = \begin{vmatrix} a_{n-1} & a_n \\ a_{n-3} & a_{n-2} \end{vmatrix} = a_{n-1}a_{n-2} - a_n a_{n-3} \tag{4.8}$$

$$\Delta_3 = \begin{vmatrix} a_{n-1} & a_n & 0 \\ a_{n-3} & a_{n-2} & a_{n-1} \\ a_{n-5} & a_{n-4} & a_{n-3} \end{vmatrix} = \Delta_2 a_{n-3} - a_{n-1}(a_{n-1}a_{n-4} - a_n a_{n-5}) \tag{4.9}$$

then the feedback system is a stable one.

Table 4.1 Composition of the determinant for Hurwitz criterion of stability

$$\begin{vmatrix} a_{n-1} & a_n & 0 & 0 & 0 & 0 & \ldots 0 \\ a_{n-3} & a_{n-2} & a_{n-1} & a_n & 0 & 0 & \ldots 0 \\ 0 & 0 & a_{n-3} & a_{n-2} & a_{n-1} & a_n & \ldots 0 \\ 0 & 0 & 0 & 0 & a_{n-3} & a_{n-2} & \ldots 0 \\ \ldots\ldots\ldots\ldots & & & & & & 0 \\ \ldots\ldots\ldots & & & & & & a_o \end{vmatrix}$$

Example 4.1

Let us investigate, with the assistance of the explained criterion, the stability of the second-order system. First we shall assemble determinant 4.1 from the coefficients a_1 in the denominators of eq. (2.12) or (2.22),

$$a_2 = 1; \quad a_1 = 2\zeta\omega_n; \quad a_o = \omega_n^2; \quad a_{-1}, a_{-2}, \ldots. = 0 \tag{4.10}$$

Computation of the first determinant reveals

$$\Delta_1 = a_{n-1} = 2\zeta\omega_n \tag{4.11}$$

Since both the damping factor and the natural frequency are positive constants, condition (4.6) is always met for the subdeterminant Δ_1. What remains is the evaluation of the determinant Δ_2.

$$\Delta_2 = a_{n-1}a_{n-2} - a_n a_{n-3} = 2\zeta\omega_n\omega_n^2 \tag{4.12}$$

Again the determinant Δ_2 is positive and we learn that PLLs of the second order are unconditionally stable.

Example 4.2

In the second example we shall investigate the stability conditions of the third-order PLLs discussed in Chapter 3 in Section 3.2.2. From relation (3.54) we have

$$P(\sigma) = \sigma^3 \frac{\mu}{2\zeta} + \sigma^2(1+\mu) + \sigma 2\zeta + 1$$

evidently

$$a_3 = \frac{\mu}{2\zeta}; \quad a_2 = 1 + \mu; \quad a_1 = 2\zeta; \quad a_o = 1$$

The subdeterminants are

$$\Delta_1 = 1 + \mu; \quad \Delta_2 = (1+\mu)2\zeta - \frac{\mu}{2\zeta}; \quad \Delta_3 = \Delta_2 a_o = \Delta_2$$

The stability condition follows from

$$\Delta_2 > 0 \quad \text{that is} \quad \zeta > \frac{1}{2}\sqrt{\frac{\mu}{1+\mu}}$$

4.2 COMPUTATION OF THE ROOTS OF THE POLYNOMIAL $P(s)$

The Hurwitz criterion of stability loses its lucidity and simplicity for PLLs of higher orders. The problem is worsened in instances in which some terms in (4.4) are parameters of other variables. Another difficulty is that we cannot gain any information about the effective damping, that is, we are not informed about the *distance* to the stability limit. As a result of this drawback, we do not know the roots of the polynomial $P_n(s)$.

Application of a suitable computer program for computation of the roots of the polynomials will provide us with numerical values. We shall inspect real parts of all roots – if all are negative, the investigated PLL system is stable.

Example 4.3

In the following example we shall investigate a third-order loop of type 2 with the additive RC section. Since

$$H(\sigma) = \frac{A(\sigma)}{A(\sigma) + \sigma^m B(\sigma)} = \frac{A(\sigma)}{P(\sigma)}$$

we get $P(\sigma)$ from eq. (3.49):

$$P(\sigma) = \sigma^3 2\zeta\kappa + \sigma^2 + \sigma 2\zeta + 1$$

With the Mathcad *Polyroots* we get (for $\zeta = 0.7$, $\kappa = 0.3$):

$$\text{polyroots }(k) = \begin{bmatrix} -1.241 \\ -0.57 + 1.262\text{j} \\ -0.57 - 1.262\text{j} \end{bmatrix}$$

Note that all the real parts of the roots are negative; consequently the loop is stable.

Example 4.4

Next we shall investigate a third-order loop of type 3, with two integrating lag-lead filters in PLL forward path. With the assistance of the relations (3.99) and (4.5), we get for $\gamma = 2$ and the second-order loop $\zeta = 0.5$

$$P_3(s) = s^3 + 2s^2 + 4s + 2 \tag{4.13}$$

By application of the *Polyroots Mathcad*, we immediately get the roots

$$s_1 = -0.639$$

$$s_{2,3} = -0.681 \pm \text{j}1.633$$

Since real parts of all roots are negative, this third-order loop is stable (see also Fig. 4.3).

4.3 EXPANSION OF THE FUNCTION $1/[1+G(s)]$ INTO A SUM OF SIMPLE FRACTIONS

Investigation of the function $1/[1+G(s)]$ reveals that it is equal to the ratio of two polynomials $R(s)/S(s)$, where for realizable networks the order of the polynomial $R(s)$ is smaller than that of $S(s)$.

4.3.1 Polynomial $S(s)$ Contains Simple Roots Only

In this case we have

$$\frac{1}{1+G(s)} = \frac{R(s)}{S(s)} = \frac{R(s)}{(s-s_1)(s-s_2)\ldots(s-s_n)} \tag{4.14}$$

where $s_1, s_2, \ldots s_n$ are roots of the polynomial $S(s)$. Next, application of the tables with Laplace transform pairs (e.g., Tab. 10.1 or [3]) provides a solution in the time domain. Since the polynomial $R(s)$ is of a lower order than $S(s)$, the above relation can be changed into a sum of simple fractions with constants in the numerators, that is,

$$\frac{R(s)}{(s-s_1)(s-s_2)\ldots(s-s_n)} = \frac{K_1}{s-s_1} + \frac{K_2}{s-s_2} + \cdots + \frac{K_n}{s-s_n} \tag{4.15}$$

With the assistance of the computed roots s_r, we can evaluate the respective constants K_r

$$K_r = \frac{R(s)(s-s_r)}{S(s)}\Big|_{s=s_r} \tag{4.16}$$

and after application of the inverse Laplace transform we get the time domain solution

$$K_r e^{s_r t} = \mathscr{L}^{-1}\left[\frac{K_r}{s-s_r}\right] \tag{4.17}$$

This procedure is also valid for a simple root in the origin, that is, $s_r = 0$. A practical solution with the assistance of a computer is presented in Chapter 5 in connection with Example 5.4.

4.3.2 Polynomial $S(s)$ Contains a Pair of Complex Roots

The situation is complicated in cases with pairs of complex conjugate roots since the respective K_r will also be complex and the inverse transform will reveal either decreasing or increasing “sine” waves.

4.3.3 Polynomial $S(s)$ Contains Multiple-order Roots

In this case, we encounter another difficulty presented with the multiple-order roots in the polynomial $S_n(s)$. By assuming that

$$(s - s_r)^m = 0 \tag{4.18}$$

expansion of $R(s)/S(s)$ has additional terms

$$\frac{K_1}{s - s_1} + \frac{K_2}{s - s_2} + \cdots + \frac{K_{r,m}}{(s - s_r)^m} + \frac{K_{r,m-1}}{(s - s_r)^m} + \cdots \frac{K_{r,1}}{s - s_r} + \cdots + \frac{K_n}{s - s_n} \tag{4.19}$$

where

$$K_{r,m} = \frac{(s - s_r)^m R(s)}{S(s)|_{s=s_r}}; \quad K_{r,m-1} = \frac{\mathrm{d}}{\mathrm{d}s} \frac{(s - s_r)^m R(s)}{S(s)|_{s=s_r}} \tag{4.20}$$

Derivation is repeated until all constants $K_{r,m} \ldots K_{r,1}$ are evaluated. For the inverse transform into the time domain we use the following relation

$$\mathcal{L}^{-1}\left[\frac{1}{(s - s_r)^m}\right] = \frac{t^{m-1} \mathrm{e}^{s_r t}}{(m - 1)!} \tag{4.21}$$

Note again that in instances in which s_r contains a negative real part, the corresponding time function approaches zero for large t.

4.4 THE ROOT-LOCUS METHOD

The root-locus method of the function $1 + G(s)$ is intended to find the location of the respective roots in the complex plane [1, 2]. The advantage of this approach is information about the location of the roots in the complex plane. In the past a set of rules was devised for finding, at least, the approximate position or direction of the position of the roots. Nowadays, our situation is much simpler since the computer solution of the polynomial of $P_n(s)$, with the changing parameter K or any other, provides us with a set of roots, which can thereafter be plotted in the complex plane.

Nevertheless, we feel that a little information about the basic definition and rules would be useful:

zero is designated by that variable s for which the gain $G(s) = 0$, that is, for which the numerator of $G(s)$ is zero;

pole is designated by that variable s for which the denominator of $G(s)$ is zero, that is, for which both $G(s)$ and $1 + G(s)$ are nearing infinity;

root is designated by that variable s for which $1 + G(s) = 0$.

Theorem 1 Branches of the root-locus plot start in each pole of $G(s)$ for the gain $K = 0$ and end in *zeros* for $K \to \infty$.

Example 4.5

The problem of the root locus will be illustrated with a simple example of the second-order loop with a lag RC filter. In Chapter 2 we have found for the loop gain (2.7)

$$G(s) = \frac{K}{s(sT_1 + 1)} \tag{4.22}$$

Since the above equation has only two poles

$$s = 0 \quad \text{and} \quad s = -1/T_1 \tag{4.23}$$

both branches of root locus must end on zeros in infinity. After introducing $G(s)$ into (4.1), we arrive at the quadratic equation

$$s^2 T_1 + s + K = 0 \tag{4.24}$$

the roots of which are

$$s_{1,2} = \frac{-1 \pm \sqrt{1 - 4KT_1}}{2T_1} \tag{4.25}$$

For $K = 0$, we get

$$s_1 = 0; \quad s_2 = -1/T_1 \tag{4.26}$$

In agreement with Theorem 1, as long as $4KT_1 < 1$, both roots are real and negative and move along the zero axis. As soon as $4KT_1 > 1$, the roots are complex with a constant real part and the locus proceeds as a vertical parallel with the imaginary axis in the distance $-1/2T_1$ from the origin (see Fig. 4.1).

Theorem 2 Root locus coincides with the zero axis where an odd number of poles plus zeros are found to the right of the point. Verify this statement with the assistance of Fig. 4.1.

Theorem 3 For large values of the gain K, the locus is asymptotic to the angles

$$\frac{(2k + 1)180^\circ}{P - Z}, \quad k = 0, 1, 2, \ldots \tag{4.27}$$

where P is the number of poles and Z is the number of zeros.

There exist other theorems for estimation of the locus plot; however, with application of computers they lose importance. Nevertheless, we want to mention one important property, that is, we can estimate the effective damping of the PLL from the distance of the operating point from the imaginary axis. We shall illustrate the problem with the assistance of the following example.

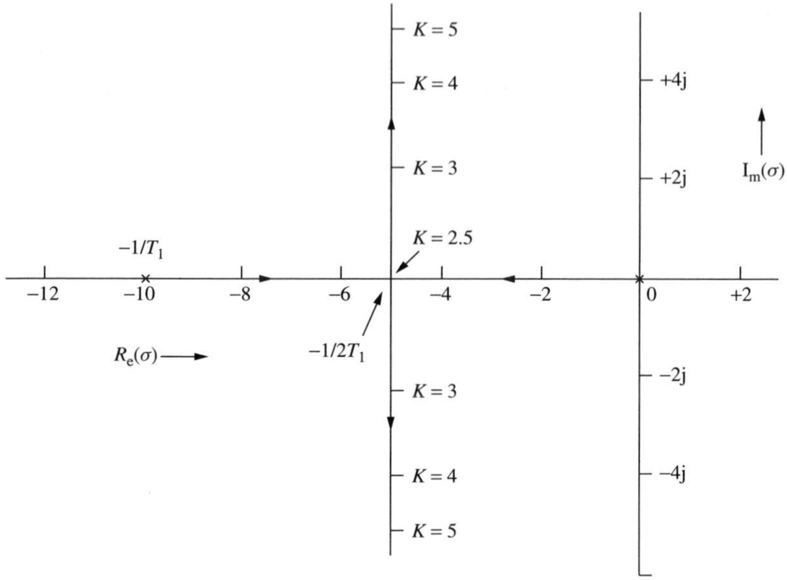

Figure 4.1 The root locus of the $1 + G(\sigma)$ for the second-order PLL with a simple RC filter.

Example 4.6

Let us plot roots of the second-order PLLs. With the open-loop gain,

$$G(s) = \frac{K_d K_A K_o(1 + sT_2)}{s(s + T_1)} \tag{4.28}$$

The system has two poles, $s = 0$ and $s = -1/T_1$, and one zero in the finite range, that is, $s = -1/T_2$. The polynomial for computation of roots is of the second order and consequently easily solved:

$$s^2 T_1 + s(1 + KT_2) + K = 0 \tag{4.29}$$

With the assistance of analytic geometry, we recognize the above equation as that of a circle with center $-1/T_1, 0$, and radius $r^2 = 1 - T_2^2 - 1/T_1 T_2$. However, we have introduced into the plot the natural frequency ω_n and the damping factor ζ instead of the gain K and the loop filter time constants (cf. (2.21)). In that case the cosine of the angle between the straight line connecting the origin with the operating point and the real axis is equal to the damping factor, as illustrated in Fig. 4.2.

$$\cos \Theta = \zeta \tag{4.30}$$

This property is important for estimation of the effective damping in instances in which we investigate PLLs of higher order (see Fig. 4.18). In this connection we point out that properties of higher-order loops might often be better appreciated from the root locus plotted in the plane $\sigma = \omega/j\omega_n$.

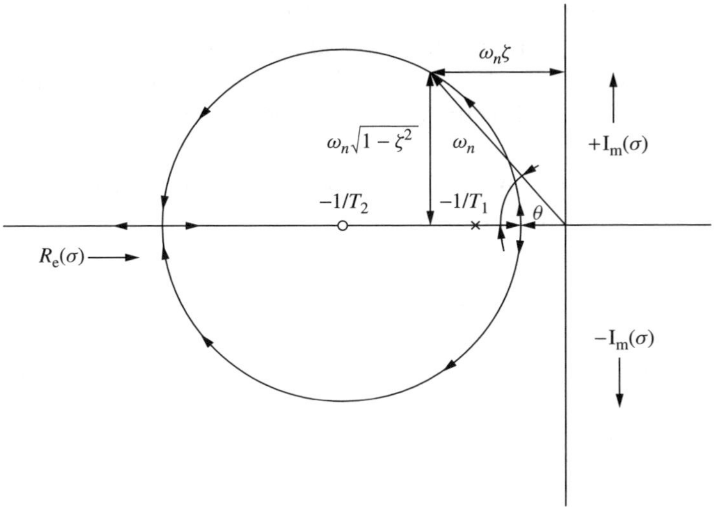

Figure 4.2 The root locus of $1 + G(\sigma)$ for the second-order PLL type 2 with the RRC loop filter.

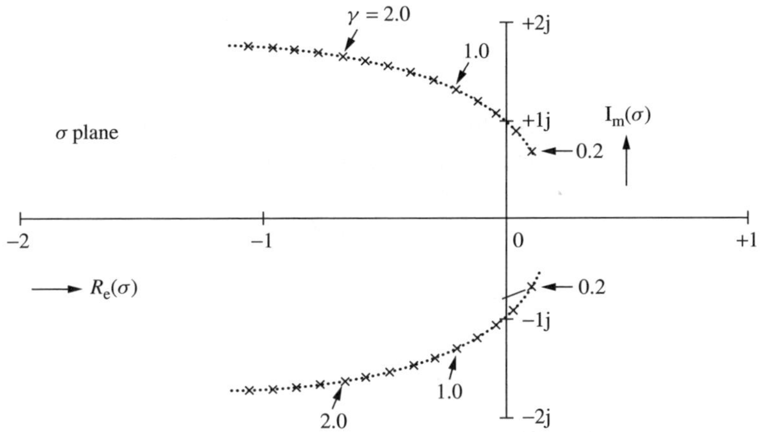

Figure 4.3 The root locus of $1 + G(\sigma)$ for the third-order PLL type 3.

Example 4.7

Another example is the root-locus plot of $1 + G(\sigma)$ for the third-order type 3 PLL, in the σ-plane with parameter γ, reproduced in Fig. 4.3. Inspection of the plot reveals that for small γs the roots are in the right-hand half plane – consequently, the system would be unstable – whereas for larger γ the loop is stable.

4.5 FREQUENCY ANALYSIS OF THE TRANSFER FUNCTIONS – BODE PLOTS

In earlier and often in contemporary literature, stability of the PLL systems is investigated with simple Bode plots in accordance with the old tradition of servo systems [2, 3]. However, now we usually replace the earlier rules with modern personal computers. They provide more insight into the problem and make possible immediate corrections at the design state. Furthermore, we can easily plot transfer functions $|H(s)|^2$ and $|1 - H(s)|^2$, even of higher-order loops, as we have seen in the previous chapter.

In the Bode plot we combine in one figure the open-loop gain in decibel measure and the respective phase shift, that is,

$$20\log(|G(\mathrm{j}x)|) \quad \text{and} \quad \frac{180}{\pi}\ \mathrm{ln}[\mathrm{I_m}(G(\mathrm{j}x)] \tag{4.31}$$

We have seen that the open-loop gain $G(s)$ or $G(\mathrm{j}x)$ consists of factors with simple transfer functions.

Generally, in respect to Tab. 4.2, the open-loop gain $G(\mathrm{j}\omega)$ or $G(\mathrm{j}x)$ is a complex number that can be changed into the following form

$$G(\mathrm{j}\omega) = KA_1A_2\ldots A_r\mathrm{e}^{-\mathrm{j}(\phi_1+\phi_2+\cdots+\phi_r+\omega\tau)} \tag{4.32}$$

4.5.1 Bode Plots

The PLL is a feedback system that will oscillate in those instances in which the denominator in the transfer function $H(s)$ is zero. In accordance with eq. (1.35) this happens when

$$1 + G(s) = 0 \tag{4.33}$$

With the assistance of (4.32), we have

$$|G(\mathrm{j}\omega)|\mathrm{e}^{\mathrm{j}\Psi} = -1 \tag{4.34}$$

To be met, the above condition requires

$$|G(\mathrm{j}\omega)| = 1 \quad \text{and} \quad \Psi = (2k+1)\pi \quad (k = 0, \pm 1, \pm 2, \ldots) \tag{4.35}$$

Table 4.2 Simple partial transfer functions encountered in PLL

1. Frequency-independent gain	$K = K_\mathrm{d}K_\mathrm{A}K_\mathrm{o}$
2. Factor with one zero in the origin	$\mathrm{j}\omega$
3. Factor with one pole in the origin	$1/\mathrm{j}\omega$
4. Factor with one zero	$1 + \mathrm{j}\omega T_0$
5. Factor with one pole	$1/(1 + \mathrm{j}\omega T_0)$
6. Time delay	$\exp(-\mathrm{j}\omega\tau)$
7. A quadratic transfer function that can be encountered both in the numerator and in the denominator	$[(\mathrm{j}\omega)^2 + 2\mathrm{j}\zeta\omega_n + \omega_n^2]^{\pm 1}$

When $|G(j\omega)| = 1$, the logarithm of the expression (4.34) results in

$$\log|G(j\omega)| = j[(2k+1)\pi - \Psi_G] \tag{4.36}$$

By introducing $k = 0$, we compute from the rhs of (4.35) the so-called *phase margin* Ψ_{pm} if the following difference is positive:

$$\Psi_{pm} = \pi - \Psi_G \tag{4.37}$$

Otherwise, for negative Ψ_{pm}, the stability condition is not met and the corresponding PLL is not a stable one. Similarly, we look for the frequency where (4.36) is equal to π and compute the so-called *gain margin*, that is, the magnitude [1, 4–6]

$$-20\log|G(j\omega)| \tag{4.38}$$

4.5.1.1 Drawing of Bode plots

We shall revert to the relation (4.32) and compute its logarithm

$$\begin{aligned}\log G(j\omega) = {} & \log K + \log A_1 + \log A_2 + \cdots + \log A_r \\ & - j(\phi_1 + \phi_2 + \cdots + \phi_r + \omega\tau)\end{aligned} \tag{4.39}$$

After plotting the rhs of the above relation in two separate graphs, we get information about the system stability. In addition, the gain characteristic $G(j\omega)$, for frequencies above ω_n, also provides information about the transfer function $H(j\omega)$ or $H(jx)$.

For construction of the Bode plots we apply the asymptotes. Here we shall repeat some basic rules for the construction.

1. Frequency-independent gain is represented by a straight line at a distance of $20\log(K)$ dB from the horizontal axis (see the plot in Fig. 4.4).
2. Factor with one zero in the origin is represented by a straight line with a slope of -20 dB/dec drawn through the zero in the horizontal axis. The phase characteristic is $+90°$.
3. Similarly, the factor with one pole in the origin is represented by a straight line with the slope of -20 dB/dec drawn through the zero in the horizontal axis. The phase characteristic is $-90°$.

   These three simple cases are illustrated in Fig. 4.4.
4. The factor with one zero in the transfer function

$$1 + j\omega T_0 \tag{4.40}$$

is composed of two asymptotes: one is the straight line in the 0-dB level and the other is the straight line with the slope $+20$ dB/dec starting from the "cut off" frequency

$$\omega_{cut,0} = 1/T_0 \tag{4.41}$$

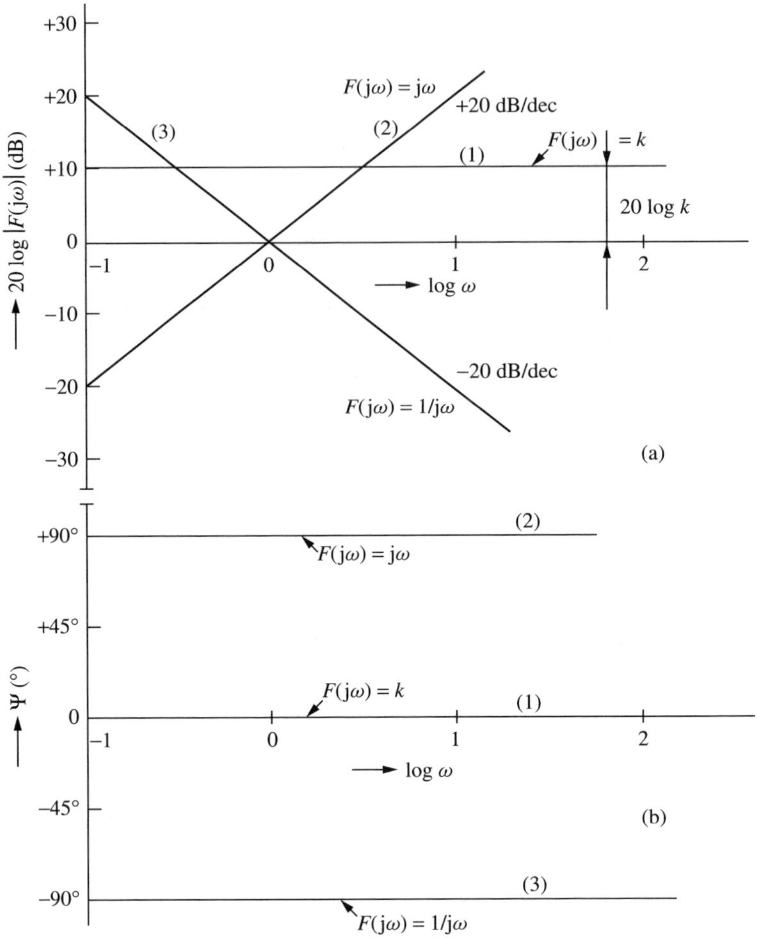

Figure 4.4 Bode plots of the first three simple transfer functions cited in Tab. 4.2: (1) frequency-independent gain $K = K_{\mathrm{d}}K_{\mathrm{A}}K_{\mathrm{o}}$, (2) characteristic with one zero in the origin, and (3) characteristic with one pole in the origin; (a) logarithm of the gain and (b) phase plots.

Note the plot in Fig. 4.5(a). The error is small with a maximum of $-3\,\mathrm{dB}$ since

$$10 \log |1 + \mathrm{j}|^2 \doteq 3 \quad [\mathrm{dB}] \tag{4.42}$$

For $\omega = 2/T_0$ or $\omega = 1/2T_0$, the error, due to the asymptotic approximations, is approximately 1 dB. The phase is

$$\Psi_0 = \frac{180}{\pi} \tan^{-1}(\omega T_0) \tag{4.43}$$

which for $\omega_{\mathrm{cut},0}$ is just $+45^\circ$. With the assistance of Taylor expansion, we find asymptotic approximation: for very low frequencies the phase is virtually zero,

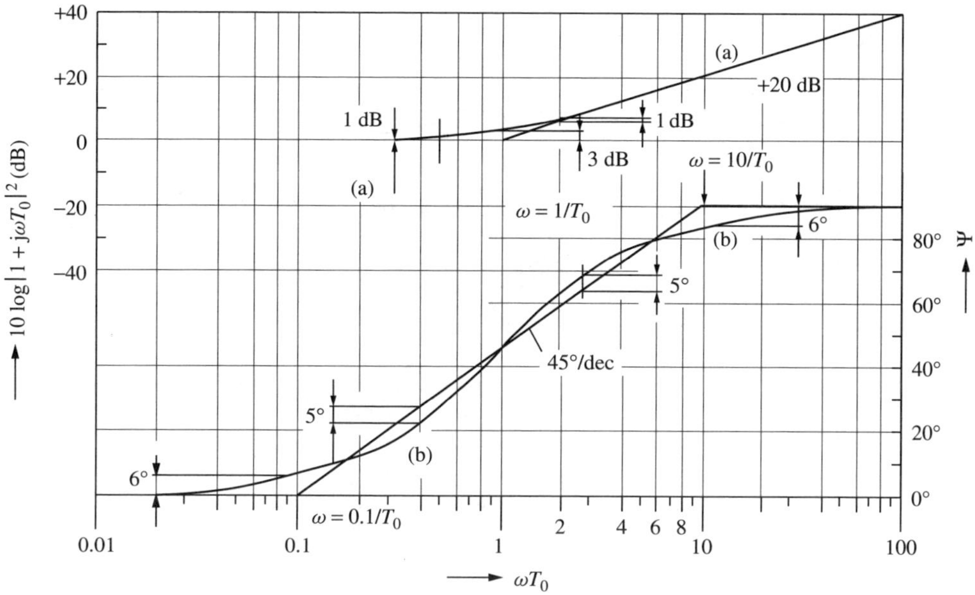

Figure 4.5 Bode plot of the transfer function $1 + j\omega T_0$: (a) amplitude characteristic and (b) phase characteristic.

whereas for very high frequencies 90° in between a straight line is plotted practically with the slope 45°/dec as shown in Fig. 4.5(b). Note that the errors are in the range of ±5° or ±6°.

5. The factor with one pole

$$\frac{1}{1 + j\omega T_p} \tag{4.44}$$

The characteristics are shown in Fig. 4.6 and are provided as mirrors of the previous cases discussed in (4).

6. The time delay τ introduces the factor

$$e^{-j\omega\tau} \tag{4.45}$$

The phase shift due to the time delay is plotted in Fig. 4.7.

7. The quadratic transfer functions are very rare in the numerator of $G(jx)$; however, they are often encountered in the denominators of $G(jx)$ of the higher-order loops discussed in Chapter 3. For drawing of the Bode plots, we shall limit our discussion to the factor

$$\frac{\omega_{nf}^2}{(j\omega)^2 + 2jd\omega\omega_{nf} + \omega_{nf}^2} = \frac{1}{1 + 2jxd - x^2} = F(jx); \quad (x = \omega/\omega_{nf}) \tag{4.46}$$

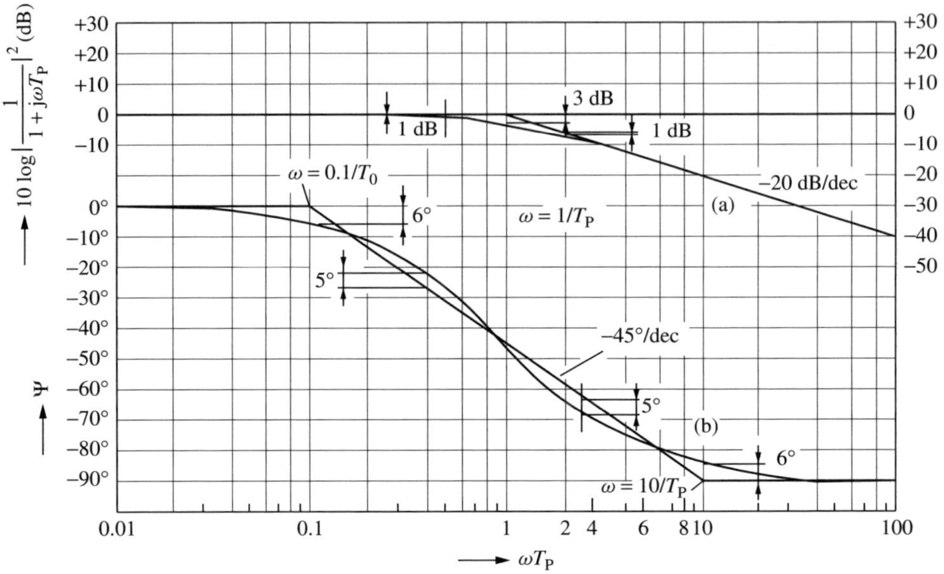

Figure 4.6 Bode plot of the transfer function $1/(1 + j\omega T_0)$: (a) amplitude characteristic and (b) phase characteristic.

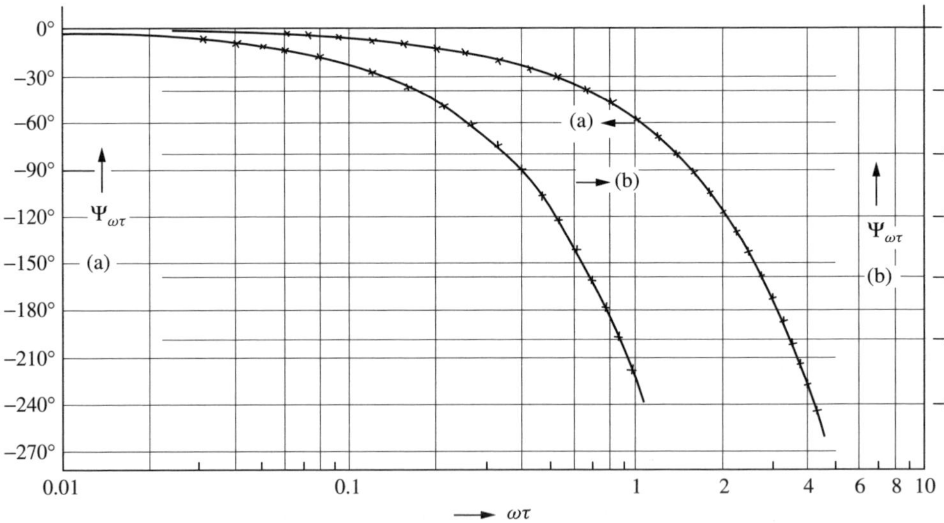

Figure 4.7 Phase shift due to the time delay.

which presents the transfer function of the active low-pass filter discussed in Chapter 3, where its transfer and phase characteristics are plotted. For the use in the Bode plot diagrams, we limit ourselves to finding the slope of $F(jx)$. For high normalized frequencies, the slope of (4.46) is -40 dB/dec and, in addition, it crosses the point $(1, 0)$

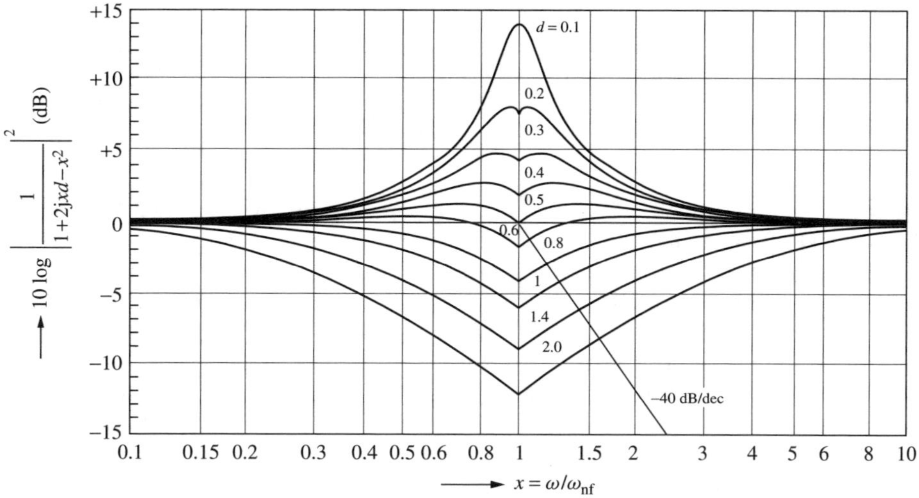

Figure 4.8 Amplitude differences of the quadratic transfer functions $|F(jx)|^2$ in eq. (4.46) for small damping factors, d (Reproduced from C.J. Savant Jr., *Basic Feedback Control System Design*. New York, Toronto, London: McGraw-Hill, 1958 by permission of McGraw Hill, 2002).

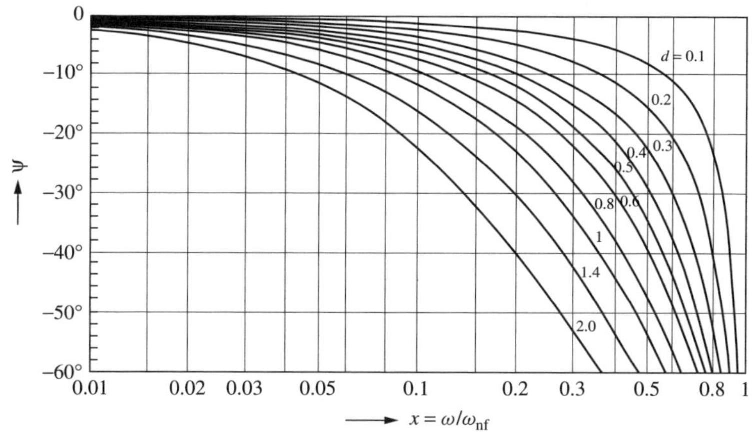

Figure 4.9 Phase characteristics of the transfer function with the quadratic term in the denominator (cf. eq. (4.46)) for different damping factors at very low normalized frequencies.

(cf. Fig. 4.8). For small damping factor, d, the amplitude characteristics of $|F(jx)|^2$ exhibit a range of overshoots, which are not negligible. In Fig. 4.8 we have plotted only differences (cf. Fig. 3.4).

We have seen, in Chapter 3, that filters additional to the basic low-pass filter in the PLL can generally introduce only small phase shifts, without any serious stability problems. Therefore, we have plotted the phase of the filter (4.46) in Fig. 4.9.

We know that nowadays the Bode plots are solved or drawn with the assistance of computers. Nevertheless, we shall provide an example for illustration.

Example 4.8

Draw the Bode plot for the popular third-order PLL with the open-loop gain

$$G(\mathrm{j}x) = \frac{1}{-x^2} \cdot \frac{1 + \mathrm{j}2\zeta x}{1 + \mathrm{j}2\zeta\kappa x} \tag{4.47}$$

where $\zeta = 0.7$ and $\kappa = 0.3$. The first factor is represented with a straight line through the point (1, 0) with the slope -40 dB/dec. The numerator line is drawn from the point $x = -\zeta$ of the first asymptote, with the slope $(-40 + 20)$ dB/dec, and finally the denominator line reverts the slope to -40 dB/dec from the point $x = +1/2\zeta\kappa$ onwards. The gain characteristic reaches zero for approximately $x \approx 1.5$. Now the phase characteristic starts as the horizontal line, $-180°$, with respect to the x-axis, it must cross the point $(-1/2\zeta, -135°)$ with the slope $+45°$/dec. Similarly, the denominator branch starts as the horizontal line, $+90°$, and crosses the point $(+1/2\zeta\kappa; -135°)$ with the slope $-45°$/dec; the difference is a parallel line with axis x in the distance of approximately $-155°$ – evidently the phase margin is about 25° (see Fig. 4.10). The computer verification of the zoomed Bode plot in Fig. 4.11 gives approximately

$$x_0 = 1.37 \quad \text{and} \quad \Psi_{\mathrm{pm}} = +33°$$

Whereas evaluation of the argument from (4.47) reveals

$$-180 + 70 - 35 = -35° \tag{4.48}$$

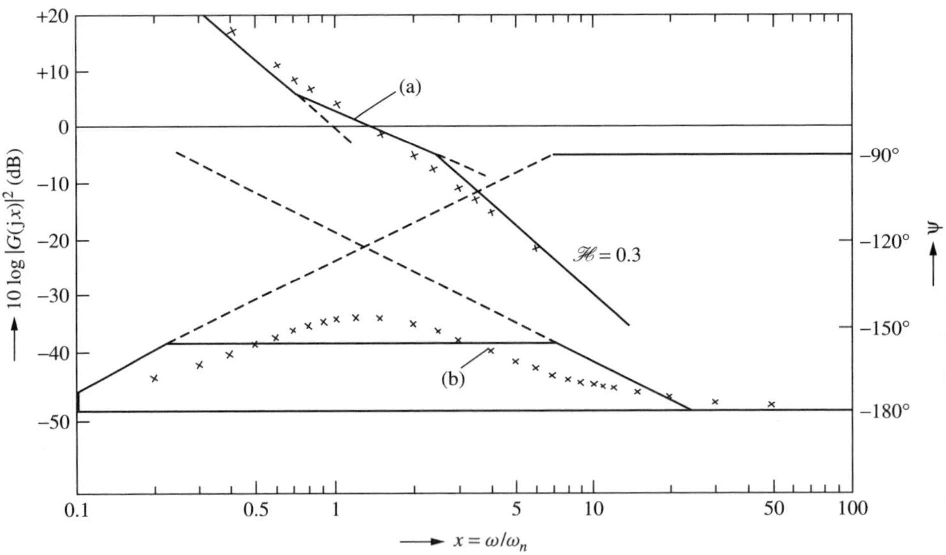

Figure 4.10 Straight line Bode plot for the popular third-order PLL with $\zeta = 0.7$ and $\kappa = 0.3$.

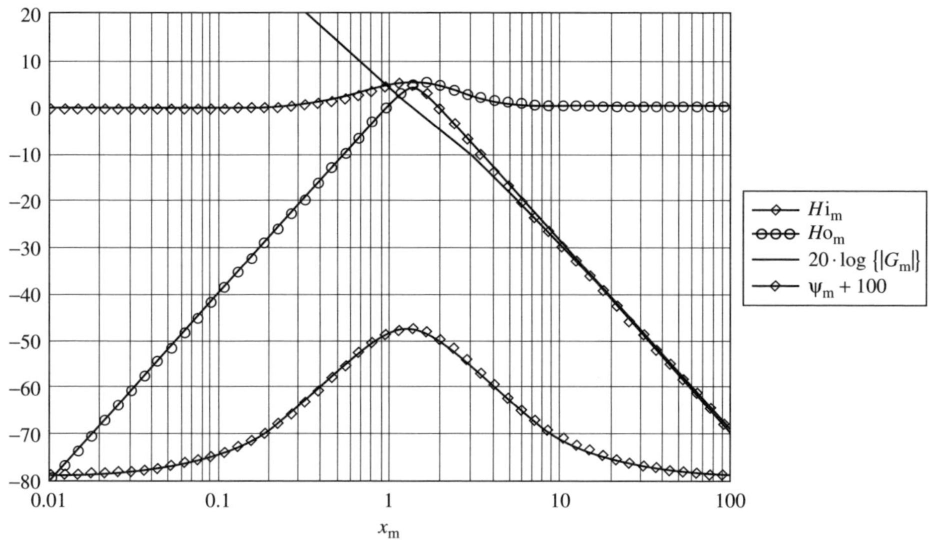

Figure 4.11 Computer plot of the open-loop gain and of the phase margin of the asymptotic version of Fig. 4.10.

4.5.2 Polar Diagrams

In Section 4.5.1, we investigated PLL stability with the asymptotic approximation of the gain and the phase of $G(\mathrm{j}x)$ in semilog diagrams. The same information provides the polar diagram in which the radius vector is the absolute value of $|G(\mathrm{j}x)|$ and the angle Ψ is plotted in the counterclockwise direction

$$G(\mathrm{j}x) = |G(\mathrm{j}x)|\mathrm{e}^{-\mathrm{j}\Psi} \tag{4.49}$$

(see the example in Fig. 4.12). We arrive at the same result by plotting, in the Cartesian coordinate system, the points (Re$[G(\mathrm{j}x)]$; $[\mathrm{I_m}G(\mathrm{j}x)]$) on the x- and y-axes. If the characteristic crosses the point -1 on the real axis, the corresponding PLL is just on the verge of stability. If the polar plot encircles the point $(-1, 0)$, the system is unstable; on the other hand, if this point remains on the lhs of the curve, the investigated PLL is stable.

The construction of the polar Bode plot starts from $-\infty$ for $x = 0$ (in the instance that the loop contains only the integrator connected with the VCO). Further, since the polynomial in the numerator of $G(\mathrm{j}x)$ is of a lower order than the polynomial in the denominator, the polar Bode plot ends in the origin. However, earlier investigation of the stability taught us that the critical range is for the normalized frequencies $x = \omega/\omega_n$

$$0.1 < x < 10 \tag{4.50}$$

To appreciate the stability, it is advantageous to draw a circle with the radius equal to 1 and the straight line rotated by -150° (see Appendix to this chapter).

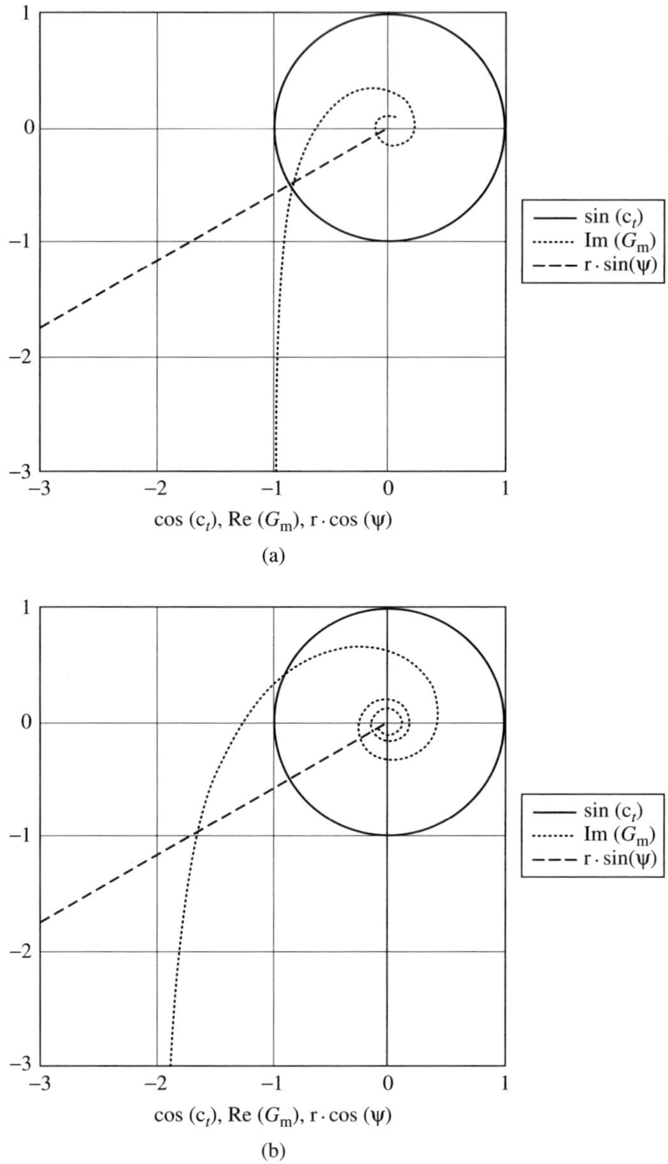

Figure 4.12 Plot of the polar diagram of the first-order PLL with a time delay: (a) $K\tau = 1$ and (b) $K\tau = 2$.

Example 4.9

Plot the polar diagram for a first-order PLL with a time delay, that is,

$$G(\mathrm{j}\omega) = \frac{K}{\mathrm{j}\omega} \cdot \mathrm{e}^{-\mathrm{j}\omega\tau} \tag{4.51}$$

normalization $\sigma = j\omega/K$ reveals

$$G(\sigma) = \frac{1}{\sigma} e^{-\sigma\tau} \tag{4.52}$$

In Fig. 4.12 we have plotted two different cases, that is, for (a) $K\tau = 1$ and (b) $K\tau = 2$. Inspection of the drawings reveals that only the first system is stable; in the second one the time delay is so large that this quasi-simple PLL system is unstable.

4.6 NYQUIST CRITERION OF STABILITY

In some cases, particularly in conditionally stable feedback systems, the Bode plots do not provide reliable information about the stability. In such instances we have to revert back to eq. (4.1), solve it, respectively solve eq. (4.5), and investigate if the signs of all real components of the roots are negative. We have discussed such a solution with an eventual plot of the root locus in Section 4.4.

However, the Nyquist criterion of stability provides a simpler solution. We shall see later that it answers the question about the existence of the roots in the rhs of the "s" plane. It is beyond the scope of this book to give a detailed derivation of this criterion; however, we shall discuss the leading ideas and explain its application with the assistance of examples.

The leading idea is a conformal mapping of the points of the "s" plane into the corresponding points of the "$G(s)$" plane. By inserting $s = j\omega$, we can easily find out that the positive half of the imaginary axis is mapped into the $G(s)$ plane as the polar frequency characteristic $G(j\omega)$, as discussed in Section 4.5. Its mirroring around the real axis is the respective negative half of the imaginary axis in the s plane (see Fig. 4.13). Next we shall investigate the mapping of the vicinity around

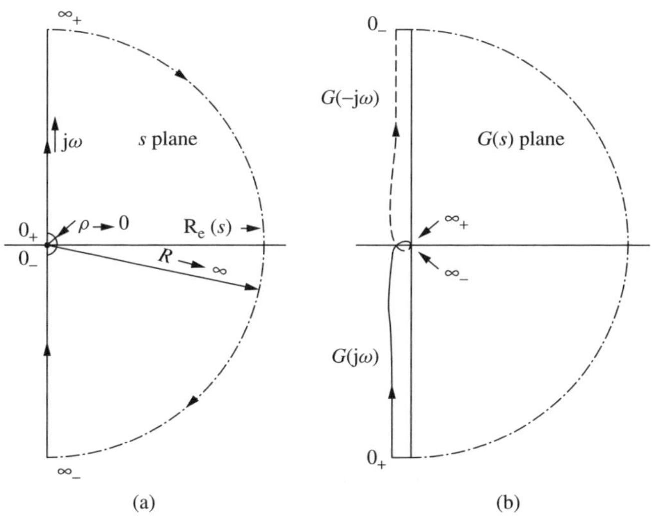

Figure 4.13 Conformal mapping of the (a) s plane (b) into the plane $G(s)$.

the origin, which in Fig. 4.13(a) is encircled with a minute radius. In accordance with relation (1.41), the corresponding vector is equal to

$$\lim_{s\to 0} G(s) = \frac{1}{s^n} R \tag{4.53}$$

where R is a complex constant not yet defined. The preceding relation can be rearranged into

$$\lim_{s\to 0} G(s) = R\left|\frac{1}{s^n}\right| \mathrm{e}^{-\mathrm{j}n\Psi} \tag{4.54}$$

Note that Ψ represents rotation around the origin in the positive sense. In accordance with Fig. 4.13(a), this rotation is just equal to π, that is, from 0_- to 0_+. This means that the limit value of $G(s)$, for $s \to 0$, in (4.54) is rotated n times, that is, $-n\pi$, clockwise by passing from 0_- to 0_+. However, the end of the vector $G(s)$ encircles the plane around a hypothetic circle in infinity, since

$$\lim_{s\to 0}\left|\frac{1}{s^n}\right| \longrightarrow \infty \tag{4.55}$$

Next we shall investigate the trajectory from $+\infty$ to $-\infty$. In the s plane, it is at such a distance from the origin that the graph in Fig. 4.13(a) encircles through "points" 0_-, 0_+, ∞_+, ∞_-, 0_- all the roots and poles with the positive real parts. From the relation (1.42) we conclude that the connecting line from $+\infty$ to $-\infty$ in the $G(s)$ plane is mapped into the origin (cf. Fig. 4.13(b)).

Now let us revert to the problem of roots of eq. (4.1) in the right-hand half of the s plane. To this end we rewrite eq. (4.1) as follows:

$$1 + G(s) = \frac{(s + s_{k1})(s + s_{k2})\ldots}{(s + s_{p1})(s + s_{p2})\ldots} \tag{4.56}$$

where $s_{k,i}$ are roots of the above relation and $s_{p,\mathrm{j}}$ are the respective poles; however, they are also the poles of the function $G(s)$. In the following we shall consider only roots and poles in the right-hand half of the s plane.

Let us choose in this plane an arbitrary point "s" (Fig. 4.14(a)), then the connecting lines $(s - s_{k,i})$ and $(s - s_{p,i})$ are complex vectors. Let us encircle one of the roots, for example $s_{k,r}$, with such a small circle that all other roots and plots remain outside. A vector with origin in this root and rotating clockwise around the circle periphery makes the whole angle of 360°; however, all other vectors connecting the remaining roots and poles after the rotation around the small circle periphery return to the original position and do not contribute to the resulting rotation.

However, mapping from the s plane into $1 + G(s_{k,r}) = 0$ maps the root $s_{k,r}$ into the origin of $1 + G(s)$ plane, and consequently the vector $1 + G(s)$ makes one rotation around the circle as illustrated schematically in Fig. 4.14(b). Similarly, we conclude that the pole after conform mapping also rotates the whole 360°, however, in the

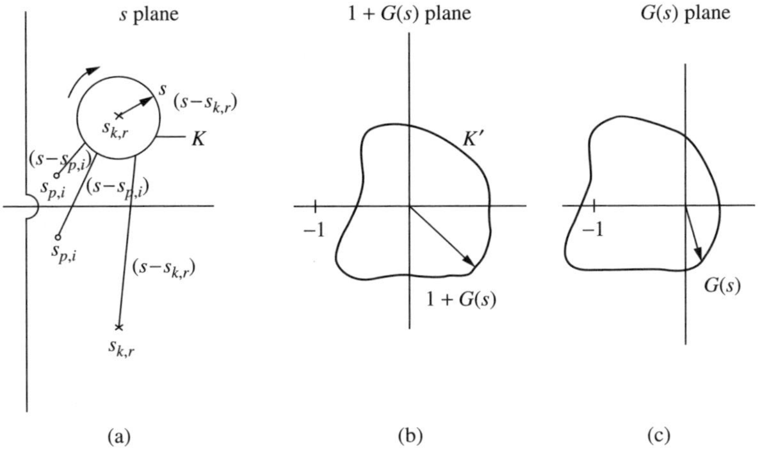

Figure 4.14 Transition from s plane to the $G(s)$ plane: (a) encircling one of the roots, for example, $s_{k,r}$, in the s plane around the curve K result into encircling; (b) of the origin in the $1 + G(s)$ plane following curve K'; and (c) finally, transition to the $G(s)$ plane (Reproduced from C.J. Savant Jr., *Basic Feedback Control System Design*. New York, Toronto, London: McGraw-Hill, 1958 by permission of McGraw Hill, 2002).

opposite sense. The result is that the number of roots, R, and poles, P, in the right-hand half of the "s" plane is equal to the number of encirclements of the origin in the $1 + G(s)$ plane, that is,

$$N = R - P \tag{4.57}$$

Transition to the $G(s)$ or $G(\mathrm{j}\omega)$ plane is provided, simply, by shifting the origin as illustrated in Fig. 4.14(c). Consequently, we arrive at the conclusion: the number of roots R with real parts is equal to the number of poles P with a real part increased about the number N of clockwise rotations of the polar characteristic $G(\mathrm{j}\omega)$ around the point (−1,0) via route, $0_+, \infty_+, \infty-, 0_-, 0_+$, that is,

$$R = N + P \tag{4.58}$$

In the introduction to this section we emphasized the fact that both numerators and denominators of the gain $G(s)$ are multiples of simple transfer functions (see Tab. 4.2). The consequence is that the number of poles with positive real parts is easily found. After plotting the transfer function $G(\mathrm{j}\omega)$ in the polar coordinates and evaluating the number of rotations, we can judge the stability in accordance with relation (4.58) without solution of $1 + G(s)$.

In Fig. 4.15 we have plotted the polar characteristic $G(\mathrm{j}\omega)$ for a stable (a) and an unstable (b) PLL of type 1 and in Fig. 4.16 for type 2. Solution of the stability of the PLL type 3 will be performed in Example 4.10.

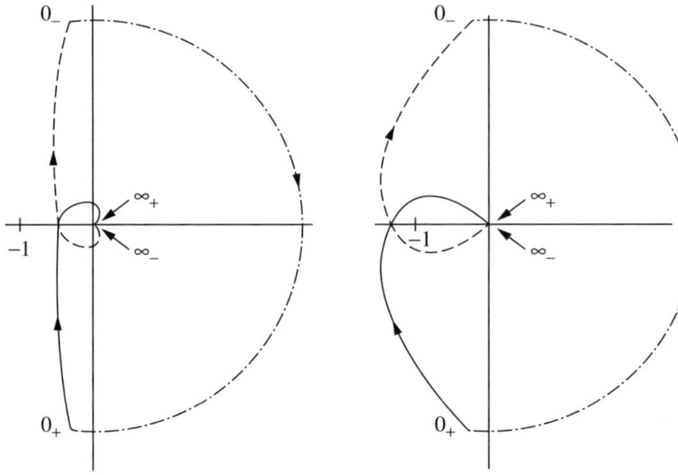

Figure 4.15 Polar characteristic $G(j\omega)$ for (a) a stable and (b) an unstable PLL of type 1.

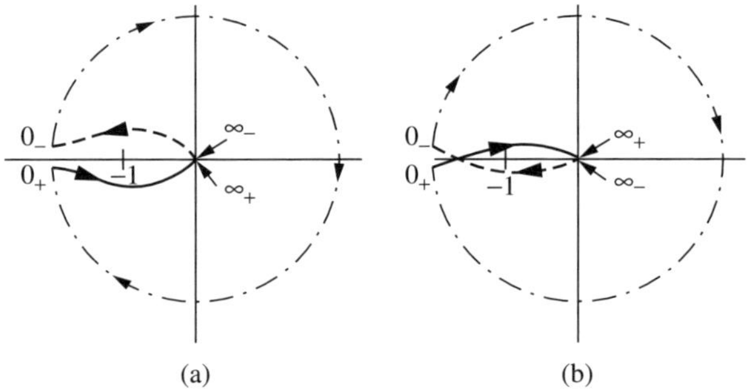

Figure 4.16 Polar characteristic $G(j\omega)$ for (a) a stable and (b) an unstable PLL of type 2.

Example 4.10

Let us examine PLLs of the third-order type 3. We shall start with the open-loop gain

$$G(s) = \frac{K(sT_2 + 1)^2}{s^3 {T_1}^2} \tag{4.59}$$

and with the assistance of normalization

$$sT_2 = \sigma = jx \tag{4.60}$$

we arrive at

$$G(jx) = -\gamma \frac{(1 + jx)^2}{jx^3} \tag{4.61}$$

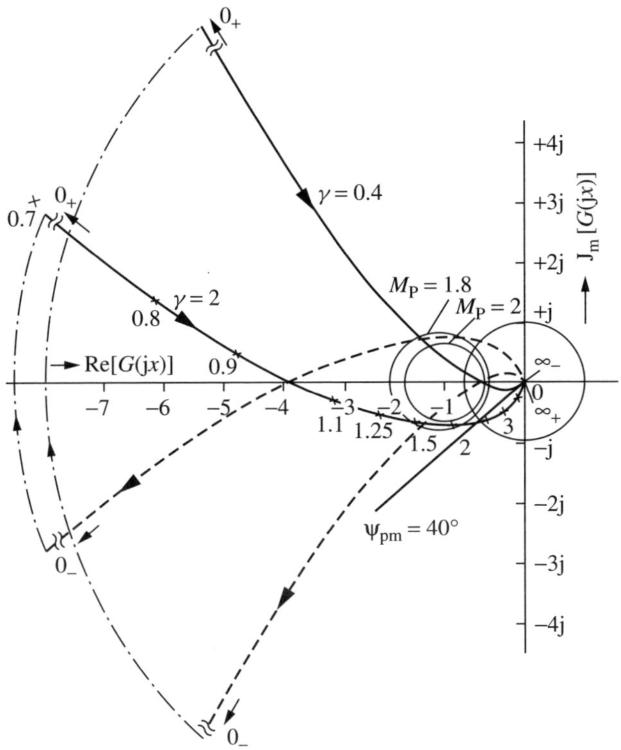

Figure 4.17 Polar characteristic of third-order PLL type 3. Note that for $\gamma = 0.4$, the characteristic encircles point $(-1, 0)$ clockwise and the feedback system is not a stable one, whereas for $\gamma = 2$, it encircles point $(-1, 0)$ anticlockwise and the system is stable (cf. Example 4.1).

where

$$\gamma = KT_2\left(\frac{T_2}{T_1}\right)^2 \tag{4.62}$$

Inspection of Fig. 4.3 reveals that the roots of eq. (4.1) are in the rhs of plane σ for a small constant γ. For $\gamma = 0.5$, the PLL would be on the verge of stability as proven, for example in the Hurwitz criterion, and for larger γ the PLL feedback system will be stable.

Example 4.11

In Fig. 4.17 the polar characteristics of $G(\mathrm{j}x)$ are plotted for two different constants γ, namely, for $\gamma = 0.4$ and $\gamma = 2$. After exclusion of the origin, the open-loop gain (4.53) has no poles, that is,

$$P = 0 \tag{4.63}$$

Inspection of the plot in Fig. 4.17 reveals that for $\gamma = 0.4$ the point $(-1, 0)$ is encircled once clockwise on the trajectory from 0_+ to ∞_+ and so on. However, after enclosing the contribution of the triple pole in the origin (i.e., -3π in accordance with (4.54) from 0_- to 0_+], we get

$$N = 2 \tag{4.64}$$

and after introduction into (4.52)

$$R = 2 + 0 \tag{4.65}$$

We shall conclude that the system must have two roots in the right-hand half of the plane in agreement with Fig. 4.3.

On the contrary, for $\gamma = 2$, the point $(-1, 0)$ is once encircled counterclockwise and once clockwise (due to the triple pole in the origin); thus $N = 0$ and also $R = 0$ and there are no roots of (4.1) in the right half of the plane and the respective PLL system is stable.

4.7 THE EFFECTIVE DAMPING FACTOR

One of the important design parameters of the PLLs of the second order is the damping constant ζ. However, with higher-order loops the situation is not so simple. Nevertheless, a good guide is inspection of two of the roots of (4.5) with the smallest negative value. By considering that the influence of all other roots might be neglected, since they are too far from the imaginary axis, we face a quasi-second-order loop, and the effective damping factor can be computed from these two roots only (compare Fig. 4.2). We find the angle Θ, and from (4.30) evaluate the expected ζ_{eff}.

Another possibility is to compare the phase margin of the investigated loop, Ψ_{pm}, with that of the second-order PLL (with the simple RC filter) plotted in Fig. 4.18 versus the damping factor ζ. Note that we arrive at good approximations in the range up to $\zeta \le 0.5$. Linearization reveals in degrees the following approximation of the damping factors for Ψ_{pm}

$$\zeta_{\mathrm{eff}} \approx \frac{\Psi_{\mathrm{pm}}}{100} \quad \text{or} \quad \zeta_{\mathrm{eff}} \approx \frac{\Psi_{\mathrm{pm}}}{360}\pi \tag{4.66}$$

Finally, we can approximate ζ_{eff} from maximum peak, M_{p}, of the transfer function $|H(\mathrm{j}x)|^2$. To this end we have plotted (in Fig. 4.19) relations between peaks of both types of second-order PLL (lag and lag-lead arrangements) and the respective damping factors. For systems with a slope of $|H(\mathrm{j}x)|^2 - 40\,\mathrm{dB/dec}$, for $x \gg 1$, we use the characteristic (Fig. 4.19(a)) derived from the PLL with a simple RC filter, whereas for $|H(\mathrm{j}x)|^2$ with slopes $-20\,\mathrm{dB/dec}$ we take the plot in Fig. 4.19(b). Note that the computer solution provides us immediately with the overshoot M_{p} (cf. Fig. 4.20).

Several examples of the evolution of ζ_{eff} are summarized in Tab. 4.3.

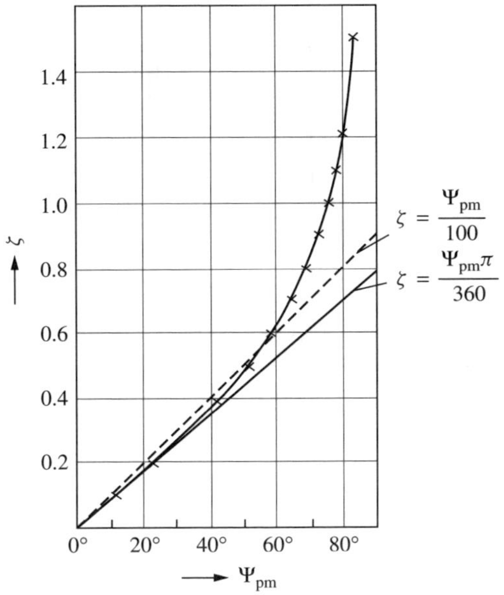

Figure 4.18 Phase margin versus damping factor ζ in PLL of the second-order with RC filter (© John Wiley & Sons, Inc., 2002).

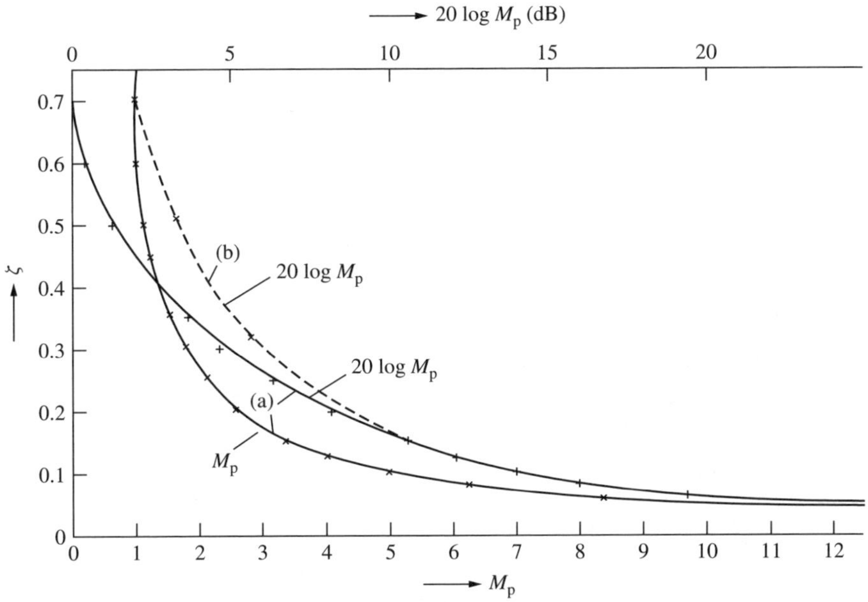

Figure 4.19 Damping factor ζ as function of the overshoots M_P for second-order PLL: (a) with simple RC filter and (b) with lag-lead or RRC filter (note the linear and decibel scales).

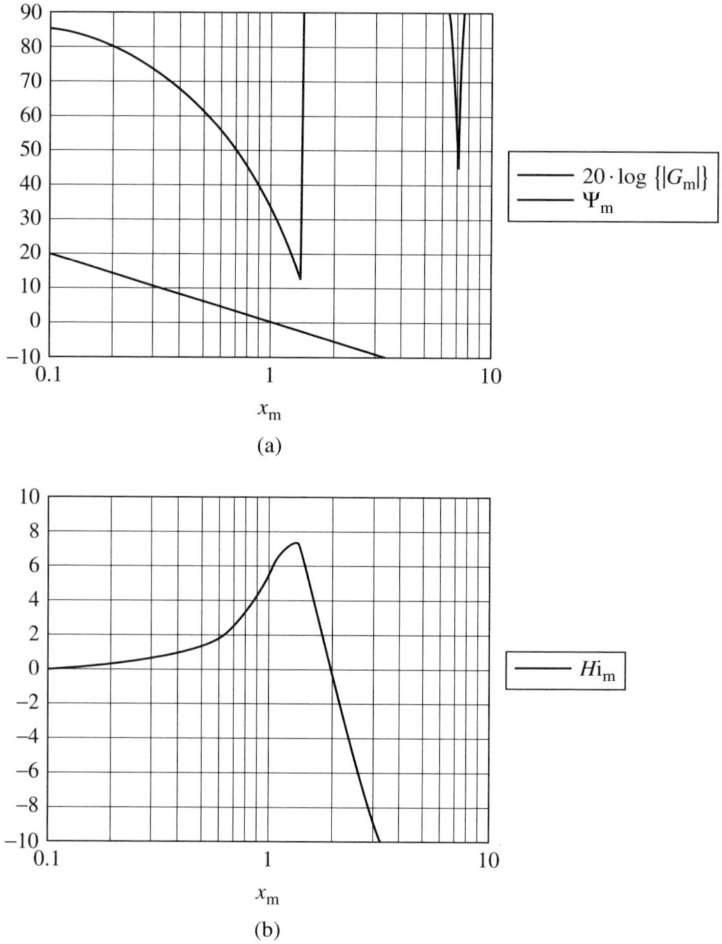

Figure 4.20 Solution of the effective damping factor ζ for the first-order PLL with a time delay: (a) with the use of phase margin and (b) with the assistance of overshoot.

Table 4.3 Comparison of the effective damping evaluated with assistance of the overshoots M_P and phase margin for three different examples.

Example			ζ_{ef}	
4.3	M_P	8[dB]	0.21	Fig. 3.12
	Ψ	24°	0.2	
4.4	M_P	5[dB]	0.25/0.27	Fig. 3.24(b)
	Ψ	21.4°	0.19	
4.12	M_P	7	0.25	Fig. 4.20
	Ψ	32°	0.28	

Example 4.12

Let us investigate the effective damping factor of the first-order loop with the time delay $K\tau = 1$ (cf. Fig. 4.12(a)). In Fig. 4.20(a) we have plotted the open-loop gain $G(\mathrm{j}x)$ and evaluated the phase margin $\Psi_{\mathrm{pm}} \approx 35^\circ$, and with the assistance of Fig. 4.18 we evaluate $\zeta_{\mathrm{eff}} \approx 0.35$. However, from Fig. 4.20(b) we have for the overshoot $M_{\mathrm{P}} \approx 7\,\mathrm{dB}$ and from Fig. 4.19 we find a smaller $\zeta_{\mathrm{eff}} \approx 0.25$.

4.8 APPENDIX

Program for computer plotting of the polar diagram with the assistance of Mathcad.

In the first line there are commands for plotting the unit circle.

In the second line we define the variable σ_{m} for $x = K/\omega$ and the effective time delay τ.

In the third line there is a definition of the first-order loop gain $G1(\sigma)$ and an additive gain of the time delay ge(σ,τ).

The program in the last line will enable us to read the phase margin by setting a slope of the straight line passing the origin and the intersection of the polar plot $G(\sigma)$ with the unit circle.

Polar diagram – Computer program (see Fig. 4.21)

$r := 0..100 \quad t := 0..1000 \quad \mathrm{c}_t := \dfrac{t}{100}$

$\mathrm{m} := 1..100 \quad x_{\mathrm{m}} := \dfrac{\mathrm{m}}{10} \quad K := 1 \quad \sigma_{\mathrm{m}} := \dfrac{\mathrm{j} \cdot x_{\mathrm{m}}}{K} \quad \tau := 1$

$G1_{\mathrm{m}} := \dfrac{1}{\sigma_{\mathrm{m}}} \quad \mathrm{ge}_{\mathrm{m}} := \mathrm{e}^{(-\sigma)_{\mathrm{m}} \cdot \tau} \quad G_{\mathrm{m}} := G1_{\mathrm{m}} \cdot \mathrm{ge}_{\mathrm{m}}$

$\psi_{\mathrm{pm}} := 30 \quad \psi\mathrm{set} := -1 \cdot \psi_{\mathrm{pm}} + 180 \quad \psi := -\pi \cdot \dfrac{\psi\mathrm{set}}{180}$

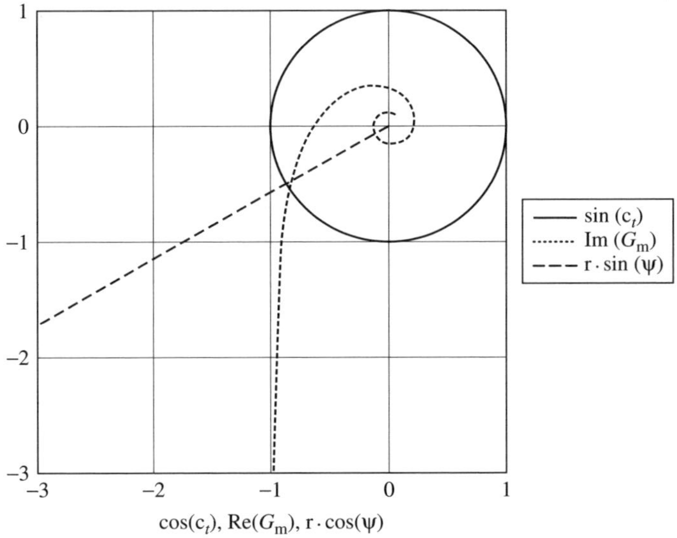

Figure 4.21 Computer program for polar plot of $G(\sigma)$.

REFERENCES

[1] C.J. Savant Jr., *Basic Feedback Control System Design*. New York, Toronto, London: McGraw-Hill, 1958.
[2] M. Gardner, *Phase-Lock Techniques*. New York: Wiley, 1966 and 1979.
[3] G.A. Korn and T.M. Korn, *Mathematical Handbook*. New York: McGraw-Hill, 1958.
[4] W.F. Egan, *Phase-Lock Basics*. New York: John Wiley and Sons, 1999.
[5] W.F. Egan, *Frequency Synthesis by Phase Lock*. 1981, 2nd ed. New York: John Wiley, 2000.
[6] V.F. Kroupa, *Theory of Phase-Locked Loops and their Applications in Electronics*. Praha: Academia 1995 (in Czech).

5 Tracking

Up to now, we have only investigated properties of PLLs in steady state conditions. However, such situations are generally not present. In practice, we encounter either wanted or unwanted frequency changes both in reference generators and more often in voltage-controlled oscillators (VCO) (mainly because of adjusting the division ratio in the feedback path). The respective changes can be divided into three major groups:

1. phase or frequency steps;
2. periodic changes (spurious phase or frequency modulations, discrete spurious signals, etc.);
3. noises accompanying both reference and VCO signals.

In this section we shall discuss the first two problems. The last one deserves special treatment and will be considered later in Chapter 9.

5.1 TRANSIENTS IN PLLs

Application of PLLs in modern communications is connected with nearly permanent carrier frequency changes. Evidently, for proper operation we need to know the duration of the switching process, how long it takes before the output frequency is settled, and how large the eventual steady state error might be. The information provides the phase difference at the output of the phase detector (PD) $\Phi_e(s)$:

$$\frac{\Phi_e(s)}{\Phi_i(s)} = 1 - H(s) \tag{5.1a}$$

or more exactly its time domain behavior $\varphi_e(t)$. To this end we shall investigate the following relation with its time response:

$$\Phi_e(s) = \Delta\Phi_i(s)[1 - H(s)] \tag{5.1b}$$

Phase Lock Loops and Frequency Synthesis V.F. Kroupa
 ISBN: 0-470-84866-9

and evaluate the time needed for setting the phase error to or below the predetermined value. To this end we shall analyze the phase lock (PL) transient effects. For the final steady state we shall evaluate (1.43).

5.1.1 Transients in First-order PLLs

Using (1.44) and (5.1b) we get for the phase step, $\Delta\phi_i$, in the Laplace notation

$$\Phi_e(s) = \frac{\Delta\phi_i}{s + K} \tag{5.2}$$

which results in the time domain in

$$\phi_{e1}(t) = \Delta\phi_i e^{-Kt} \tag{5.3}$$

Evidently for $t \to \infty$ the time error is zero.

For the frequency step we have the Laplace transform

$$\Phi_e(s) = \frac{\Delta\omega_i}{s(s + K)} \tag{5.4}$$

and the corresponding time dependance

$$\phi_{e2}(t) = \frac{\Delta\omega_i}{K}(1 - e^{-Kt}) \tag{5.5}$$

which would result in a correcting phase for $t \to \infty$

$$\phi_{e2}(t) = \frac{\Delta\omega_i}{K} \tag{5.6}$$

with the tacit assumption that $\phi_{e2}(t)$ does not exceed $\pi/2$ at any time.

Linear change of the frequency will result in agreement with (1.44)

$$\Phi_e = \frac{\Delta\dot{\omega}}{s^3} \cdot \frac{s}{s^2(s + K)} \tag{5.7}$$

The reverse transform reveals

$$\phi_{e3}(t) = \frac{\Delta\dot{\omega}_i}{K^2}(e^{-Kt} - 1 + Kt) \tag{5.8}$$

We conclude that the phase error, $\phi_{e3}(t)$, increases linearly with time for larger values of Kt and in steady state condition remains a constant frequency error.

5.1.2 Transients in Second-order PLLs

PLLs of the second order, particularly with a high gain, form the fundamental building blocks of many communication systems, and therefore we shall investigate their transient properties due to the phase, frequency, and step of acceleration (frequency ramp $\Delta\dot{\omega}_\text{i}$ (radians/s^2)) changes in some detail.

5.1.2.1 Phase step

First we shall investigate the influence of the phase step $\Delta\phi_\text{i}$ at the input of the PD $(\Delta\phi_\text{i}(s) = \Delta\Phi_\text{i}/s)$.

$$\Phi_\text{e}(s) = \frac{\Delta\phi_\text{i}}{s}[1 - H(s)] \tag{5.9}$$

To get the most general solution we shall provide a solution in the normalized form. Substituting for $[1 - H(\sigma)]$ from (2.24) in (5.9) we get

$$\Phi_\text{e}(\sigma) = \frac{\Delta\phi_\text{i}}{\sigma} \cdot \frac{\sigma(\sigma + \omega_n/K_v)}{\sigma^2 + 2\zeta\sigma + 1} \tag{5.10}$$

Solution of the quadratic equation in the denominator reveals roots

$$\sigma_{1,2} = -\zeta \pm \sqrt{\zeta^2 - 1} \tag{5.11}$$

With them we can rearrange (5.9) into

$$\Phi_\text{e}(\sigma) = \Delta\phi_\text{i}\frac{\sigma + \omega_n/K_v}{(\sigma - \sigma_1)(\sigma - \sigma_2)} \tag{5.12}$$

Application of the Laplace transform tables (see Tab. 10.1 or [1, D6–1.6]) reveals the time domain transient

$$\phi_\text{el}(t) = \frac{\Delta\phi_\text{i}}{\sigma_1 - \sigma_2}[(\sigma_1 + \omega_n/K_v)\text{e}^{(\sigma_1\omega_n t)} - (\sigma_2 + \omega_n/K_v)\text{e}^{(\sigma_2\omega_n t)}] \tag{5.13a}$$

Note that owing to the normalized frequency σ we must introduce the natural frequency ω_n into the exponents of the exponential functions.

$$\phi_\text{el}(t) = \Delta\phi_\text{i}[(K_1\text{e}^{\sigma_1\omega_n t} + K_2\text{e}^{\sigma_2\omega_n t})] \tag{5.13b}$$

Example 5.1

Another solution of the relation provides expansion into a sum of simple fractions as discussed in Section 4.3.1. With the assistance of (4.15) we get for K_1 and K_2 in (5.13b)

$$K_1 = \frac{\sigma_2 + \omega_n/K}{\sigma_2 - \sigma_1} \quad \text{and} \quad K_2 = \frac{\sigma_1 + \omega_n/K}{\sigma_1 - \sigma_2} \tag{5.14}$$

Further, we shall proceed to a solution with the assistance of (4.17).

To get a deeper insight into the problem we shall proceed with the rearrangement and finally we arrive at

$$\phi_{\mathrm{el}}(t) = \Delta\phi_{\mathrm{i}} \mathrm{e}^{-\zeta\omega_n t}\left[\cosh(\omega_n t\sqrt{\zeta^2-1}) - \frac{\zeta - \omega_n/K_v}{\sqrt{\zeta^2-1}}\sinh(\omega_n t\sqrt{\zeta^2-1})\right] \quad (5.15)$$

For the critical damping, that is, $\zeta = 1$, we get

$$\phi_{\mathrm{el}}(t) = \Delta\phi_{\mathrm{i}} \mathrm{e}^{-\omega_n t}[1 - \omega_n t(1 - \omega_n/K_v)] \quad (5.16)$$

In the case of $\zeta < 1$, the hyperbolic functions will have imaginary arguments and will change into the ordinary trigonometric functions with the result that eq. (5.15) will change into

$$\phi_{\mathrm{el}}(t) = \Delta\phi_{\mathrm{i}} \mathrm{e}^{-\zeta\omega_n t}\left[\cos(\omega_n t\sqrt{1-\zeta^2}) - \frac{\zeta - \omega_n/K_v}{\sqrt{1-\zeta^2}}\sin(\omega_n t\sqrt{1-\zeta^2})\right] \quad (5.17)$$

Normalized transients due to the phase steps in the second-order high-gain PLLs are plotted in Fig. 5.1(a) for several damping factors ζ. Note that for $\zeta < 1$ they exhibit damped oscillations that decay below 10% for about $\omega_n\tau \approx 2.5$ – (cf. [2]).

Example 5.2

Investigate the transient due to the phase step in the second-order loop with the simple RC filter. In this case we have to introduce relation (2.16) into (5.9) with the consequence that in (5.10) to (5.17) we must replace the term ω_n/K_v with 2ζ. The normalized transients $\Delta\phi_{\mathrm{el}}(t)/\Delta\phi_{\mathrm{i}}$ are plotted in Fig. 5.1(b). Note that for the larger damping factors the decay is much longer than that for the lag-lead high-gain loop.

5.1.2.2 Frequency steps

Frequency step $\Delta\omega_{\mathrm{i}}$ at the input of the PD originates the phase error that, in the Laplace notation, is

$$\Phi_{\mathrm{e}}(s) = \Delta\omega_{\mathrm{i}}/s^2 \frac{\Delta\omega_{\mathrm{i}}/\omega_n^2}{\sigma^2} \quad \text{or} \quad \Phi_{\mathrm{e}}(\sigma) = \frac{\Delta\omega_{\mathrm{i}}}{\sigma^2} \quad (5.18)$$

After introduction into (2.24) we have

$$\Phi_{\mathrm{e}}(\sigma) = \frac{\Delta\omega_{\mathrm{i}}}{\sigma^2}[1 - H(\sigma)] = \frac{\Delta\omega_{\mathrm{i}}}{\sigma^2}\cdot\frac{\sigma(\sigma+\omega_n/K)}{\sigma^2+2\zeta\sigma+1} = \frac{\Delta\omega_{\mathrm{i}}}{\sigma}\cdot\frac{\sigma+\omega_n/K}{\sigma^2+2\zeta\sigma+1} \quad (5.19)$$

The corresponding time domain transient at the PD output phase, $\phi_{\mathrm{e}}(t)$, is computed with the assistance of the roots found earlier in (5.11). With the assistance of the Laplace transform tables we find [1, D6–1.8]

$$\Phi_{\mathrm{e2}}(t) = \frac{\Delta\omega_{\mathrm{i}}}{\sigma_1-\sigma_2}\left[\frac{(\sigma_1+\omega_n/K)}{\sigma_1}\mathrm{e}^{(\sigma_1\omega_n t)} - \frac{(\sigma_2+\omega_n/K)}{\sigma_2}\mathrm{e}^{(\sigma_2\omega_n t)}\right] + \frac{\Delta\omega_{\mathrm{i}}}{K} \quad (5.20)$$

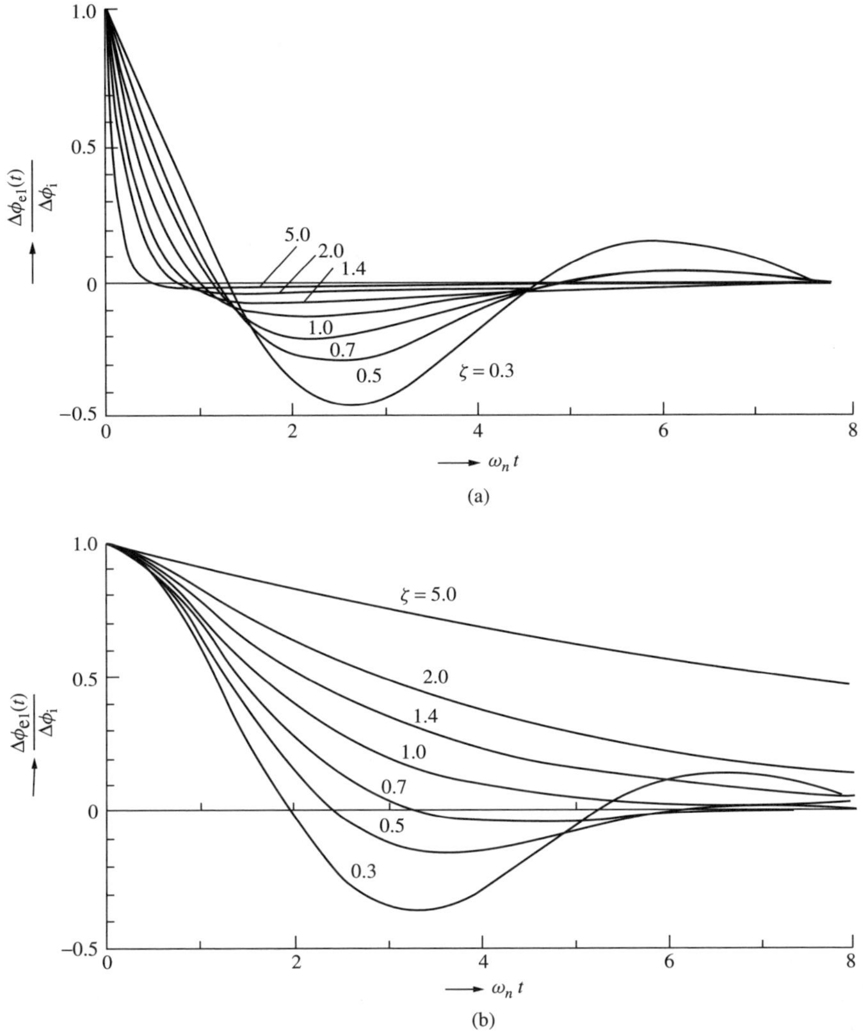

Figure 5.1 Normalized transients $\Delta\phi_{e1}(t)/\Delta\phi_i$ due to the phase step $\Delta\phi_i$ for different damping factors ζ; (a) for a high-gain loop with lag-lead RC filter and (b) for simple RC loop filter (© John Wiley & Sons, Inc., 2002).

which simplifies for very high-gain type 2 loops

$$\phi_{e2}(t) = \frac{\Delta\omega_i}{\omega_n} \cdot \frac{e^{(-\zeta+\sqrt{\zeta^2-1})\omega_n t} - e^{(-\zeta-\sqrt{\zeta^2-1})\omega_n t}}{2\sqrt{\zeta^2-1}} + \frac{\Delta\omega_i}{K_v} \tag{5.21}$$

For very long times we get from (5.20) the constant phase correction (the velocity constant) in accordance with (1.45)

$$\phi_{e2}(t) = \Delta\omega_i/K_v \tag{5.22}$$

The velocity constant is very small for high-gain PLLs and nearly zero for type 2 loops because of the very large DC gain in the active lag-lead RC filter (cf. Fig. 2.8).

After reverting to relation (5.21) we get for $\zeta > 1$

$$\begin{aligned}\phi_{e2}(t) &= \frac{\Delta\omega_i}{K_v} - e^{-\zeta\omega_n t}\left[\frac{\Delta\omega_i}{K_v}\cosh(\omega_n t\sqrt{\zeta^2-1})\right. \\ &\quad \left. - \frac{\Delta\omega_i}{\omega_n}\cdot\frac{1-\zeta\omega_n/K_v}{\sqrt{\zeta^2-1}}\sinh(\omega_n t\sqrt{\zeta^2-1})\right] \\ &\approx \frac{\Delta\omega_i}{\omega_n}\cdot\frac{1}{\sqrt{\zeta^2-1}}\sinh(\omega_n t\sqrt{\zeta^2-1})e^{-\zeta\omega_n t}\end{aligned} \tag{5.23}$$

for $\zeta = 1$

$$\begin{aligned}\phi_{e2}(t) &= \frac{\Delta\omega_i}{K_v} + e^{-\omega_n t}\frac{\Delta\omega_i}{\omega_n}\left[1 - \frac{\omega_n}{K_v}\right]t \\ &\approx \frac{\Delta\omega_i}{\omega_n}e^{-\omega_n t}\omega_n t\end{aligned} \tag{5.24}$$

and for $\zeta < 1$

$$\begin{aligned}\phi_{e2}(t) &= \frac{\Delta\omega_i}{K_v} - e^{\zeta\omega_n t}\left[\frac{\Delta\omega_i}{K_v}\cos(\omega_n t\sqrt{1-\zeta^2})\right. \\ &\quad \left. - \frac{\Delta\omega_i}{\omega_n}\cdot\frac{1-\zeta\omega_n/K_v}{\sqrt{1-\zeta^2}}\sin(\omega_n t\sqrt{1-\zeta^2})\right]\end{aligned} \tag{5.25}$$

These transients for several values of ζ are plotted in Fig. 5.2. Note that for $\zeta < 1$, the plot again exhibits damped oscillations that decay below 10% for $\omega_n\tau \approx 2.5$ (cf. [2]). Furthermore, for type 2 loops the effective *velocity error* is practically zero (cf. Fig 5.2(b)).

5.1.2.3 Frequency ramp

The Laplace transform of steady state changes of the frequency (frequency ramp) is

$$\Delta\Phi_e(\sigma) = \frac{\Delta\dot{\omega}}{s^3} = \frac{\Delta\dot{\omega}_i}{s^3} \tag{5.26}$$

In this case we get for the second-order high-gain or type 2 loops ($\omega_n/K_v \to 0$)

$$\Phi_e(\sigma) = \frac{\Delta\dot{\omega}_i}{\sigma^3}[1 - H(\sigma)] = \frac{\Delta\dot{\omega}_i/\omega_n^3}{\sigma^3}\cdot\frac{\sigma^2}{\sigma^2+2\zeta\sigma+1} = \frac{\Delta\dot{\omega}_i}{\sigma(\sigma^2+2\zeta\sigma+1)} \tag{5.27}$$

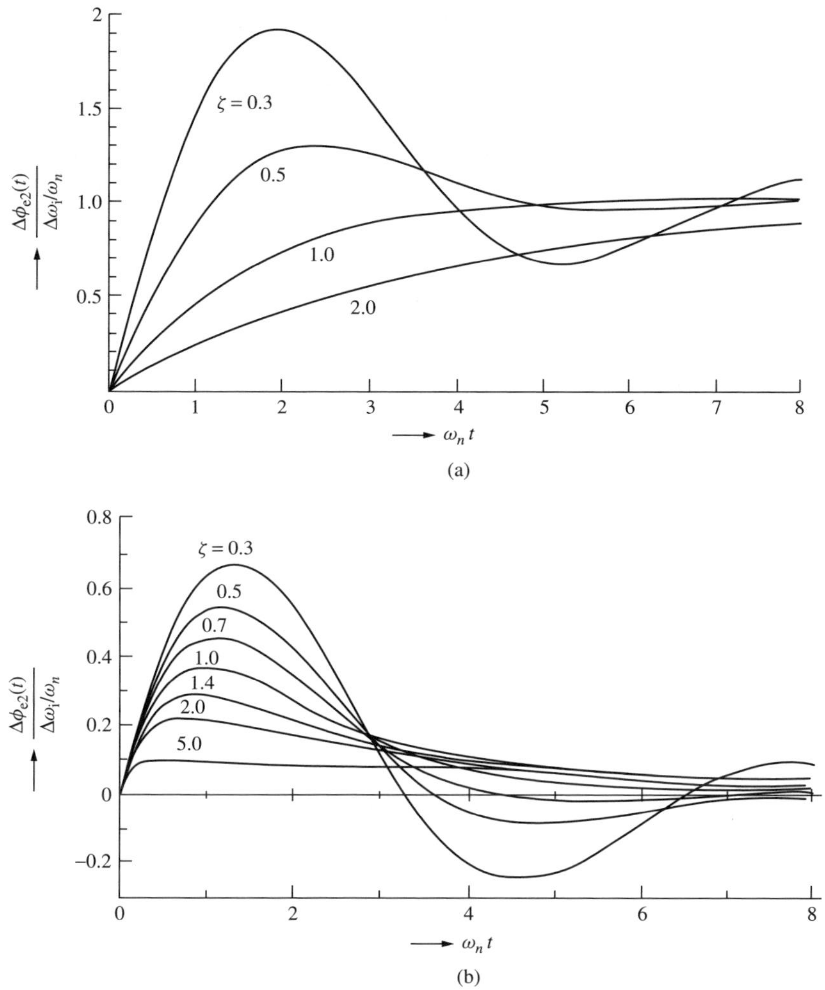

Figure 5.2 Normalized transients $\Delta\phi_{e2}(t)/(\Delta\omega_i/\omega_n)$ due to the frequency step $\Delta\omega_i$ for different damping factors ζ for high-gain loop; (a) for simple RC loop filter and (b) for high-gain loop with lag-lead RC filter (© John Wiley & Sons, Inc., 2002).

For performing the inverse Laplace transform we again have the roots (5.11) and with the assistance of [1, D6–1.7] we get

$$\phi_{e3}(t) = \left[\frac{1}{\sigma_1(\sigma_1 - \sigma_2)} e^{\sigma_1 \omega_n t} + \frac{1}{\sigma_2(\sigma_2 - \sigma_1)} e^{\sigma_2 \omega_n t} + \frac{1}{\sigma_1 \sigma_2}\right] \frac{\Delta\dot{\omega}}{\omega_n^3} \tag{5.28}$$

After inserting the roots from (5.11)

$$\phi_{e3}(t) = \frac{\Delta\dot{\omega}_i}{\omega_n^2}\left[1 + \frac{e^{(-\zeta+\sqrt{\zeta^2-1})\omega_n t} - e^{(-\zeta-\sqrt{\zeta^2-1})\omega_n t}}{2\sqrt{\zeta^2 - 1}}\right] \tag{5.29}$$

In cases in which we do not neglect the term ω_n/K_v in the numerator, we have

$$\Phi_e(\sigma) = \frac{\Delta\dot{\omega}/\omega_n{}^2}{\sigma^2}\frac{\sigma + \omega_n/K_v}{\sigma^2 + 2\zeta\sigma + 1} \tag{5.30}$$

With the assistance of [1, D6–3.14] we get for the time domain solution

$$\begin{aligned}\phi_{e3}(t) = {} & \frac{\Delta\dot{\omega}_i t}{K_v} + \frac{\Delta\dot{\omega}_i}{\omega_n{}^2}\left(1 - \frac{2\zeta\omega_n}{K_v}\right) - \frac{\Delta\dot{\omega}_i}{\omega_n{}^2}e^{-\zeta\omega_n t} \\ & \times\left[\left(1 - \frac{2\zeta\omega_n}{K_v}\right)\cosh(\omega_n t\sqrt{\zeta^2 - 1}) \right. \\ & \left. + \frac{\zeta - (\omega_n/K_v)(2\zeta^2 - 1)}{\sqrt{\zeta^2 - 1}}\sinh(\omega_n t\sqrt{\zeta^2 - 1})\right]\end{aligned} \tag{5.31}$$

For the damping factor, $\zeta < 1$, the hyperbolic functions again change into the trigonometric function and the steady state is approached with damped oscillation (see Fig. 5.3). For the critical damping, (5.28) simplifies into

$$\begin{aligned}\phi_{e3}(t) = {} & \frac{\Delta\dot{\omega}_i t}{K_v} + \frac{\Delta\dot{\omega}_i}{\omega_n{}^2}\left(1 - 2\frac{\omega_n}{K_v}\right) \\ & - \frac{\Delta\dot{\omega}_i}{\omega_n{}^2}e^{-\omega_n t}\left[\left(1 - 2\frac{\omega_n}{K_v}\right) + \left(1 - \frac{\omega_n}{K_v}\right)\omega_n t\right]\end{aligned} \tag{5.32}$$

The final steady state error is

$$\phi_{e3}(t) = \frac{\Delta\dot{\omega}t}{K_v} + \frac{\Delta\dot{\omega}_i}{\omega_n^2} \tag{5.33}$$

Example 5.3

In this example we shall provide the computer solution of the above investigated transient effects in the high-gain second-order PLLs. First, we shall solve the roots of the denominator in eq. (5.10) for the chosen damping factor. Next, we evaluate the partial fraction expansion of normalized phase errors in the σ notation and eventually compute the K-factors. Note that this time we have designated the normalized Laplace variable σ by s. The variable t_r in the plot is actually equal to $\omega_n\tau$ – see Fig. 5.4.

In instances in which $\zeta > 1$, both roots are real and the K-factors are easily evaluated by computer. For $\zeta < 1$, we must use individual evaluations (see Sections 5.1.2.1 and 5.1.2.2).

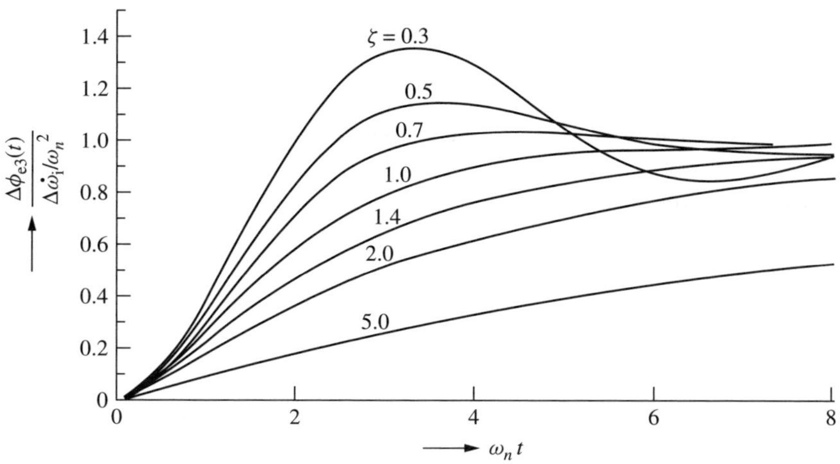

Figure 5.3 Normalized transients $\Delta\phi_{e3}(t)/(\Delta\dot{\omega}_i/\omega_n^2)$ due to the frequency ramp $\Delta\dot{\omega}_i$ for different damping factors ζ for a high-gain loop (DC phase error is retained) (© John Wiley & Sons, Inc., 2002).

5.1.3 Transients in Higher-order Loops

The pure second-order PLLs are rather rare. Either we change the order intentionally to enhance attenuation in the stop band of the transfer function $H(s)$ or $H(\sigma)$ as explained in Chapter 4, or the PLL order is increased involuntarily because of the spurious RC sections (mostly in connection with the VCO tuning). Another PLL order increase is caused by sampling or digital operation. We have already seen that this process can be respected, in the first-order approximation, with an RC section (cf. relation (3.42)).

In the following example we shall investigate properties of the fourth-order high-gain loop with the computer assistance (by application of Mathcad 8). Note that by choice of very small values of the normalized time delay δ we may investigate the behavior of the transients in the third-order loops, or even the fourth-order loops.

Example 5.4

We shall investigate transients in the fourth-order loop with two additional independent RC sections. We start with the relation (3.61), with the second RC constant η equal to the normalized time delay δ. We shall start with the third-order loop with the constant $\kappa = 0.3$ and the normalized time delay $\delta = 0.105$. The Mathcad solution is presented in Fig. 5.5.

5.2 PERIODIC CHANGES

In these instances we are interested in the settled or steady states that are easily evaluated with the assistance of the loop transfer functions $H(s)$ and $1 - H(s)$.

Polyroots 2nd order $\zeta := 1.5 \quad \omega n := 1 \quad K := 10 \quad r := 0..200 \quad t_r := r \cdot 0.1$

$c0 := 1 \quad c1 := 2\cdot\zeta \quad c2 := 1$

$$kk := \begin{bmatrix} c0 \\ c1 \\ c2 \end{bmatrix} \quad \text{Polyroots}(kk) = \begin{bmatrix} -2.618 \\ -0.382 \end{bmatrix} \quad n := 0..1 \quad b2_n := \text{polyroots}(kk)_n \quad b2 = \begin{bmatrix} -2.618 \\ -0.382 \end{bmatrix}$$

The phase step: Partial fractions

$$\frac{s + \frac{\omega n}{K}}{s^2 + 2\cdot\zeta\cdot s + 1} \left| \begin{array}{l} \text{convert, parfrac, s} \\ \text{float, 3} \end{array} \right. \rightarrow \frac{1.13}{(s + 2.62)} - \frac{.126}{(s + .382)}$$

$$K1 := \frac{b2_0 + \frac{\omega n}{K}}{b2_0 - b2_1} \quad K1 := 1.126 \quad K2 := \frac{b2_1 + \frac{\omega n}{K}}{b2_1 - b2_0} \quad K2 := -0.126$$

$$\phi 1_r := K1 \cdot e^{t_r \cdot b2_0} + K2 \cdot e^{t_r \cdot b2_1}$$

The frequency step: Partial fractions

$$\frac{1}{s^2 + 2\cdot\zeta\cdot s + 1} \quad K21 := \frac{1}{b2_0 - b2_1} \quad K21 = -0.447 \quad K22 := \frac{1}{b2_1 - b2_0} \quad K22 = 0.447$$

$$\phi 2_r := K21 \cdot e^{t_r \cdot b2_0} + K22 \cdot e^{t_r \cdot b2_1}$$

The frequency ramp: Partial fractions

$$\frac{1}{s\cdot(s^2 + 2\cdot\zeta\cdot s + 1)} \left| \begin{array}{l} \text{convert, parfrac, s} \\ \text{float, 3} \end{array} \right. \rightarrow \frac{1.00}{s} + \frac{.171}{(s + 2.62)} - \frac{1.17}{(s + .382)}$$

$$\phi 3_r := K32 \cdot e^{t_r \cdot b2_0} + K33 \cdot e^{t_r \cdot b2_1} + K31$$

(a)

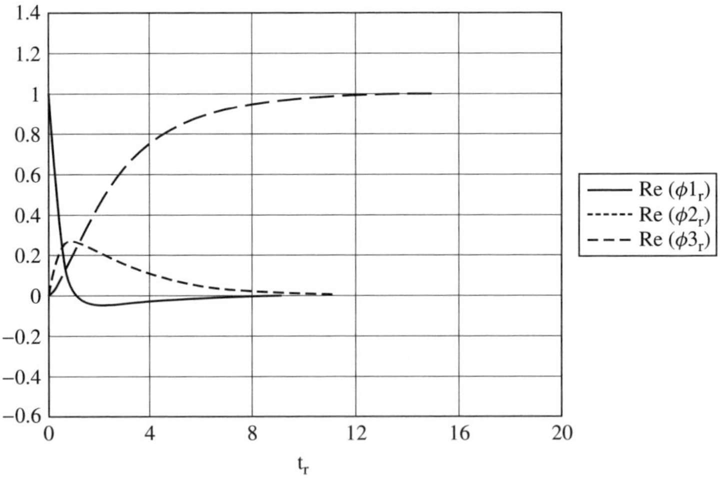

(b)

Figure 5.4 Computer evaluation of the transients in the second-order high-gain PLL. (a) The Mathcad program; (b) the transients' characteristics for the phase step (full line), for the frequency step (points), and for the frequency ramp (dashed).

$$HO_m := \frac{\left[(\sigma_m)^2 \cdot 2 \cdot \zeta \cdot \kappa \cdot \delta + \sigma_m\,(2.\zeta \cdot \kappa + \delta) + 1\right] \cdot (\sigma_m)^2}{\left[(\sigma_m)^4 \cdot 2 \cdot \zeta \cdot \kappa \cdot \delta + (\sigma_m)^3 . (2 \cdot \zeta \cdot \kappa + \delta) + (\sigma_m)^2 + 2 \cdot \zeta \cdot \sigma_m\right] + 1}$$

$$\frac{s \cdot \left[s^2 \cdot (2 \cdot \zeta \cdot \kappa \cdot \delta) + s \cdot (2 \cdot \zeta \cdot \kappa + \delta) + 1\right]}{s^4 \cdot (2 \cdot \zeta \cdot \kappa \cdot \delta) + s^3 \cdot (2 \cdot \zeta \cdot \kappa + \delta) + s^2 + 2 \cdot \zeta \cdot s + 1} \left| \begin{matrix}\text{convert, parfrac, s} \\ \text{float, 3}\end{matrix} \right. \rightarrow \frac{(4.22 \cdot 10^{-2})}{(s + 10.0)} - \frac{.285}{(s + 3.21)} + 39.8 \cdot \frac{(2.63 \cdot 10^{-3} + 3.12 \cdot 10^{-2} \cdot s)}{(s^2 + .481 \cdot s + 1.24)}$$

Polyroots 4th order $\zeta = 0.4$ $\kappa = 0.3$ $\delta = 0.105$

$a0 := 1$ $a1 := 2 \cdot \zeta$ $a2 := 1$ $a3 := (\delta + 2 \cdot \zeta \cdot \kappa)$ $a4 := 2 \cdot \zeta \cdot \kappa \cdot \delta$ $n := 0..3$

$$k := \begin{bmatrix} a0 \\ a1 \\ a2 \\ a3 \\ a4 \end{bmatrix} \quad \text{polyroots}(k) = \begin{bmatrix} -10.024 \\ -3.211 \\ -0.24 + 1.085i \\ -0.24 - 1.085i \end{bmatrix} \quad b4_n := \text{polyroots}(k)_n \quad b4 = \begin{bmatrix} -10.024 \\ -3.211 \\ -0.24 + 1.085i \\ -0.24 - 1.085i \end{bmatrix}$$

$$K41 := \frac{b4_0\left[(b4_0)^2 \cdot a4 + b4_0 \cdot a3 + 1\right]}{(b4_0 - b4_1) \cdot (b4_0 - b4_2) \cdot (b4_0 - b4_3)} \qquad \frac{K41}{a4} = 0.042 - 2.108i \cdot 10^{-10}$$

$$K42 := \frac{b4_1\left[(b4_1)^2 \cdot a4 + b4_1 \cdot a3 + 1\right]}{(b4_1 - b4_0) \cdot (b4_1 - b4_2) \cdot (b4_1 - b4_3)} \qquad \frac{K42}{a4} = -0.285 + 4.557i \cdot 10^{-9}$$

$$K43 := \frac{b4_2\left[(b4_2)^2 \cdot a4 + b4_2 \cdot a3 + 1\right]}{(b4_2 - b4_0) \cdot (b4_2 - b4_1) \cdot (b4_2 - b4_3)} \qquad \frac{K43}{a4} = 0.622 + 0.089i$$

$$K44 := \frac{b4_3\left[(b4_3)^2 \cdot a4 + b4_3 \cdot a3 + 1\right]}{(b4_3 - b4_0) \cdot (b4_3 - b4_2) \cdot (b4_3 - b4_1)} \qquad \frac{K44}{a4} = 0.622 - 0.089i$$

$$f41_r := \left(K41 \cdot e^{t_r \cdot b4_0} + K42 \cdot e^{t_r \cdot b4_1} + K43 \cdot e^{t_r \cdot b4_2} + K44 \cdot e^{t_r \cdot b4_3}\right) \cdot \frac{1}{a4} \quad \zeta = 0.4 \;\; \kappa = 0.3 \;\; \delta = 0.105$$

(a)

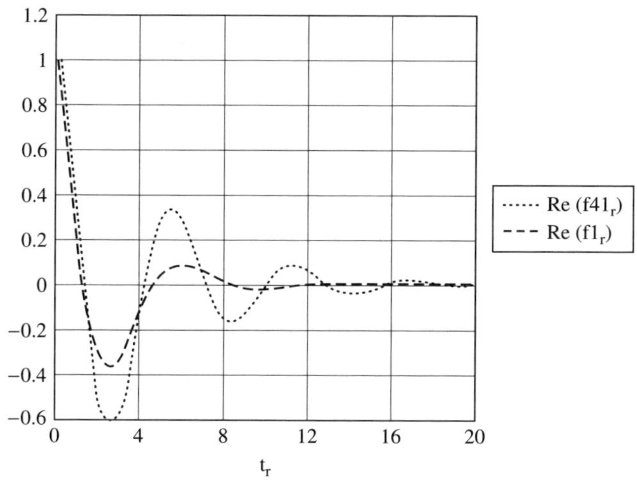

(b)

Figure 5.5 Computer evaluation of the transients in the fourth-order high-gain PLL. (a) The Mathcad program and (b) the transients' characteristics for the phase step of the basic second-order PLL (dashed) and for the fourth-order PLL (points).

5.2.1 Phase Modulation of the Input Signal

For simplicity we shall consider modulation with a single sine wave

$$\phi_i(t) = \Delta\phi_i \sin \Omega t \tag{5.34}$$

Consequently, the output modulation would remain sinusoidal,

$$\phi_o(t) = |H(j\Omega)| \Delta\phi_i \sin(\Omega t + \Psi_o) \tag{5.35}$$

however, shifted by

$$\Psi_o = \arctan \frac{\mathrm{Im}[H(j\Omega)]}{\mathrm{Re}[H(j\Omega)]} \tag{5.36}$$

Inspection of the PLL transfer functions reveals that in the $H(s)$ pass band the modulation index $\Delta\phi_i$ remains unaltered but attenuated for frequencies $\Omega > \omega_n$.

$$\Delta\phi_o = |H(j\Omega/\omega_n)| \Delta\phi_i \tag{5.37}$$

In instances in which the PLL is used as a PD, the desired information must be recovered at the output of the loop detector, however, this only works for frequencies outside of the pass band, that is, for $\Omega > \omega_n$.

$$\phi_e(t) = |1 - H(j\Omega)| \Delta\phi_i \sin(\Omega t + \Psi_e) \tag{5.38}$$

As an example, see Section 9.4.8.1.

5.2.2 Frequency Modulation of the Input Signal

By starting again with the sinusoidal modulation we get for the modulation phase index

$$\Delta\phi_i(t) = \Delta\omega_i \int_0^t \cos \Omega t \, dt = \frac{\Delta\omega_i}{\Omega} \sin \Omega t \tag{5.39}$$

which remains unaltered for modulation frequencies $\Omega < \omega_n$, however, only for PLLs of type 1. The amplitude of the normalized phase at the output of the loop detector in the instances of the PLLs of type 2

$$\frac{\Delta\phi_e}{\Delta\omega_i/\omega_n} = \frac{1 - H(jx)}{x}; \quad \text{where} \quad x = \frac{\Delta\omega_i}{\omega_n} \tag{5.40}$$

for $\Delta\omega_i/\omega_n = 1$ has a maximum (see Fig. 5.6)

$$\frac{\Delta\phi_e}{\Delta\omega_i/\omega_n} = \frac{1}{2\zeta} \tag{5.41}$$

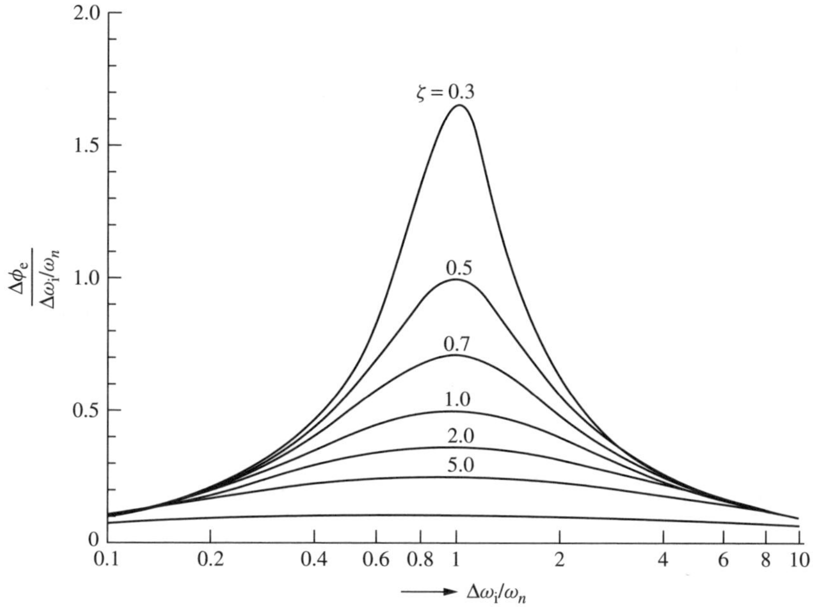

Figure 5.6 Steady state normalized phase error due to frequency modulation of the high-gain second-order PLL as a function of the damping factor ζ (© John Wiley & Sons, Inc., 2002).

Note that the phase shift between the input signal to the PD and the modulation signal is zero. This property and relation (5.41) are often used for experimental evaluation of the normalized frequency ω_n in the second-order high-gain PLLs.

5.3 DISCRETE SPURIOUS SIGNALS

Small discrete spurious signals found at the PD output are of different origins: leakage of the reference signal and of its harmonics, signals superimposed to the reference or VCO voltages, and different mixing products. Furthermore, the PD itself can enhance already present signals or generate new ones owing to inherent nonlinearities. The situation is schematically depicted in Fig. 5.7.

5.3.1 Small Discrete Spurious Signals at the Input

In this case the input voltage is composed of the desired and spurious signal or signals:

$$v_i(t) = V_i \sin \omega_i t + V_{sp} \sin \omega_{sp} t. \tag{5.42}$$

After replacing ω_{sp} in the above equation with

$$\omega_{sp} t = \omega_i t - (\omega_i - \omega_{sp})t = \omega_i t - \Omega t \tag{5.43}$$

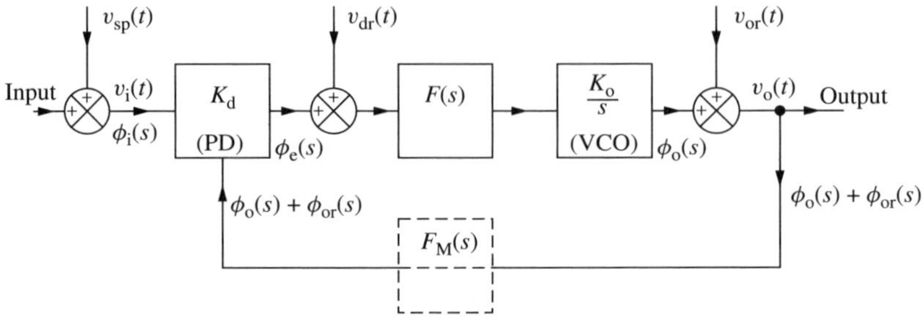

Figure 5.7 Block diagram of the PLL with indicated spurious signals: at the input of the PD, at the output of the PD, and at the output of the PLL.

we can rearrange (5.42) as

$$v_i(t) = V_i \sin \omega_i t + \frac{V_{sp}}{2} \sin \omega_i t \cos \Omega t - \frac{V_{sp}}{2} \cos \omega_i t \sin \Omega t \tag{5.44}$$

Summation of all three terms, with the assistance of the trigonometric rules, reveals, when $V_{sp} \ll V_i$ (cf. Chapter 9, Section 9.3.5.1),

$$v_i(t) \doteq V_i \left(1 + \frac{V_{sp}}{V_i} \cos \Omega t\right) \sin \left(\omega_i t - \frac{V_{sp}}{V_i} \sin \Omega t\right). \tag{5.45}$$

Evidently, we find the reference input signal to be simultaneously amplitude- and phase-modulated with modulation indexes

$$m_a = m_p = \frac{V_{sp}}{V_i} \ll 1. \tag{5.46}$$

Introduction of the input voltage $v_i(t)$, (5.42), together with the output voltage of the VCO – by assuming the locked state

$$v_O(t) = -V_O \cos[\omega_i t + \phi_o(t)] \tag{5.47}$$

into (1.4) reveals for the output voltage of the PD

$$\begin{aligned} v_d(t) &= -K_m V_i (1 + m_a \cos \Omega t)\ V_O \sin(\omega_i t - m_p \sin \Omega t) \\ &\quad \times \cos[\omega_i t + \phi_o(t)] \\ &= -K_d (1 + m_a \cos \Omega t) \sin[-m_p \sin \Omega t - \phi_o(t)]. \end{aligned} \tag{5.48}$$

In cases in which the PLL is locked, the phase $\phi_o(t)$ is small, particularly for a high-gain PLL, and if (5.46) is met the above equation simplifies into

$$v_d(t) = K_d (1 + m_a \cos \Omega t)[m_p \sin \Omega t + \phi_o(t)] \tag{5.49a}$$

and further into

$$v_{\mathrm{d}}(t) \approx K_{\mathrm{d}}[m_{\mathrm{p}} \sin \Omega t + \phi_{\mathrm{o}}(t)] \tag{5.49b}$$

since the contribution by the "amplitude modulation" is of the second order. Consequently, the final effect of a small spurious signal at the input of the PLL is the spurious phase modulation with the index m_{p} and frequency $\Omega = \omega_{\mathrm{i}} - \omega_{\mathrm{sp}}$. Evidently, we may apply results found in Section 5.2.1 for the pure phase modulation of the input signal. The knowledge of the output phase modulation is important for the researcher. With the assistance of (5.35) we find

$$m_{\mathrm{po}} = m_{\mathrm{p}} |H(\mathrm{j}\Omega)|. \tag{5.50}$$

This is an important result for designers.

5.3.2 Small Spurious Signal at the Output of the Phase Detector

At the PD output, we encounter spurious AC signals with the reference frequency and its harmonics. In addition, some leakage of the power frequency and other signals, particularly, intermodulation products in the IC system, cannot be excluded. Investigation proceeds with the assistance of Fig. 5.7. For simplicity, we assume the situation of one spurious voltage $v_{\mathrm{dr}}(t)$ superimposed on the useful output voltage of the PD. As always, we leave the time domain and after application of the Laplace symbology we have the following relation for the PLL:

$$([\Phi_{\mathrm{i}}(s) - \Phi_{\mathrm{o}}(s) F_M(s)] K_{\mathrm{d}} + V_{\mathrm{dr}}(s)) \frac{F(s) K_0}{s} = \Phi_{\mathrm{o}}(s) \tag{5.51}$$

Without any loss of generality we shall assume $\phi_{\mathrm{i}}(t) = 0$ and the above relation remains a function of two variables V_{dr} and $\Phi_{\mathrm{o}}(s)$

$$\Phi_{\mathrm{o}}(s) = \frac{V_{\mathrm{dr}}(s)}{K_{\mathrm{d}}} \cdot \frac{KF(s)/s}{1 + KF(s) F_M(s)/s} \tag{5.52}$$

Using (1.35) we find a spurious modulation index, in the case of the sinusoidal voltage $v_{\mathrm{dr}}(t)$:

$$v_{\mathrm{dr}}(t) = V_{\mathrm{dr}} \sin(\omega_{\mathrm{dr}} t + \phi_{\mathrm{dr}}), \tag{5.53}$$

$$m_{\mathrm{po}} = \frac{V_{\mathrm{dr}}}{K_{\mathrm{d}}} |H(\mathrm{j}\omega_{\mathrm{dr}})| \tag{5.54}$$

In instances in which the natural frequency of the loop, ω_n, is small compared with the frequency of the disturbing signal, which is generally the case for the remainder of the reference signal, the above relations can be further reduced to

$$m_{\text{po}} \doteq \frac{V_{\text{dr}}}{K_{\text{d}}} \left| \frac{KF(\text{j}\omega_{\text{dr}})}{\text{j}\omega_{\text{dr}}} \right|, \tag{5.55a}$$

This equation points to the possibility of reducing the spurious modulation index m_{po} in instances in which $\omega_n \ll \omega_{\text{dr}}$

$$m_{\text{po}} = V_{\text{dr}} K_0 \left| \frac{F(\text{j}\omega_{\text{dr}})}{\text{j}\omega_{\text{dr}}} \right| \tag{5.55b}$$

In order to achieve this

1. the choice of phase detector (PD) is very important since, for example, sampling PD (see Section 8.4.3) generates a much smaller leakage of the spurious signals in respect to simpler arrangements;
2. the gain of the voltage-controlled oscillators (VCO) should not be excessive, the magnitude guaranteeing that the correct operation in the desired temperature and frequency range might be sufficient;
3. finally, we use loop filters with high attenuation around the spurious frequency $\omega_{\text{d,sp}}$ (additional RC section notch filter, etc.).

As long as $\text{m}_{\text{po}} \ll 1$, a condition nearly always met in practice, the signal-to-noise ratio (SNR) is

$$\frac{\text{signal}}{\text{one spurious side band}} = \frac{\text{S}}{\text{SSB}_{\text{dr}}} = \frac{2}{m_{\text{po}}} \tag{5.56}$$

After introduction into (5.55b) we compute the spurious level in decibels:

$$\frac{(\text{SSB})_{\text{dr}}}{\text{S}} = 20 \log \frac{V_{\text{dr}} K_0}{\omega_n} + 20 \log \left| \frac{F(\text{j}\omega_{\text{dr}}/\omega_n)}{\text{j}\omega_{\text{dr}}/\omega_n} \right| - 6 \text{ [dB]} \tag{5.57}$$

When we would like to use normalized transfer functions, in accordance with Figs. 2.2, 2.5(a), and 2.10(a), we must take into account that they have been plotted for the transfer function with $F_M(\sigma) = 1$ (as done in the case of the primed transfer functions – cf. (1.47) and (1.48)). In this instance the spurious sideband (SSB) noise in respect to the signal is

$$\begin{aligned} \frac{(\text{SSB})_{\text{d,sp}}}{\text{S}} &= 20 \log(V_{\text{d,sp}}/K_{\text{d}}) + 20 \log |H'(\text{j}x_{\text{d,sp}})| \\ &\quad - 20 |\log F_M(\text{j}x_{\text{d,sp}})| - 6 \text{ [dB]} \end{aligned} \tag{5.58}$$

where $x_{\text{d,sp}} = \omega_{\text{d,sp}}/\omega_n$.

As long as the feedback path is only a simple frequency divider, which is often the case, with the division factor N,

$$F_M(\mathrm{j}x_{\mathrm{d,sp}}) = \frac{1}{N} \tag{5.59}$$

Its introduction into (5.58) simplifies it into

$$\frac{(\mathrm{SSB})_{\mathrm{d,sp}}}{\mathrm{S}} = 20\log(V_{\mathrm{d,sp}}/K_{\mathrm{d}}) + 20\log|H'(\mathrm{j}x_{\mathrm{d,sp}})| + 20\log N - 6 \text{ [dB]} \tag{5.60}$$

5.3.3 Small Spurious Signals at the Output of the PLLs

Proceeding as in Section 5.3.1 we find that the major contribution is provided by the phase modulation generated by superposition. The only difference is caused by the presence or the absence of the feedback path "filter" block.

First, we shall consider a situation in which $F_M(s) = 1$ is valid for all the investigated frequency ranges. With the assistance of Fig. 5.7 we find for the PD output phase

$$\Phi_{\mathrm{e}}(s) = \Phi_{\mathrm{i}}(s) - \Phi_{\mathrm{o}}(s) - \Phi_{\mathrm{o,sp}}(s) \tag{5.61}$$

After the introduction of

$$\Phi_{\mathrm{o}}(s) = \Phi_{\mathrm{e}}(s)K/s \tag{5.62}$$

we eliminate $\Phi_{\mathrm{o}}(s)$ and arrive at the result

$$\Phi_{\mathrm{e}}(s) = \frac{\Phi_{\mathrm{i}}(s) - \Phi_{\mathrm{o,sp}}(s)}{1 + KF(s)/s}, \tag{5.63a}$$

$$\Phi_{\mathrm{e}}(s) = \frac{\Phi_{\mathrm{i}}(s) - \Phi_{\mathrm{o,sp}}(s)}{1 + G(s)} \tag{5.63b}$$

By comparing the above result with relation (5.1b), we perceive the equivalence of the input and the output spurious phase signals. Consequently, we can in Sections 5.3.1 and 5.3.2 leave out the labeling of "i" as is usually the case (cf. Gardner [2], Kroupa [3], and [4] to [10]).

The situation is difficult in cases in which the transfer function $F_M(s)$ cannot be neglected. The output is given by

$$\Phi_{\mathrm{e}}(s) = \frac{\Phi_{\mathrm{i}}(s) - \Phi_{\mathrm{o,sp}}(s)F_M(s)}{1 + G(s)} \tag{5.64}$$

In the PLLs of the first and second orders, the feedback transfer block will be a simple frequency divider. The investigation will reveal after introduction of (5.59) into (5.64) that the influence of the output disturbances reduces N times the contribution supplied by the output.

REFERENCES

[1] G.A. Korn and T.M. Korn, *Mathematical Handbook for Scientists and Engineers*. London: McGraw-Hill, 1961.
[2] F.M. Gardner, *Phaselock Techniques*. 2nd ed. New York: John Wiley, 1979.
[3] V.F. Kroupa, *Frequency Synthesis: Theory, Design et Applications*. London: Ch. Griffin, New York: John Wiley, 1973.
[4] V. Manassewit, *Frequency Synthesizers, Theory and Design*. New York: Wiley, 1976 and 1980.
[5] W.F. Egan, *Frequency Synthesis by Phase Lock*. New York: Wiley, 1981.
[6] U.L. Rohde, *Digital PLL Frequency Synthesizers, Theory and Design*. Englewood Cliffs: Prentice Hall, 1983.
[7] J.A. Crawford, *Frequency Synthesizer Design Handbook*. Boston/London: Artech House, 1994.
[8] G. Bar-Giora, *Digital Techniques in Frequency Synthesis*. New York: McGraw-Hill, 1996.
[9] U.L. Rohde, *Microwave and Wireless Synthesizers, Theory and Design*. Chichester: John Wiley, 1997.
[10] V.F. Kroupa, ed. *Direct Digital Frequency Synthesizers*. New York: IEEE Press 1999.

6

Working Ranges of PLLs

An important problem, encountered while designing the phase lock loop (PLL), is the question of a reasonable difference between the input frequency, ω_i, and the true free-running frequency of the voltage-controlled oscillator (VCO), ω_c.

The maximum frequency difference before losing lock of the PLL system is called the *hold-in range*. Another criterion might be the frequency difference $|\omega_i - \omega_c|$, for which the phase lock will take place in all circumstances, the so-called *pull-in range*. However, as the ideal state we can design a situation in which the lock is achieved without any cycle slipping, that is, without any loss of lock after switching on. This state between the input and the output frequencies of the PLLs is described as the *lock-in range*.

The final question while investigating working ranges is the problem of tolerating the progressive frequency difference before the lock breakdown, the so-called *pull-out frequency*. A graphical representation of the above discussed parameters is schematically plotted in Fig. 6.1.

6.1 HOLD-IN RANGE

Proceeding with the definition given above, the *hold-in range* $\Delta\omega_H = |\omega_i - \omega_c|$ is the largest frequency that can be tolerated in the PLLs before losing lock without any serious event. In these circumstances we can define $\Delta\omega_H$ as a function of the phase detector output as long as it is valid.

$$\frac{d\Psi_e(t)}{dt} = 0 \tag{6.1}$$

After introducing the above relation into (1.10), we have

$$\Delta\omega_H = K_0 v_{2,\max} \tag{6.2}$$

However, for different types of phase detectors (PD) we find different limits.

Phase Lock Loops and Frequency Synthesis V.F. Kroupa
 ISBN: 0-470-84866-9

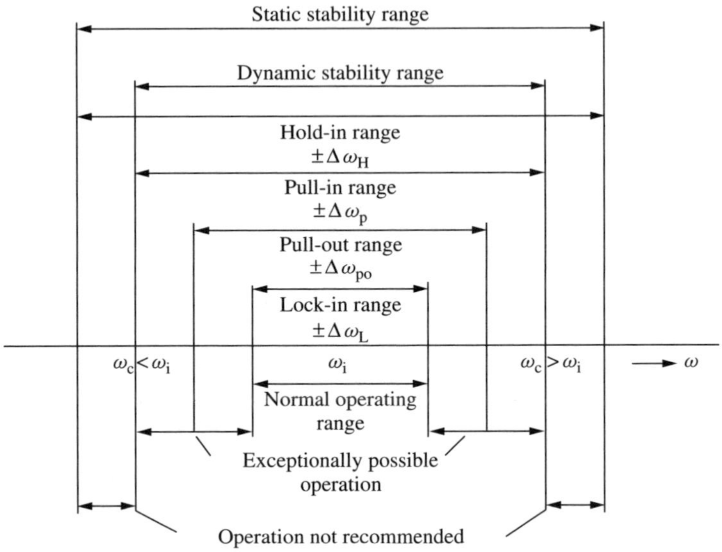

Figure 6.1 Working ranges of a PLL.

6.1.1 Phase Detector with the Sine Wave Output

From eq. (1.5) or (1.31) we get for $v_{2,\max}$

$$v_{2,\max} = K_d F(0) \sin(\pi/2) = K_d F(0) \tag{6.3}$$

After introducing this relation into (6.2) and by taking into account that Ψ_e can change in the range from $-\pi/2$ to $+\pi/2$, we get for the hold-in range

$$\Delta\omega_H = \pm K_v \tag{6.4}$$

However, this theoretical hold-in range may be limited by different causes to smaller values: by the maximum output voltage of the amplifier, the maximum frequency change tolerated by the VCO, and others.

6.1.2 The PD with Triangular Output

After linearization of the sine wave in (6.3) into a triangular one we get

$$\Delta\omega_H = \pm K_v \frac{\pi}{2} \tag{6.5}$$

6.1.3 The PD with a Sawtooth Wave Output

Similarly, as mentioned above we get

$$\Delta\omega_H = \pm K_v \pi \tag{6.6}$$

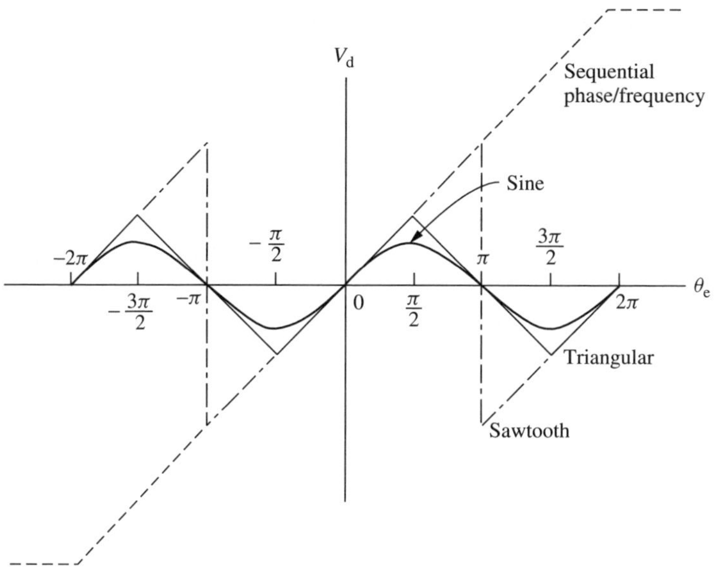

Figure 6.2 Phase detector characteristic (Reproduced from F.M. Gardner, *Phaselock Techniques*. 2nd ed. New York: John Wiley, 1979 by permission of John Wiley & Sons, Inc, © 2002).

6.1.4 Sequential Phase Frequency Detectors

Figure 8.19 reveals that

$$\Delta\omega_H = \pm K_v 2\pi \tag{6.7}$$

Note that all definitions of the hold-in range enclose the velocity constant K_v, which is generally very large, in some instances indicated as infinite; in these cases the maximum, $v_{2,max}$, or the detuning range of the VCO is decisive for the actual, $\Delta\omega_H$. The above definitions of the *hold-in range* are illustrated with the PD characteristics plotted in Fig. 6.2 in which no nonlinearities are considered.

6.2 THE PULL-IN RANGE

The pull-in range presents a much more complicated task. In the following paragraphs we shall discuss the procedure for second-order PLLs of type 1 and 2.

The problem was solved in the past by many authors from different points of view. Here, we will provide a solution that is as simple as possible but accurate enough for practical applications.

For investigation of the PLLs pull-in range $\Delta\omega_p$, we may use two methods:

1. In the first approach we will compute the time needed for a given frequency difference $|\omega_i - \omega_c|$, then increase this difference and note $|\omega_i - \omega_c| = \Delta\omega_p$ for which the pull-in time starts to be nearly infinite.

2. In the second approach we will proceed the other way, that is, by choosing the difference, $\Delta\omega = |\omega_i - \omega_c|$, so large that no locking takes place. Then we shall reduce the difference till the beat note at the PD output starts to be constant; in other words as long as the differential equation of the system has a periodic solution. The smallest detuning, $\Delta\omega = \Delta\omega_p$, is the sought pull-in range.

Let us assume a sinusoidal PD with the gain K_d. In the asynchronous steady state the PD output voltage, in accordance with relation (1.4) and Figs. 1.1, 1.5(b), and 3.7, will be

$$v_d(t) = K_1 e^{j[(\omega_i-\omega_o)t-\phi_{on}(t)]} + K_{-1} e^{-j[(\omega_i-\omega_o)t-\phi_{on}(t)]} \tag{6.8}$$

where

$$K_1 = K_d/2j \quad \text{and} \quad K_{-1} = -K_d/2j = K^*{}_1 \tag{6.9}$$

Further, in the asynchronous steady state the VCO is phase-modulated with a beat frequency, ν,

$$\begin{aligned} \nu &= |\omega_i - \omega_o| = |\omega_i - \omega_c| - \Delta\omega \\ &= \nu_c - \Delta\omega \end{aligned} \tag{6.10}$$

After expressing the free-running frequency ω_c as a function of ν we have

$$\nu_c = \nu + \Delta\omega(\nu) \tag{6.11}$$

It is clear that a slowly varying VCO phase error ϕ_o(t) can be expanded in the Fourier series

$$\phi_o(t) = \sum_{n=-\infty}^{\infty} \phi_n e^{jn\nu t} \quad (n \neq 0) \tag{6.12}$$

Similarly, in instances in which a block with the transfer function $F_M(s)$ is in the feedback path, the above relation will change to

$$\phi_{on} = \sum_{n=-\infty}^{\infty} F_M(jn\nu)\phi_n e^{jn\nu t} \quad (n \neq 0) \tag{6.13}$$

Now we shall revert to the PD output voltage $v_d(t)$, which after passing the loop filter $F(s)$ changes into the VCO steering voltage $v_2(t)$, the DC component of which causes the $\Delta\omega$ shift of the VCO frequency with a phase modulation $\varphi_o(t)$. All three components are connected with a differential equation

$$\Delta\omega + \frac{d\phi_o(t)}{dt} = K_o v_2(t) \tag{6.14}$$

For solution of the above differential equation we shall first simplify $v_d(t)$ after applying the Taylor expansion on (6.8)

$$v_d(t) \approx K_1 e^{j\nu t}[1 - j\phi_{on}(t)] + K_{-1} e^{-j\nu t}[1 + j\phi_{on}(t)] \tag{6.15}$$

After introduction of (6.13) we arrive at[1]

$$\begin{aligned}\Delta\omega + j\sum_n{}' n\nu\phi_n e^{jn\nu t} &= K_1 K_o[F(j\nu)e^{j\nu t} - j\sum_n{}' F[j(n+1)\nu]F_M(jn\nu)\phi_n e^{j(n+1)\nu t}] \\ &\quad + K_{-1}K_o[F(-j\nu)e^{-j\nu t} + j\sum_n{}' F[j(n-1)\nu] \\ &\quad \times F_M(jn\nu)\phi_n e^{j(n-1)\nu t}]\end{aligned} \tag{6.16}$$

The next step is to compare terms with the same frequency on the left- and right-hand sides

$$\Delta\omega \doteq -jK_1 K_o F(0) F_M(-j\nu)\phi_{-1} + jK_{-1}K_o F(0) F_M(j\nu)\phi_1 \tag{6.17}$$

$$\begin{aligned}\phi_1 &\doteq \frac{K_o F(j\nu)}{j\nu}[K_1 + jK_{-1}F_M(j2\nu)\phi_2] \\ &= -\frac{KF(j\nu)}{2\nu}[1 - jF_M(j2\nu)\phi_2]\end{aligned} \tag{6.18}$$

$$\phi_{-1} = \phi^*{}_1 \tag{6.19}$$

$$\phi_2 \doteq \frac{KF(j2\nu)}{j4\nu}[F_M(j\nu)\phi_1 + F_M(j3\nu)\phi_3] \tag{6.20}$$

$$\phi_{-2} = \phi^*{}_2 \tag{6.21}$$

and so on.

In instances in which the beat frequency is large, or more exactly in which

$$\nu \gg \frac{K}{2}|F(j\nu)| \tag{6.22}$$

we easily verify that the following condition is met:

$$1 \gg \phi_1 \gg \phi_2 \gg \cdots \tag{6.23}$$

Unfortunately, the above condition is not fulfilled in PLL type 1 but practically always in PLL type 2. In the latter case the beat frequency, outside of the pull-in range, is much larger compared with the former case. The consequence is that the maximum

[1] The primed summation does not incorporate terms with $n = 0$.

amplitude of the phase fluctuation $|\varphi_o(t)|$ is small compared with *one* and application of the simplified version of $v_d(t)$ in (6.15) is justified. Similarly from relation (6.23) it follows that the second harmonic of the phase error, that is, φ_2, in (6.18) can be neglected. In such a situation, the DC frequency change in VCO may be simplified to

$$\Delta\omega = \mathrm{j}\frac{K_v}{4}[G(\mathrm{j}\nu) - G(-\mathrm{j}\nu)] = \frac{K_v}{2}|G(\mathrm{j}\nu)|\sin\Psi \tag{6.24}$$

The above equation can be rearranged in two ways, either with the assistance of the gain of the open loop $G(\mathrm{j}\nu)$ or with the velocity constant $K_v = KF(0)$ into

$$\Delta\omega = \mathrm{j}\frac{KF(0)}{4}\left[\frac{K}{\mathrm{j}\nu}F(\mathrm{j}\nu)F_M(\mathrm{j}\nu) - \frac{K}{-\mathrm{j}\nu}F(-\mathrm{j}\nu)F_M(-\mathrm{j}\nu)\right] \tag{6.25}$$

Further, by introduction of the operation amplifier gain $A = F(0)$, we get for the real part of the loop filter, gain

$$\Phi(\nu) = \mathrm{Re}[F(\mathrm{j}\nu)F_M(\mathrm{j}\nu)] \tag{6.26}$$

And by combination of both relations (6.25) and (6.26), we get

$$\Delta\omega = \frac{AK^2}{2\nu}\Phi(\nu) \tag{6.27}$$

After inserting the above result into (6.11), we finally get ν for the beat frequency, as a function of the free-running frequency ν_c,

$$\nu_c = \nu + \frac{AK^2}{2\nu}\Phi(\nu) \tag{6.28}$$

Reduction of the original detuning will finally lead to a minimum beat note frequency

$$\nu = \nu_m \tag{6.29}$$

where the asynchronous state is not to be held any further and the pull-in starts. The respective $\nu_{c,min}$ is the sought $\Delta\omega_p$, that is, the pull-in range frequency. To this end we shall differentiate the rhs of the relation (6.28) and make it equal to zero:

$$1 + \frac{AK^2}{2\nu^2}\left[\frac{\mathrm{d}\Phi(\nu)}{\mathrm{d}\nu}\nu - \Phi(\nu)\right] = 0 \tag{6.30}$$

Its solution reveals ν_m and after its introduction into (6.28) we get the upper limit of the pull-in range, that is, $\Delta\omega_p$. In instances in which $\Phi(\nu)$ is nearly independent of the beat frequency ν, that is, in which

$$\mathrm{d}\Phi(\nu)/\mathrm{d}\nu \approx 0 \tag{6.31}$$

the solution of (6.30) is simplified to

$$\nu_m \doteq K\sqrt{A\Phi(\nu_m)/2} \tag{6.32}$$

After introduction of the above relation into (6.28) we get for the pull-in range

$$\Delta\omega_P \doteq 2\nu_m \doteq K\sqrt{2A\Phi(\Delta\omega_P/2)} \tag{6.33}$$

At this stage we are prepared for application of the above theory on practical examples.

6.2.1 PLLs of the First Order

Phase lock loops (PLLs) of the first order have a pull-in range equal to the hold-in range:

$$\Delta\omega_P = \Delta\omega_H \tag{6.34}$$

This statement is easily verified by inspecting Fig. 1.2 in which we have plotted relation (1.10) in the phase plane. As long as $\Delta\omega/K < \pi$ is valid the derivative $\dot{\Psi}_e/K$ passes the zero level twice. For a smaller change of $\Delta\Psi_e$ in the neighborhood of $\dot{\Psi}_e = 0$, the Taylor expansion reveals

$$\frac{\mathrm{d}\Delta\Psi_e}{\mathrm{d}t} = -K\,\Delta\Psi_e \cos\Psi_{eo} \tag{6.35}$$

After integration of the above relation we have

$$\Delta\Psi_e(t) \approx \exp(-Kt\cos\Psi_{eo}) \tag{6.36}$$

and inspection reveals that $\Delta\Psi_e(t)$ will increase as long as $\cos\Psi_{eo}$ is negative and vice versa and will converge to zero where $\cos\Psi_{eo}$ is positive, which is valid in the interval

$$-\pi/2 < \Psi_{eo} < \pi/2 \tag{6.37}$$

We conclude that this zero crossing will result in a fast phase locking without cycle slipping. Another confirmation of the relation (6.34) follows from the knowledge that for $|\Delta\omega/K| > \pi$ no zero crossing exists and the lock is impossible.

Behavior of the PLLs of the first order for a larger frequency difference, exceeding the hold-in range, is of assistance in understanding the pull-in process in PLLs of higher orders.

Let us investigate the sawtooth wave PD in situations in which $|\Delta\omega/K| > \pi$. In this case the phase error, $\Psi_e(t)$, will change periodically in the range from $-\pi$ to

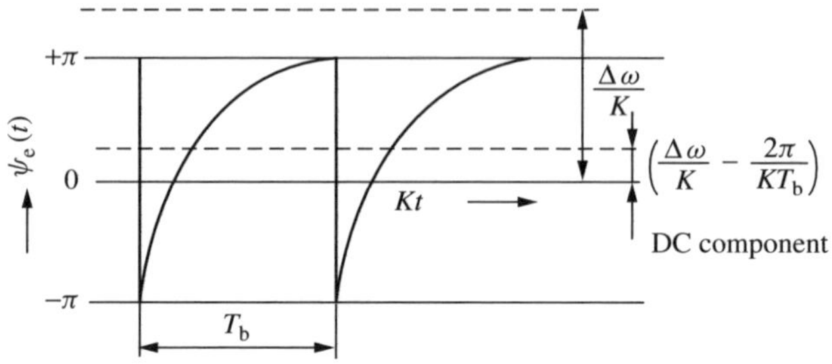

Figure 6.3 Plot of the beat note $\Psi_e(t)$ in the first-order loop with a sawtooth PD for $\Delta\omega/K = 1.4\pi$ (© Kroupa [1]).

$+\pi$. After introduction of these conditions into (1.20) we find the beat note, as shown in Fig. 6.3, and its period as

$$T_b = \frac{2\pi}{\nu_b} = -\frac{\log(\pi - \Delta\omega/K) - \log(\pi + \Delta\omega/K)}{K} \tag{6.38}$$

Note the unsymmetrical shape of the beat wave due to a DC component reducing the original frequency difference $\Delta\omega = |\omega_i - \omega_o|$. The lesson learnt here is that even PLLs of the first order are pulled to the lock, which is, however, beyond reality, when $\Delta\omega$ is larger than $\Delta\omega_H$.

6.2.2 PLLs of the Second Order

The situation is different with second-order PLLs having a memory in the filter capacity. The consequence is that the pull-in process is not immediate but may take some time.

6.2.2.1 PLLs of the second order of type 1

We shall solve the pull-in range in these cases with the assistance of examples.

Example 6.1

First we shall consider a simple RC filter and perform the pull-in solutions. After inserting (2.6) into (6.26), we find for the square of the loop filter transfer function

$$\Phi(\nu) = \frac{1}{1 + (\nu T_1)^2} \tag{6.39}$$

To arrive at a more general solution we shall introduce ω_n and ζ from (2.10) and arrive at

$$\Phi(x) = \frac{4\zeta^2}{x^2 + 4\zeta^2} \tag{6.40a}$$

which can be simplified for small values of ζ to

$$\Phi(x) \approx (2\zeta/x)^2 \tag{6.40b}$$

Application of (6.28) and (6.30) reveals for the normalized pull-in range

$$x_p = 1.475 \tag{6.41a}$$

or in the frequency notation

$$\Delta\omega_p = 1.475\omega_n \tag{6.41b}$$

The performed solution agrees with a more exact one provided by Greenstein [2] and is recalled in Fig. 6.4.

Example 6.2

Let us perform solution of the pull-in range for the second-order loop with lag-lead or RRC filter. For large beat frequencies we have from (2.17)

$$\Phi(\nu) \approx T_2/T_1 \tag{6.42}$$

After application of (6.32) and (6.33), we immediately arrive at the known formula

$$\Delta\omega_p = K\sqrt{2T_2/T_1} \tag{6.43a}$$

which for high-gain loops simplifies with the assistance of the basic PLL parameters (cf. (2.20)) to

$$\Delta\omega_p \approx 2\sqrt{\zeta\omega_n K} \tag{6.43b}$$

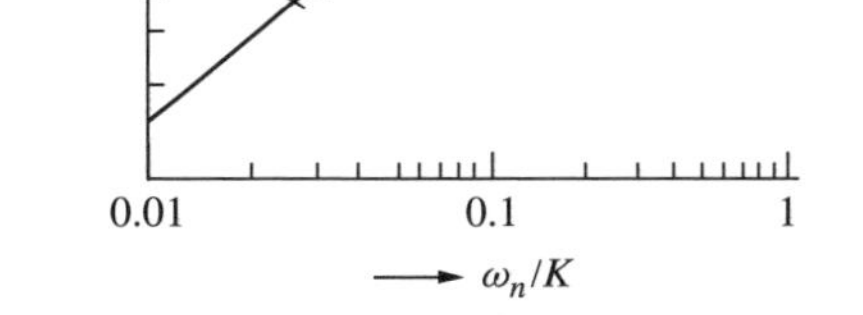

Figure 6.4 Plot of the normalized pull-in range for the second-order PLL with a simple RC loop filter ($F(0) = 1$) as a function of $x = \omega_n/K$; (+) (Data by L.J. Greenstein, "Phase-locked loop pull-in frequency", *IEEE Trans. Commun.*, **COM-22**(8), 1005–1013, 1974).

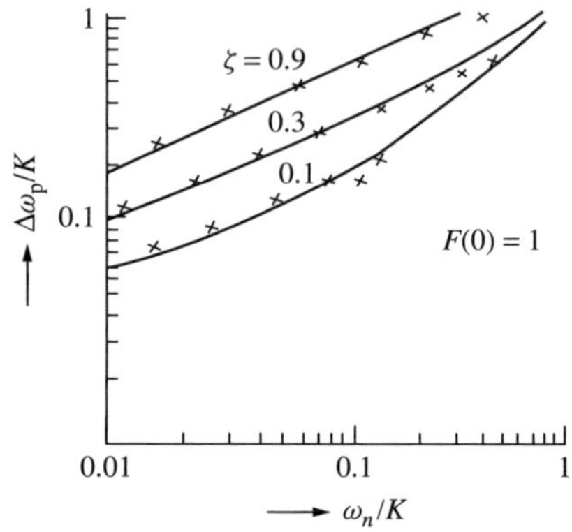

Figure 6.5 Plot of the normalized pull-in range for the second-order high-gain PLLs with a lag-lead RRC loop filter as a function of $x = \omega_n$/K; (+) (Data by L.J. Greenstein, "Phase-locked loop pull-in frequency", *IEEE Trans. Commun.*, **COM-22**(8), 1005–1013, 1974).

All that remains is to investigate the validity of the above solution. To this end we would calculate the normalized frequency x_{m} (cf. (6.27) to (6.29)):

$$x_{\mathrm{m}} = \nu_{\mathrm{m}}/\omega_n = \sqrt{\zeta K/\omega_n} \tag{6.43c}$$

For the high-gain loops we have $\zeta \gg \omega_n/K$, and consequently $x_{\mathrm{m}} > 1$ and the above simplification (6.42) is justified. This also follows from Fig. 6.5 in which we compared solutions (6.43) with those by Greenstein in [2].

Example 6.3

Investigation of the pull-in range for PLLs with a sawtooth PD.

Let us revert to the relation (6.38). We find that for the duration of the beat note period, $T_{\mathrm{b}} \to \infty$, we must put

$$\frac{\Delta\omega}{K} = \pi \tag{6.44}$$

Since for high frequencies the gain K in the PLLs with RRC filters is reduced to K_r (cf. (2.17) and (6.42)), we get for the pull-in range

$$\Delta\omega_{\mathrm{p}} < \pi K_r = \pi \frac{T_2}{T_1} < \pi\sqrt{K K_r} \tag{6.45}$$

Byrne [3] found in this case

$$\Delta\omega_{\mathrm{p}} \approx \pi K \frac{2}{\sqrt{3}} \sqrt{T_2/T_1} \quad (T_2/T_1 < 0.5;\ T_2 K \gg 1) \tag{6.46}$$

which is not much different from relation (6.45). In Fig. 6.6 we see the slowly increasing beat note period on the scope.

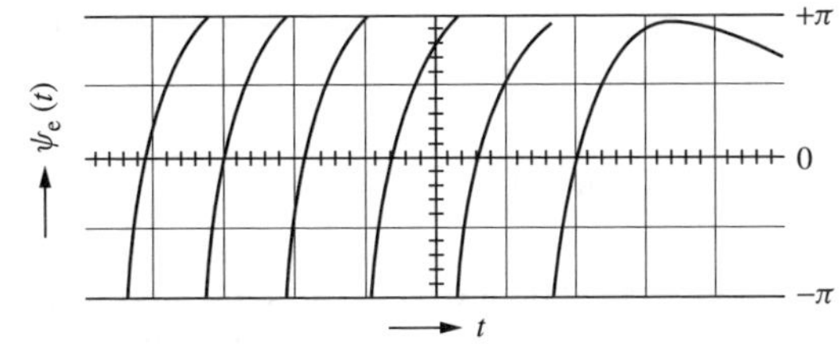

Figure 6.6 Pull-in process of the PLLs with a sawtooth PD; note the slowly increasing beat note (adapted from [1]).

6.2.2.2 PLLs of the second-order of type 2

We have seen in the preceding chapters the advantages of the PLLs of type 2. Computation of the pull-in range will again be illustrated with several examples.

Example 6.4

We shall investigate second-order PLLs of type 2 with a lag-lead or RRC filter. Evidently, relation (6.42) still holds. However, we must introduce the gain A of the operation amplifier into (6.43) in accordance with (6.24) or (6.25). Thus,

$$\Delta\omega_{\mathrm{P}} = K\sqrt{2AT_2/T_1} \tag{6.47a}$$

and by introducing ω_n, ζ, and K_v

$$\Delta\omega_{\mathrm{P}} \approx 2\sqrt{\zeta\omega_n K_v} \tag{6.47b}$$

Finally, in the normalized form we have

$$x_{P,1} = 2\sqrt{A\zeta x_k} = 2\sqrt{A'}, \quad (x_k = K/\omega_n) \tag{6.48}$$

where we have introduced the effective amplification

$$A' = A\zeta x_k \tag{6.49}$$

Example 6.5

In this example we shall investigate PLLs with a lag-lead filter and an additive time delay. We assume the filter transfer function

$$F(\mathrm{j}\nu) = \frac{1 + \mathrm{j}\nu T_2}{\mathrm{j}\nu T_1}\mathrm{e}^{-\mathrm{j}\nu t} \tag{6.50}$$

the real part of which is

$$\Phi(\nu) = \frac{\nu T_2\cos(\nu\tau) - \sin(\nu\tau)}{\nu T_1} \tag{6.51a}$$

First, we shall investigate cases in which the time delay is small. In this instance we can simplify the above equation to

$$\Phi(\nu) \approx \frac{T_2}{T_1}\cos(\nu\tau) \tag{6.51b}$$

By assuming $\cos(\nu\tau)$ nearly independent of ν, that is, the argument $\nu\tau$ is close to zero, we can use the earlier results (6.32), (6.33), and (6.47a), with which we get

$$\nu_{\mathrm{m}} = K\sqrt{\frac{AT_2}{2T_1}\cos(\nu_{\mathrm{m}}\tau)} \tag{6.52a}$$

$$\Delta\omega_{\mathrm{P},2}{}' \approx \Delta\omega_{\mathrm{P},1}\sqrt{\cos(\nu_{\mathrm{m}}\tau)} = \Delta\omega_{\mathrm{P},1}\sqrt{\cos(\Delta\omega_{\mathrm{P},1}\tau/2)} \tag{6.52b}$$

After expanding the cos function into the Taylor series and retaining the first two terms only, we find for the pull-in range

$$\Delta\omega_{\mathrm{P},2}{}' \approx \Delta\omega_{\mathrm{P},1}\sqrt{1 - A\frac{T_2}{T_1}\left(\frac{K\tau}{2}\right)^2} \tag{6.53}$$

or, in the normalized form,

$$x'_{\mathrm{P},2} = 2\sqrt{A'\left[1 - \frac{1}{2}A'(\omega_n\tau)^2\right]} \tag{6.54}$$

Inspection of both relations reveals that with the increasing gain A of the operation amplifier the pull-in range decreases. Or the larger the gain A, the smaller the time delay τ that may be tolerated. The computer solution in Fig. 6.7, by taking harmonics to the eighth order, confirms this conclusion. Here, we have plotted relations between

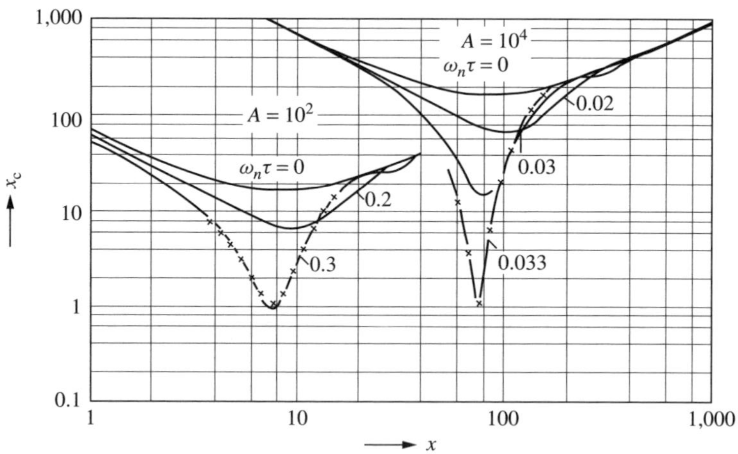

Figure 6.7 The computer solution of the pull-in range in PLL type 2 loops ($\zeta = 0.7$) for three different additional time delays $\omega_n\tau$ and two OP gains A. We have plotted relations between the normalized free-running frequency x_c and the beat note frequency x in accordance with (6.54) (© Kroupa [4]).

the normalized free-running frequency x_c and the beat note frequency x. We use this figure to call attention to the oscillating behavior of x_c for larger values of the time delay τ. We shall discuss this problem later while investigating false locks in PLLs in Section 6.5.1.

Further, analysis of eqs. (6.52) and (6.54) reveals the existence of a critical time delay reducing the pull-in range to zero. This state would requires frequency ν_c in (6.28) to be zero and a negative difference frequency of DC frequency shift $\Delta\omega$ for a nonzero beat frequency ν. The other condition seen from Fig. 6.7 is that x_c touches the x-axis at a tangent. We find the critical ν from eq. (6.28) after introduction of (6.51a):

$$x_c = x + \frac{A'}{x}\left[\cos(x\omega_n\tau) - \frac{\sin(x\omega_n\tau)}{2\zeta x}\right] \tag{6.55}$$

To make the rhs of the above equation equal to zero, the argument, $x\omega_n\tau$, must be in the second quadrant. By estimating $x\omega_n\tau \approx \pi$, we get

$$x_{\text{m,crit}} \approx \sqrt{A'} \tag{6.56}$$

After inserting the above result into (6.55), we find the contribution from the sine term small and consequently negligible. By assuming the argument $(x_{\text{m,crit}}\omega_n\tau)$ to be somewhere in the middle of the second quadrant, which seems reasonable, and applying the shortened Taylor expansion into (6.55), we arrive at

$$x_c = x - \frac{A'}{x}\frac{1}{\sqrt{2}}\left(1 + x\omega_n\tau - \frac{3\pi}{4}\right) \tag{6.57}$$

By performing the derivation of x_c in respect to x and applying relation (6.30), we find

$$x_m = \sqrt{\frac{A'}{\sqrt{2}}\left(\frac{3\pi}{4} - 1\right)} = 0.98\sqrt{A'} \approx \sqrt{A'} \tag{6.58}$$

and after its introduction into (6.57) we get for the pull-in range

$$\begin{aligned} x''_{\text{P},2} &= 1.96\sqrt{A'}\left(1 - \frac{\omega_n\tau\sqrt{A'}}{2.77}\right) \\ &\approx 2\sqrt{A'}\left(1 - \frac{\omega_n\tau\sqrt{A'}}{2\sqrt{2}}\right) \end{aligned} \tag{6.59}$$

After plotting relations (6.59) and (6.54) in Fig. 6.8, we find that (6.59) is a better approximation than (6.54). However, application of (6.54) assures a higher confidence limit since

$$(\omega_n\tau)'_{\text{crit}} = \sqrt{2/A'} \tag{6.60}$$

in contradiction with (6.59)

$$(\omega_n\tau)''_{\text{crit}} = 2\sqrt{2/A'} \tag{6.61}$$

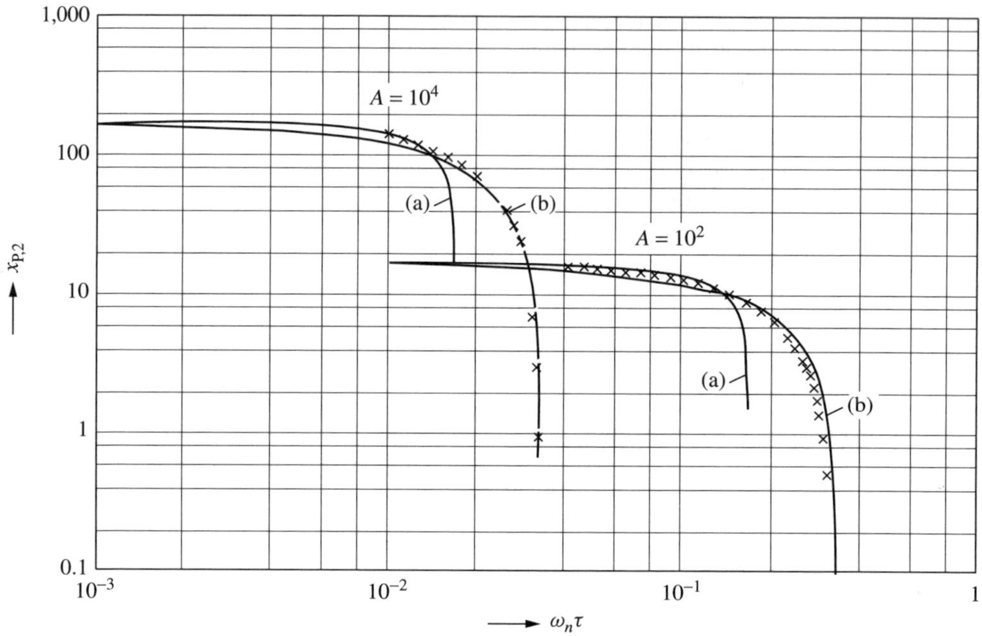

Figure 6.8 The normalized pull-in range $x_{P,2}$ in PLL type 2 loops ($\zeta = 0.7$) as a function of the time delay $\omega_n\tau$ and two OP gains A: (a) in accordance with (6.54) and (b) in accordance with (6.59) (© Kroupa [4]).

Example 6.6

In this example we shall examine a very important case, namely, PLLs of the third order, type 1 and 2 with lag-lead loop filters. It is the RRC high-gain loop with an additional RC section.

From eq. (3.46), we have

$$F(\mathrm{j}\nu) = \frac{1 + j\nu T_2}{\mathrm{j}\nu T_1} \cdot \frac{1}{1 + \mathrm{j}\nu T_3} \tag{6.62}$$

the real part of which, for $T_3 = \kappa T_2$, is

$$\Phi(\nu) = \frac{T_2 - T_3}{T_1} \cdot \frac{1}{1 + \nu^2 T_3^2} = \frac{T_2}{T_1} \cdot \frac{1 - \kappa}{1 + \nu^2\kappa^2 T^2} \tag{6.63}$$

For a very small constant κ, that is, for $\nu_\mathrm{m} T_3 < 1$, we can apply solution (6.31) to (6.33) and arrive at

$$\Delta\omega_{\mathrm{P},3}{}' = \Delta\omega_{\mathrm{P},1}\sqrt{\frac{1 - \kappa}{1 + (\Delta\omega_{\mathrm{P},3}{}' T_3/2)^2}} \tag{6.64}$$

After normalization in respect to the natural frequency ω_n of the original second-order loop, we get

$$x_{P,3}' = x_{P,1}\sqrt{\frac{1-\kappa}{1+(x_{P,3}'\zeta\kappa)^2}} \approx x_{P,1}\sqrt{1-\kappa} \tag{6.65}$$

However, in instances in which $\nu_m T_3 > 1$ we must start with the function

$$\Phi(\nu) \approx \frac{T_2 - T_3}{\nu^2 T_1 T_3^2} = \frac{1}{\nu^2 T_1 T_2} \cdot \frac{1-\kappa}{\kappa^2} \tag{6.66}$$

We arrive at

$$x_c \approx x + \frac{Ax_k(1-\kappa)}{4\zeta\kappa^2 x^3} \tag{6.67}$$

and consequently at

$$x_{m,3} = \sqrt[4]{\frac{3}{4} \cdot \frac{Ax_k}{\zeta} \cdot \frac{1-\kappa}{\kappa^2}} \tag{6.68}$$

$$x_{P,3}'' = \frac{4}{3}x_{m,3} \doteq 1.24\sqrt[4]{\frac{Ax_k}{\zeta} \cdot \frac{1-\kappa}{\kappa^2}} \tag{6.69}$$

$$\Delta\omega_{P,3}'' \doteq 1.48\sqrt[4]{\frac{\kappa^2 A}{T_1 T_2} \cdot \frac{1-\kappa}{\kappa^2}} \tag{6.70}$$

Note a dramatic reduction of the pull-in range in respect to the amplification gain of the operation amplifier (see Fig. 6.9).

We believe that the examples discussed are sufficient for an introduction to the problem. Nevertheless, we will mention, in short, cases solved in [4], namely, the PLLs of the third-order type 2 with an additional time delay. The results are reproduced in Fig. 6.10. Another example solved in [4] is the fourth-order loop with the open-loop gain in accordance with (3.58). The influence of the increasing constant η is illustrated in Fig. 6.11. Up to now we have only investigated systems with sine wave PD. PLLs with other types of PD were investigated, for example, by Hasan [5].

6.3 LOCK-IN RANGE

As we mentioned in the introduction to this chapter, the lock-in range is defined as a frequency difference between the input reference frequency ω_i and the free-running frequency ω_c, for which after closing the loop the phase difference $\phi_e(t)$ converges monotonically to a steady-state value, that is,

$$\Delta\omega_L = |\omega_i - \omega_{cL}| \tag{6.71}$$

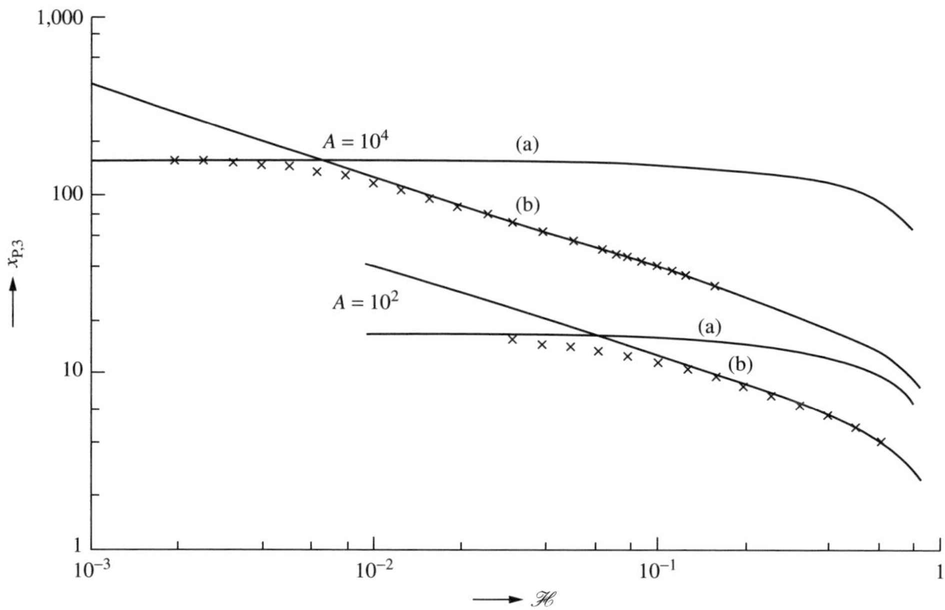

Figure 6.9 The normalized pull-in range $x_{P,3}$ in PLL type 2 loops ($\zeta = 0.7$) as a function of $\kappa = T_2/T_3$ and two OP gains A: (a) in accordance with (6.64) and (b) in accordance with (6.70); crosses indicate the minima computed from (6.28) (© Kroupa [4]).

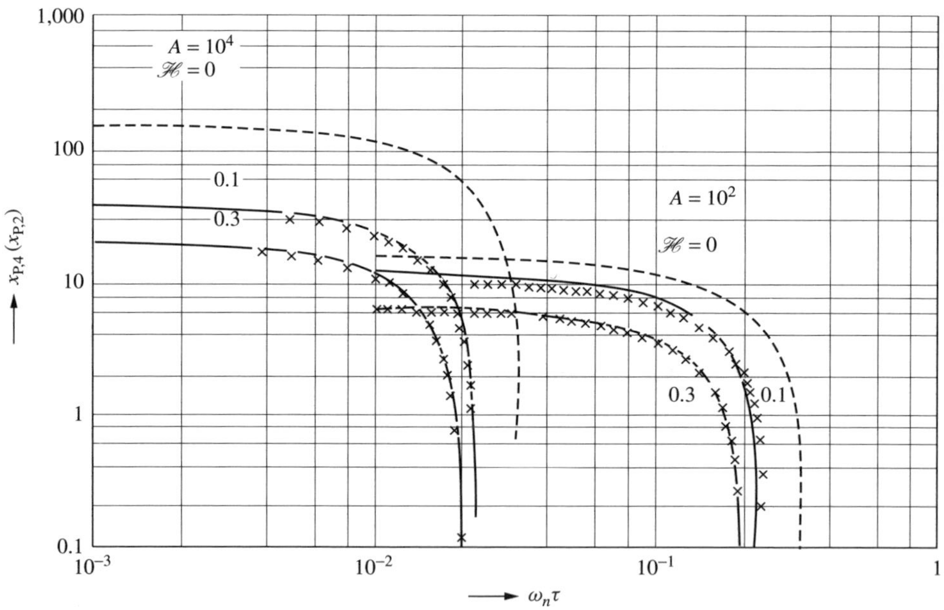

Figure 6.10 The normalized pull-in range $x_{P,4}$ and $x_{P,2}$ in PLL type 2 loops ($\zeta = 0.7$) as a function of the time delay $\omega_n\tau$ and two OP gains A for three different values of κ in accordance with [4] (© Kroupa [4]).

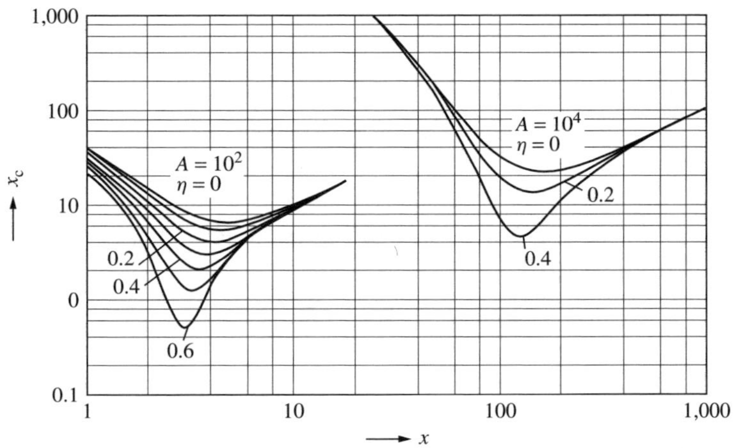

Figure 6.11 The computer solution of the pull-in range in fourth-order PLL type 2 loops ($\zeta = 0.7$, $\kappa = 0.3$) for three different additional constants η and two OP gains A (© Kroupa [4]).

By investigating ω_L, we shall start from the Byrne consideration [3], starting with the worst situation, namely, where the zero crossing of the reference and the VCO signals coincide. By closing the PLL at this moment, the PD output phase $\phi_e(t)$ is formed with two components: $\phi_{e1}(t)$, caused with an effective step change (in the worst case either $\pi/2$ or π), and $\phi_{e2}(t)$, generated with the frequency difference, $\Delta\omega_L$. The undesired cycle skipping of the beat frequency is prevented if the PD outputs are zero. This condition is met for

$$\frac{d\phi_{e1}(t)}{dt} + \frac{d\phi_{e2}(t)}{dt} = 0 \tag{6.72}$$

6.3.1 PLLs of the First Order

Investigation of the lock-in range, ω_L, in first-order loops deserves attention since for the solution of ω_L many second-order loops can be reduced into the first-order system. To provide the desired insight we should use the following example.

Example 6.7

Compute the lock-in range for the first-order loop. Combination of eq. (6.72) with (1.30) reveals

$$\frac{d}{dt}\mathscr{L}^{-1}\left[\left(\frac{\Delta\phi_i}{s} + \frac{\Delta\omega_i}{s^2}\right)\frac{s}{s+K}\right] = 0 \tag{6.73}$$

Application of the Laplace transform pairs (e.g., Tab. 10.1) results in

$$\frac{d}{dt}\left[\Delta\phi_i e^{-Kt} - \Delta\omega_i\left(\frac{1}{K}e^{-Kt} - \frac{1}{K}\right)\right] = 0 \tag{6.74}$$

After derivation we get

$$-\frac{\Delta\phi_i}{K}e^{-Kt} + \frac{\Delta\omega_i}{K^2}e^{-Kt} = 0 \tag{6.75}$$

By assuming the sawtooth PD with the starting conditions $\Delta\phi_i = \pi$ and $\Delta\omega_i = \Delta\omega_L$, we readily find from (6.75)

$$\Delta\omega_L = \pi K \tag{6.76}$$

for the triangular PD

$$\Delta\omega_L = \pi K/2 \tag{6.77}$$

and for the sine wave PD

$$\Delta\omega_L = K \tag{6.78}$$

The above result also follows immediately from Fig. 1.2.

6.3.2 PLLs of the Second Order

A major difficulty is that the Byrne condition (6.72) cannot always be used.

6.3.2.1 PLLs of the second-order lag-lead filters

We should proceed in the same way as explained for first-order loops and we shall perform a solution in the next example.

Example 6.8

After combination of (6.72), (5.13), and (5.20), we have to evaluate the following relation:

$$\frac{d}{dt}\left(\frac{\Delta\phi_i}{\sigma_1-\sigma_2}[(\sigma_1+\omega_n/K_v)e^{(\sigma_1\omega_n t)} - (\sigma_2+\omega_n/K_v)e^{(\sigma_2\omega_n t)}]\right.$$
$$\left. + \frac{\Delta\omega_i}{\sigma_1-\sigma_2}\left[\frac{(\sigma_1+\omega_n/K)}{\sigma_1}e^{(\sigma_1\omega_n t)} - \frac{(\sigma_2+\omega_n/K)}{\sigma_2}e^{(\sigma_2\omega_n t)}\right] + \frac{\Delta\omega_i}{K}\right) \tag{6.79}$$

First, we perform the derivation and then the reduction by $(\sigma_1-\sigma_2)^{-1}\omega_n$

$$\Delta\phi_i[(\sigma_1+\omega_n/K)\sigma_1\omega_n t e^{\sigma_1\omega_n t} - (\sigma_2+\omega_n/K)\sigma_2\omega_n e^{\sigma_2\omega_n t}]$$
$$+ \Delta\omega_i[(\sigma_1+\omega_n/K)\omega_n e^{\sigma_1\omega_n t} - (\sigma_2+\omega_n/K)\omega_n e^{\sigma_2\omega_n t}] = 0 \tag{6.80}$$

Inspection of Figs. 5.1 and 5.2 reveals that the maxima of the tangents, both for the phase and frequency steps, are for $t = 0$. Consequently, (6.80) simplifies to

$$\Delta\omega_{\rm i}[\sigma_1 - \sigma_2] + \Delta\phi_{\rm i}[\sigma_1{}^2 - \sigma_2{}^2 + \omega_n/K(\sigma_1 - \sigma_2)] = 0 \tag{6.81}$$

After a second reduction by $(\sigma_1 - \sigma_2)$, we arrive at

$$\Delta\omega_{\rm i} = \Delta\phi_{\rm i}\omega_n(2\zeta - \omega_n/K) \tag{6.82}$$

By introducing starting conditions, $\Delta\phi_{\rm i} = \pi$ and $\Delta\omega_{\rm i} = \Delta\omega_{\rm L}$, we readily find out

$$\Delta\omega_{\rm L} = \pi\omega_n(2\zeta - \omega_n/K) \tag{6.83}$$

By considering PDs with a sine wave output and PLLs with a lag-lead filter and a very high gain, K, we arrive at

$$\Delta\omega_{\rm L} = 2\zeta\omega_n \tag{6.84}$$

or at

$$\Delta\omega_{\rm L} = K\frac{T_2}{T_1} = K_r \tag{6.85}$$

From the above relation it follows that the lock-in range is equal to the reduced gain $K_{\rm r}$ (cf. (2.17)) at very high frequencies as suggested by Gardner [8]. He states that this conclusion is generally valid for systems with filters having the same number of nulls and zeros. After normalization, $\omega/\omega_n = x$, we have

$$x_{\rm L} = 2\zeta \tag{6.86}$$

6.3.2.2 PLLs of the second order with simple RC filters

The above procedure cannot be applied on PLL systems with simple RC filters since the introduction of $\omega_n/{\rm K} = 2\zeta$ in agreement with relation (2.10) into (6.82) would reveal the lock-in range to be zero, which is certainly not the case, and would testify the problems we encounter with the solution of PLLs in the presence of large signals. A crude estimation of the lock-in range for this type of PLLs shall be found in the following example.

Example 6.9

We shall start with the open-loop gain (2.7) and assume

$$sT_1 \ll 1 \tag{6.87}$$

In this case

$$G(s) \approx \frac{K}{s} \tag{6.88}$$

and eq. (2.9) will simplify into

$$\frac{\Phi_e(s)}{\Phi_i(s)} = T_1 \frac{s(sT_1 + 1)}{s^2 T_1 + s + K} \approx T_1 \frac{s}{s + K} \tag{6.89}$$

which is the first-order loop. The solution proceeds as in Example 6.7, that is,

$$\Delta\omega_L \approx K \tag{6.90}$$

With the assistance of (2.10) we get

$$\Delta\omega_L \approx \frac{\omega_n}{2\zeta} \tag{6.91}$$

What remains is to evaluate the validity of condition (6.91). From (2.10) we have

$$T_1 = \frac{1}{2\zeta\omega_n} \quad \text{and} \quad K = \frac{\omega_n}{2\zeta} \tag{6.92}$$

To meet condition (6.87) we evaluate

$$\omega_L T_1 = K \frac{1}{2\zeta\omega_n} = \frac{1}{4\zeta^2} < 1 \tag{6.93}$$

from which

$$\zeta > 0.5 \tag{6.94}$$

The result was also found by Hasan and Brunk [9] who suggested the form

$$\Delta\omega_L \doteq \frac{\omega_n}{\beta} = K \quad (\text{for } \zeta \geq 0.5 \quad \text{and} \quad \beta = \omega_n/K) \tag{6.95}$$

This result is surprising since the lock-in range is equal to the gain K as for PLLs of first-order loops.

6.4 PULL-OUT FREQUENCY

There are instances in which one of the output frequencies fed to the PD experiences a frequency step (as an example we mention PLL frequency synthesizers with digital dividers in the feedback path). At a certain magnitude, designed as $\Delta\omega_{PO}$ (pull-out range), the PLL loses lock, often only for a short time. The situation is the same as discussed for transients in Chapter 5 (relations (5.23) to (5.25) and Fig. 5.2). We see that the phase error, due to the frequency step, increases from zero to a maximum and then dies out either periodically or aperiodically. In instances in which the maximum phase error, $\phi_2(t)$, is equal to or exceeds the working range of the PD, the phase lock is lost. For this maximum we find from the relation (5.23), in instances of very large or infinite gain K_v, the respective normalized time $(\omega_n t)_{max}$:

For the damping factor $\zeta > 1$ we get

$$(\omega_n t)_{max} = \frac{\operatorname{arctanh}(\sqrt{1-\zeta^2}/\zeta)}{\sqrt{1-\zeta^2}} \tag{6.96}$$

In instances in which $\zeta = 1$ we get

$$(\omega_n t)_{max} = 1 \tag{6.97}$$

And for cases in which $\zeta < 1$ we get

$$(\omega_n t)_{max} = \frac{\arctan(\sqrt{1-\zeta^2}/\zeta)}{\sqrt{1-\zeta^2}} \tag{6.98}$$

From the above relations we conclude that the normalized time is a function of the damping factor, ζ, only, and consequently for the maximum phase error we can write

$$\phi_{e2\,max} = \frac{\Delta\omega}{\omega_n} f(\zeta) \tag{6.99}$$

For a sawtooth wave PD with a maximum tolerated error of π, we get for the pull-out range $\Delta\omega_{PO}$ in the normalized form

$$\Delta\omega_{PO}/\omega_n = x_{PO} = \pi/f(\zeta) \tag{6.100}$$

for the triangular wave PD

$$x_{PO} = \pi/2f(\zeta) \tag{6.101}$$

and for the sine wave PD

$$x_{PO} \approx \frac{1}{f(\zeta)} \tag{6.102}$$

In Fig. 6.12 we have plotted the inverse of $f(\zeta)$, that is, $1/f(\zeta)$, as a function of ζ. Since the plot is nearly a straight line, we have computed (with the assistance of the mean square approach) the slope and the intersection with the $1/f(\zeta)$ axis and arrived at

$$1/f(\zeta) = 1.83(\zeta + 0.53) \tag{6.103}$$

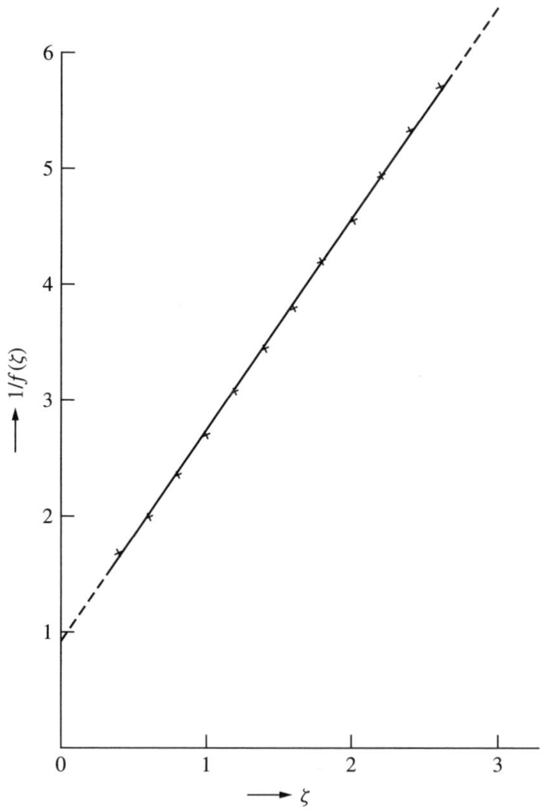

Figure 6.12 Plot of the pull-out frequency function, *1/f*(ζ), versus ζ, the damping factor.

Gardner [8] found a slightly different result, that is,

$$x_{PO} = 1.8(\zeta + 1) \tag{6.104}$$

Hasan and Brunk [5] investigated the problem in the phase plane and arrived at

$$x_{PO} = 1.84(\zeta + 1.1) \tag{6.105}$$

We see that the results of all three authors are practically the same.

6.5 HIGHER-ORDER PLLs AND DIFFICULTIES WITH LOCKING

We have seen that finding the pull-out frequency, $\Delta\omega_{PO}$, even for PLLs of the second order is not an easy process, particularly in instances with nonlinear PDs. In PLLs of higher orders, the number of variables increases and finding a closed solution is nearly impossible. Some information provides evaluation of the maximum phase error,

$\phi_{e2}(t)$, due to the frequency step $\Delta\omega$. PLL systems with sawtooth wave detectors were investigated by Hasan [6, 7]. He tabulated computer results for cases of additional time delays and one or two RC sections. His results are surprising since for practical values of $\omega_n\tau, \kappa$, and η the reduction of $\Delta\omega_{PO}$ is not important at a maximum of about 30%.

6.5.1 False Locking

In the neighborhood of the upper bound, close to the $\Delta\omega_P$, the pull-in process is slow before the lock is realized. A closer investigation of the process reveals that the PD output phase signal seems to be fed into two paths, that is, in an AC with the transfer function, T_1/T_2, and a quasi DC path with the integrator. The situation is illustrated with an approximating block diagram in Fig. 6.13. For the slowly varying component, $\Delta\omega(t)$, we have found in Section 6.2 (cf. eq. (6.25) or (6.27)):

$$\Delta\omega = \frac{KF(0)}{2}|G(j\nu)| \sin\Psi \tag{6.106}$$

In accordance with the block diagram in Fig. 6.13 we must introduce $F(0) = 1$, and the above relation is changed into

$$\Delta\omega = \frac{K^2}{2\nu}|F(j\nu)| \cos\Psi \tag{6.107}$$

However, additional low-pass filters, introduced willingly or unwillingly, in the forward path (additive pole due to the operational amplifier or to the VCO tuning connection) inclusive of the time delay due to the IF filters or digital operation, may generate the phase shift $\Psi > \pi/2$ for some error frequencies. The consequence is that the slowly varying component, $\Delta\omega$ (cf. relation (6.107)) changes sign, that is, $v_{2,DC}(t)$ will be negative and starts to pull the PLL the other way from the correct frequency. In Fig. 6.14(a) we have plotted the normalized component $\Delta\omega/\omega_n$ for the

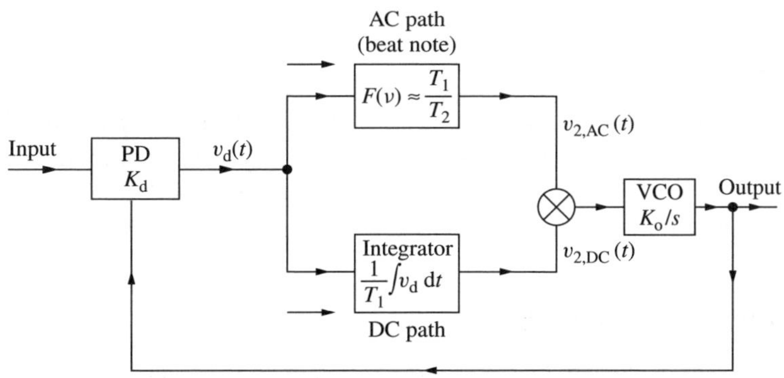

Figure 6.13 Approximating block diagram of the second-order PLL type 2 with a frequency difference $\Delta\omega = |\omega_i - \omega_o|$, only slightly smaller compared with the pull-in frequency $\Delta\omega_P$.

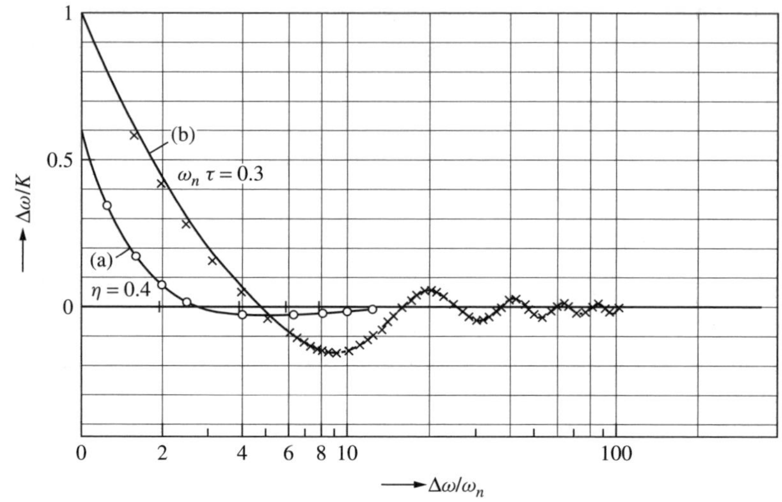

Figure 6.14 The plot of the normalized components $\Delta\omega/K$ and $\Delta\omega/\omega_n$: (a) for a fourth-order loop with two additional RC filter sections and (b) for a second-order PLL with time delay equal to $\omega_n\tau = 0.3$.

fourth-order loop with two additional RC filter sections. The normalized open-loop gain, $G(j\nu)$, is introduced from eq. (3.58).

In accordance with (3.62) and Example 3.4, the phase shift does not exceed -180° and the normalized frequency error is only once equal to zero and this zero is unstable. The other situation plotted in Fig. 6.14(b) is for the time delay $\omega_n\tau = 0.3$. Investigation reveals a characteristic with several zero crossings. We easily find out that zeros with a positive slope are stable (cf. (6.35) and (6.36)). The consequence is that the steering voltage $v_2(t)$ will be pushed, from both sides, to the respective Fourier frequency where it will be zero, and without any action from the outside, the PLL will be locked to the respective false frequency. Only detection of the beat note with an oscilloscope (or by another means) will discover this undesired situation; however, one must be careful of noise or spurious signals in such situations.

6.5.2 Locking on Sidebands

In instances in which the steering signal is periodically modulated, the VCO might lock on sidebands. If we know about this possibility, we should provide countermeasures, for example, introduce an auxiliary branch with a frequency discriminator, reduce the eventual pull-in range, introduce a coarse pretuning of the VCO, and so on.

6.5.3 Locking on Harmonics

In instances in which the VCO frequency is much higher than that of the reference frequency, there is a possibility of locking on an undesired harmonic. However, some sample PDs intentionally use this property, particularly at higher frequencies [10].

6.5.4 Locking on Mirror Frequencies

In instances in which the PLL has a mixer in the feedback path, the long-term instabilities of the input frequencies $\pm\Delta f_\mathrm{i}$ and $\pm\Delta f_\mathrm{a}$ with the $\pm\Delta f_\mathrm{o}$ of the VCO are decisive for the upper bound of the mixing range q, as discussed in [1], and are

$$q = f_\mathrm{a}/f_\mathrm{i}; (f_\mathrm{a} > f_\mathrm{i}) \tag{6.108}$$

However, we must avoid locking on the mirror frequency. The necessary conditions are shown in Fig. 6.15. After summation of individual differences one finds the lower band of the pull-in range

$$f_{\mathrm{P,min}} \geq \Delta f_\mathrm{o} + \Delta f_\mathrm{i} + \Delta f_\mathrm{a} \tag{6.109}$$

However, for the maximum we get

$$f_{\mathrm{P,max}} \leq 2 f_\mathrm{i} - \Delta f_\mathrm{o} - \Delta f_\mathrm{i} - \Delta f_\mathrm{a} \tag{6.110}$$

From the above inequalities we find the condition for the minimum of the input frequency f_i

$$f_\mathrm{i} \geq \Delta f_\mathrm{o} + \Delta f_\mathrm{i} + \Delta f_\mathrm{a} \tag{6.111}$$

and after dividing the above relation by f_0 we get for the mixing range

$$\left|\frac{1}{q+1}\right| \geq \frac{\Delta f_\mathrm{o} + \Delta f_\mathrm{i} + \Delta f_\mathrm{a}}{f_\mathrm{o}} \tag{6.112}$$

In the case in which all frequencies are supplied by crystal oscillators, the mixing range, with assistance of the PLL, can be very high.

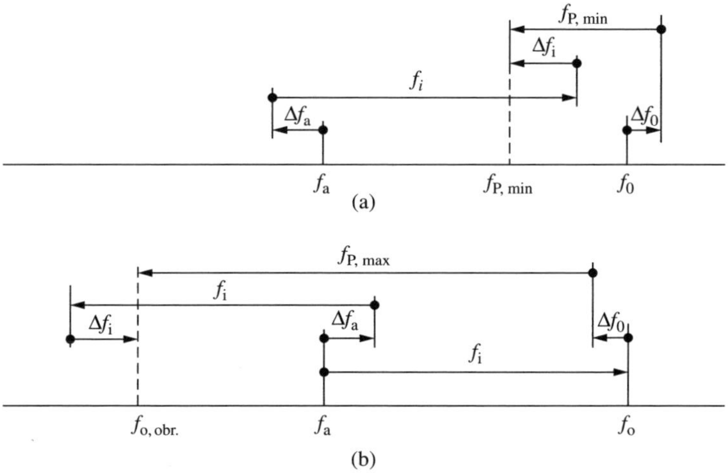

Figure 6.15 Spectral diagrams illustrating pull-in problems in PLL mixers: (a) with respect to the tolerances of the input frequencies and (b) with respect to the image block (© Kroupa [1]).

REFERENCES

[1] V.F. Kroupa, *Frequency Synthesis: Theory, Design et Applications*. London: Ch. Griffin; New York: John Wiley, 1973.

[2] L.J. Greenstein, "Phase-locked loop pull-in frequency", *IEEE Trans. Commun.*, **COM-22**(8), 1005–1013, 1974.

[3] C.J. Byrne, "Properties and design of the phase controlled oscillators with a sawtooth comparator", *Bell Syst. Tech. J.*, **41**, 559–602, 1962.

[4] V.F. Kroupa, "Pull-in range of phase lock loops of the type two", *AEÜ*, **39**(1), 37–44, 1985.

[5] P. Hasan, "Computer-aided pull in range study of phase-lock loops", *AEÜ*, **39**(2), 135–140, 1985.

[6] P. Hasan, "Pull-out frequency of second-order PLL with sawtooth phase detector", *AEÜ*, **37**(7/8), 281, 282, 1983.

[7] P. Hasan, "Pull-out frequency of phase-lock loops with sawtooth phase detector", *AEÜ*, **39**(6), 402, 403, 1985.

[8] F.M. Gardner, *Phaselock Techniques*. 2nd ed. New York: John Wiley, 1979.

[9] P. Hasan,M. Brunk, "Exact calculation of phase-locked loop lock-in frequency", *Electron. Lett.*, **22**(25), 1340, 1341, 1986.

[10] W.F. Egan, *Frequency Synthesis by Phase Lock*. 2nd ed. New York: Wiley, 1981 and 2000.

7
Acquisition of PLLs

The time needed for setting up phase locked loop (PLL) systems is another important parameter to be considered by the PLL designer. In instances in which reference and VCO frequencies are close to the pull-in limit, but not exceeding it, the process might be a slow one with step-by-step decreases of the original beat note. The duration is called the *pull-in time* T_p; however, as soon as the detuning reaches the lock-in range, the phase difference, $\phi_e(t)$, approaches, rather quickly, the steady state value with the time control being inversely proportional to $\omega_n\zeta$ (cf. Chapter 6), and the duration is the so-called *settling time*, which is generally much shorter than the pull-in time.

7.1 THE PULL-IN TIME

In the case in which the frequency difference is close to the pull-in range

$$|\nu_{VCO} - \nu_i| = \nu_c \approx \Delta\omega_P \tag{7.1}$$

the PLL is in the so-called *hang-up* state [1]. This situation, discussed in Chapter 6, might also be designated as *sticking*, and particularly the state of sticking provides a good starting point for a solution of the pull-in time process. As in Chapter 6, we shall start with the mathematical model in Fig. 6.13. The DC path causes only a small decrease of the frequency difference $|\omega_i - \omega_{VCO}|$ and consequently the AC path is decisive for the quasi-stationary behavior of the PLL and for effective detuning ν_c. As long as the loop has the same number of zeros and poles, its transfer function is constant for large frequencies and we have a first-order loop with a reduced gain; for second-order lag-lead PLLs we have

$$K_r = K\frac{T_2}{T_1} \tag{7.2}$$

Phase Lock Loops and Frequency Synthesis V.F. Kroupa
 ISBN: 0-470-84866-9

7.1.1 The Pull-in Time of a PLL with Sine Wave PD

To compute the period of the "instantaneous" beat note with a sine wave PD, we shall use relation (1.12) and perform its integration in one period in the limits from 0 to 2π. After their introduction into (1.13), we find for the beat period T_b (by taking into account that $\arctan(0) = 0, \pm\pi$, etc.)

$$T_\mathrm{b} = \frac{2\pi}{\sqrt{\nu_\mathrm{c}^2 - K_\mathrm{r}^2}} \tag{7.3}$$

From the above relation, we get for the beat note frequency

$$\nu_\mathrm{b} = \sqrt{\nu_\mathrm{c}^2 - K_\mathrm{r}^2} \tag{7.4}$$

and for the mean value of the frequency difference during one beat note

$$\Delta\nu = \nu_\mathrm{c} - \nu_\mathrm{b} \tag{7.5}$$

The phase due to this small frequency step (for the first-order loop) is

$$\phi_\mathrm{e}(t) = \frac{\Delta\nu}{K_\mathrm{r}}(1 - \mathrm{e}^{-K_\mathrm{r}t}) \tag{7.6}$$

In the instance where the time $t \to 0$, the starting phase difference is $\phi_\mathrm{e,o}$

$$\phi_\mathrm{e,o} = \Delta\nu / K_\mathrm{r} \tag{7.7}$$

After introduction of (7.5), we get for a slowly varying phase ϕ_e

$$\phi_\mathrm{e} \approx \frac{\nu_\mathrm{c} - \nu_\mathrm{b}}{K_\mathrm{r}} = \frac{\nu_\mathrm{c}}{K_\mathrm{r}} - \sqrt{\left(\frac{\nu_\mathrm{c}}{K_\mathrm{r}}\right)^2 - 1} \tag{7.8}$$

This relation was derived by Richman [2] some 50 years ago. Note that at the input of the integrator there is a voltage

$$v_\mathrm{d} = \phi_\mathrm{e} K_\mathrm{d} \tag{7.9}$$

and at the output

$$v_\mathrm{2,DC}(t) = \frac{1}{T_1}\int_0^t v_\mathrm{d}\,\mathrm{d}t \tag{7.10}$$

This voltage causes a difference of the VCO frequency by $K_\mathrm{o} v_\mathrm{2,DC}$ (cf. eq. (1.7)). After its insertion into (1.24), we find a frequency change

$$\nu = \nu_\mathrm{c} - K_\mathrm{o} v_\mathrm{2,DC} = \nu_\mathrm{c} - \frac{K_\mathrm{d} K_\mathrm{o}}{T_1}\int_0^t \phi_\mathrm{ep}\,\mathrm{d}t \tag{7.11}$$

After derivation of this relation and after the introduction of ϕ_e, we arrive at

$$dt = \frac{d\nu}{\dfrac{K_o K_d}{T_1}\left[\dfrac{\nu}{K_r} - \sqrt{\left(\dfrac{\nu}{K_r}\right)^2 - 1}\right]} \tag{7.12}$$

Integration of this differential equation will follow after multiplication of the numerator and the denominator by

$$\frac{\nu}{K_r} + \sqrt{\left(\frac{\nu}{K_r}\right)^2 - 1} \tag{7.13}$$

Using formula (112) from [3], we get for the pull-in time

$$T_P = T_2 \left| \frac{1}{2}\left(\frac{\nu}{K_r}\right)^2 + \frac{1}{2}\sqrt{\left(\frac{\nu}{K_r}\right)^2 - 1} - \frac{1}{2}\ln\left(\frac{\nu}{K_r} + \sqrt{\left(\frac{\nu}{K_r}\right)^2 - 1}\right)\right|_{K_r}^{\nu_c} \tag{7.14}$$

For the lower bound of the settling range, we may choose $\Delta\omega_L$, which is equal to K_r in accordance with (6.78). The upper bound will be $\Delta\omega$, which must be slightly smaller than the pull-in range $\Delta\omega_P$. In this situation, it is true that

$$K_r < \Delta\omega < \Delta\omega_P \tag{7.15}$$

and the normalized pull-in time follows as

$$\frac{T_P}{T_2} \approx \left(\frac{\Delta\omega}{K_r}\right)^2 \tag{7.16a}$$

After inserting the natural frequency ω_n and after evaluating K_r, we finally get

$$T_P 2\zeta\omega_n \approx \left(\frac{\Delta\omega}{\omega_n}\right)^2 \tag{7.16b}$$

These results may be considered as reliable in instances in which $\Delta\omega$ is neither close to the lock-in range $\Delta\omega_L$ nor to the pull-in range $\Delta\omega_P$. An asymptotic solution of the normalized pull-in time is shown in Fig. 7.1.

Rosenkranz [4] found a more accurate solution of T_P for cases in which $\Delta\omega$ was close to the upper bound $\Delta\omega_P$, that is,

$$\omega_n T_P = \frac{1}{\beta}\cdot\frac{2x}{\sqrt{1-x^2}}\arctan\frac{x}{\sqrt{1-x^2}} \tag{7.17}$$

where $\beta = \omega_n/K$ and $x = \Delta\omega/\Delta\omega_P$. In Fig. 7.1 Rosenkranz's solution is indicated by crosses (+).

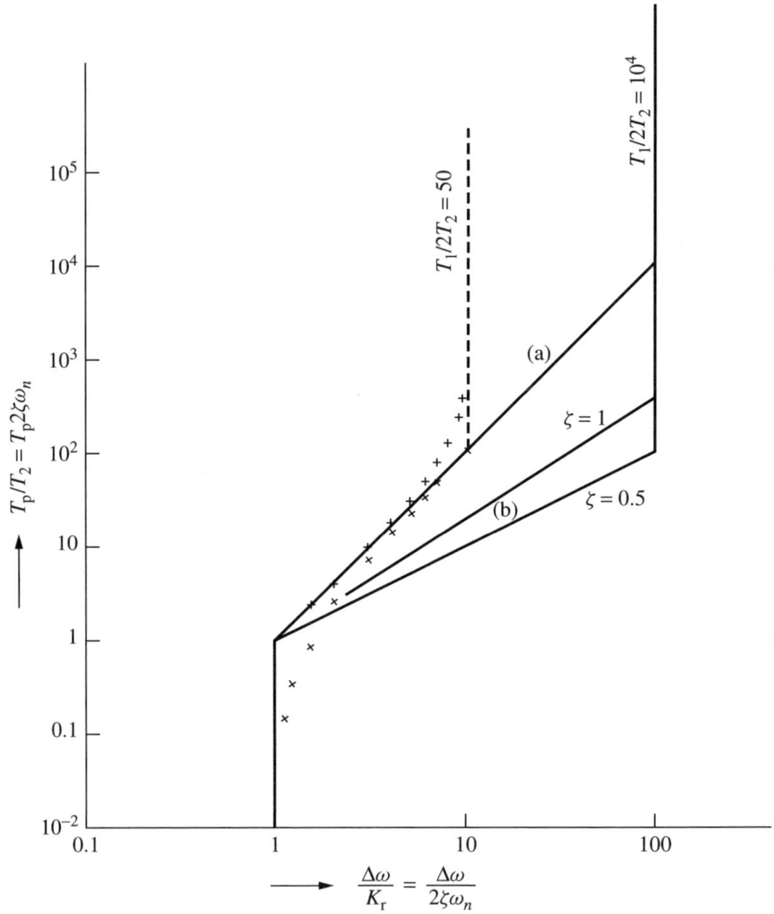

Figure 7.1 Asymptotic solution of the normalized pull-in time of the second-order PLL for the frequency difference $\Delta\omega = |\omega_i - \omega_{VCO}|$: (a) for a simple sine wave PD, ($\times$) were computed for small detuning from relation (7.14), whereas (+) were computed using (7.15) for large normalized detuning and (b) for a phase frequency detector for two different damping factors ζ.

We arrive at a more accurate solution in the vicinity of the lower bound after calculation of T_P from (7.14) for a set of different ratios $\Delta\omega/K_r$ (see points ($\times$) in Fig. 7.1).

7.1.2 The Pull-in Time of a PLL with Sawtooth Wave PD

In the case of a sawtooth wave PD, we shall start with (1.19) and after its integration into the range of one period we get the duration of the beat note T_b already computed in the previous chapter in relation (6.38). After its expansion into a Taylor series, from which we retain only the first two terms, and after introduction of ν_b from (7.5), we arrive at a frequency change during one beat note

$$\Delta\nu = \frac{1}{3}\frac{(K_r\pi)^2}{\nu_c} \tag{7.18}$$

Proceeding as in the former case (cf. relations (7.6) through (7.11)), we find out after integration for the normalized pull-in time

$$T_P 2\zeta\omega_n \approx \frac{1.5}{\pi^2}\left(\frac{\Delta\omega}{\omega_n}\right)^2 \tag{7.19}$$

7.1.3 The Pull-in Time of a PLL with Triangular Wave PD

Similarly, for the PD with a triangular output wave, we again start with eq. (1.19); however, we find that the pull-in time is four times longer, that is,

$$T_P 2\zeta\omega_n \approx \frac{6}{\pi^2}\left(\frac{\Delta\omega}{\omega_n}\right)^2 \tag{7.20}$$

As in the last example, we shall consider the popular phase frequency PD, the analog output of which is illustrated in Fig. 8.19. We easily find that outside of the lock-in range $\Delta\omega_L$ the phase is either $+\pi$ or $-\pi$. Integration reveals a much shorter pull-in time [5]

$$T_P\omega_n \approx \frac{\Delta\omega}{\omega_n} \tag{7.21}$$

The normalized solution for two different damping factors is illustrated in Fig. 7.1.

7.2 THE LOCK-IN TIME

We may define the lock-in time as the time that the PLL needs to reduce the phase error $\phi_e(t)$ to one-tenth of its maximum without cycle-skipping between $+\phi_{e,max}$ and $-\phi_{e,min}$.

For the first-order loop with sawtooth or triangular output wave PD, we arrive with the assistance of (1.20) at the relation

$$T_{L,1} \approx 2.3/K \tag{7.22}$$

Similarly, for sine wave PD, using (1.16), we find

$$T_{L,1} \approx \frac{2.3}{K\sqrt{1-(\Delta\omega/K)^2}} \tag{7.23}$$

As soon as $|\Delta\omega/K| < 0.9$, we can simplify the above relation to

$$T_{L,1} < 5/K \tag{7.24}$$

The lock-in time for PLLs of the second order with sawtooth or triangular wave PDs can be easily found from (5.15) or (5.23)

$$T_{L,2} \approx \frac{2.3}{\zeta\omega_n} \approx \frac{4.6T_1}{KT_2} \approx \frac{4.6}{K_r} \tag{7.25}$$

Several authors quote values for the lock-in time which are twice as long than that indicated by relations (7.22) and (7.23).

7.3 AIDED ACQUISITION

A quick look at Fig. 7.1 is sufficient for the statement that the pull-in time may be intolerably long for some applications of PLLs, particularly in frequency synthesis or communications applications. The situation is more favorable in instances in which phase frequency detectors are used; however, even here the pull-in time could be longer than the system will tolerate. In these cases, we generally solve the problem with a temporary change of the loop parameters or with auxiliary circuits to speed up PLL locking.

7.3.1 Pretuning of the VCO

One of the possibilities is pretuning the VCO, schematically shown in Fig. 7.2(a). This approach is advantageous in devices with large division factors in the feedback path or when a large frequency step is required (e.g., by switching from receiving to transmitting states, etc.). In some PLL systems containing a microprocessor, its use may solve the problem.

7.3.2 Forced Tuning of the VCO

A forced change of the VCO frequency can be achieved by charging or discharging of the integrating capacitor in the low-pass PLL filter (Fig. 7.2(b)); with the assistance of an auxiliary low-frequency oscillator, the output voltage of which is superimposed on the tuning element (varactor) of the PLL VCO (Fig. 7.2(c)), the forced sweeping of the VCO is suppressed as soon as the PLL is locked, either by compulsory means (by changing the impedance in the feedback path of the auxiliary oscillator) or by other means.

In accordance with relation (5.31), the steady state phase difference ϕ_e with sinusoidal PD is given by

$$\sin\phi_a = \frac{\Delta\dot{\omega}}{\omega_n^2} \tag{7.26}$$

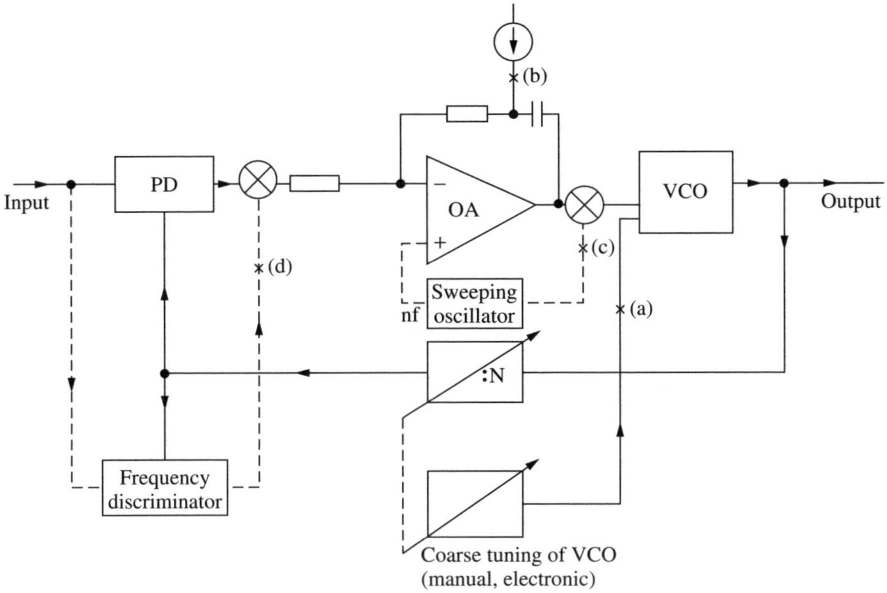

Figure 7.2 Principles of different networks for reduction of the pull-in and settling processes in PLLs: (a) by coarse pretuning of the VCO; (b) by charging or discharging of the integrating capacitor; (c) by forced sweeping of the VCO; and (d) with the assistance of a frequency discriminator.

Consequently, the speed of the frequency change cannot exceed ω_n^2. However, a practical recommendation for $\mathrm{d}(\Delta\omega)/\mathrm{d}t$ is only one half of ω_n^2 since the probability of the locking, for any starting phase condition, is high (≈ 1) (see Fig. 7.3(a)) [Gardner, 1]. By taking into account the $\mathrm{SNR_L}$ (the signal-to-noise ratio loop), the suggested value for the preliminary design should be

$$\Delta\dot{\omega} = \tfrac{1}{2}\omega_n^2[1 - 2(SNR_{\mathrm{L}})^{-1/2}] \tag{7.27}$$

Furthermore, Gardner recommends higher damping factors that enable locking, in the range from 0.5 to 2 (see Fig. 7.3(b)).

7.3.3 Assistance of the Frequency Discriminator

Another possibility for speeding up the PLL is the bridging of the PD with a frequency discriminator (see Fig. 7.2(d)), the DC output of which gradually decreases the frequency difference $\Delta\nu = |\omega_{\mathrm{i}} - \omega_{\mathrm{VCO}}|$. With the assistance of Fig. 6.13 we find for a large difference $\Delta\nu$ the relation

$$\Delta\nu(t) = \frac{K_{\mathrm{fd}}K_{\mathrm{o}}}{T_1}\int_{\omega_{\mathrm{L}}}^{\Delta\omega} \Delta\nu(t)\,\mathrm{d}t \tag{7.28}$$

where K_{fd} is the gain of the frequency discriminator.

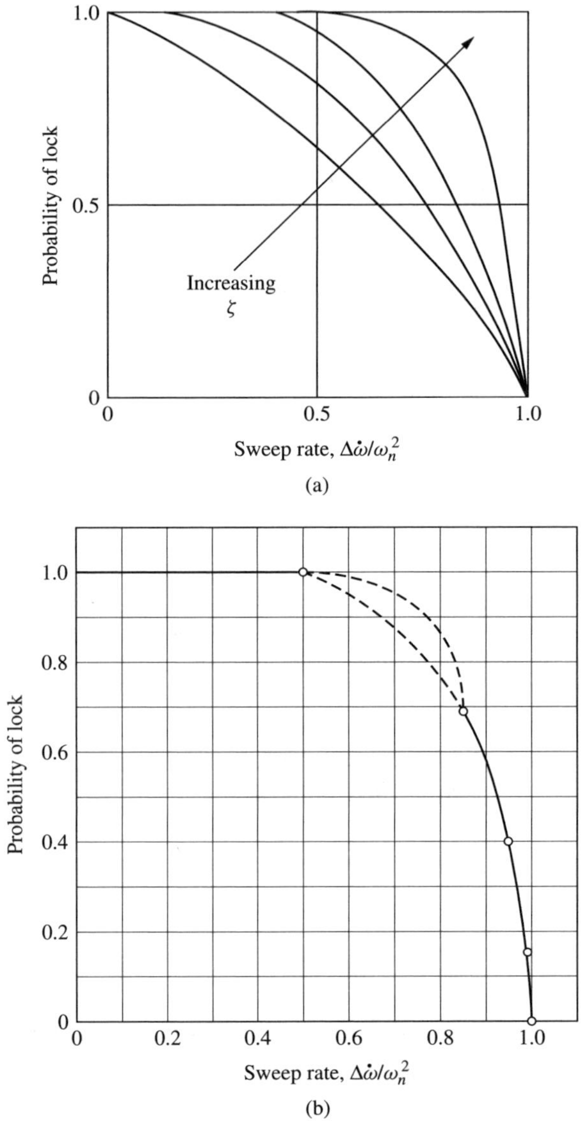

Figure 7.3 Probability of sweep acquisition: (a) second-order loop with $\zeta = 0.7$ – no noise; (b) effect of damping (Reproduced from F.M. Gardner, *Phaselock Techniques*. 2nd ed. New York: John Wiley, 1979 by permission of John Wiley & Sons, Inc., 2002).

After derivation of the above relation, separation of variables, and the succeeding integration, we arrive at the conclusion that the pull-in time is proportional to the logarithm of the detuning

$$\frac{T_P}{T_1} \approx K_{fd} K_o \ln \frac{\Delta\omega}{\omega_L} \tag{7.29}$$

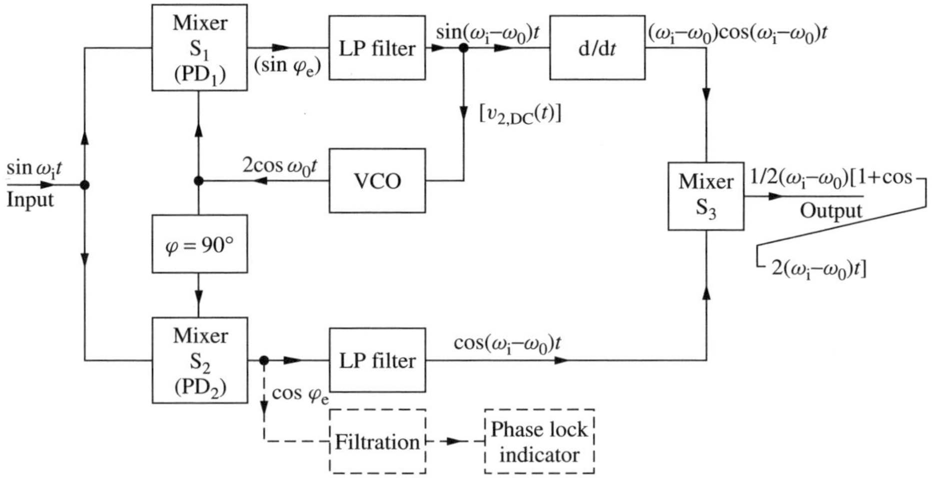

Figure 7.4 Block diagram of the quadricorrelator.

By comparing the above result with relations (7.14) and (7.19), we find that the application of the frequency discriminator provides the shortest pull-in time T_P. As soon as the PLL lock takes place, the output of the frequency discriminator is nearly zero and there is no necessity for its disconnection. The difficulty is that the input SNR must be at least 6 dB, otherwise no useful information will be gained.

Frequency discriminators with output voltage proportional only to the frequency difference are of advantage. One arrangement designated as a *quadricorrelator* is shown in Fig. 7.4. Its function is easily appreciated with the assistance of mathematical relations inserted in the figure. At first, even a superficial inspection of the network might seem very complicated. However, taking into account that one mixer with the following low-frequency filter can be part of the loop PD and that the second branch may serve for lock indication, the only addition is formed by one mixer and one differentiation with HF filter transfer function

$$G_d(j\omega) \approx \frac{j\omega/\omega_1}{1 + j\omega/\omega_1} \approx j\frac{\omega}{\omega_1} \quad (\text{for } \omega \ll \omega_1) \tag{7.30}$$

where ω_1 is smaller than the expected frequency difference. Note that circuits for indication of the phase lock are shown in dashed lines in Fig. 7.4. For the zero frequency difference, the output of PD_2 is at a maximum.

7.3.4 Increasing the PLL Bandwidth

We have seen that the pull-in time is inversely proportional to the square of the natural frequency ω_n. Evidently, its temporary increase may help to reduce T_P. Inspection

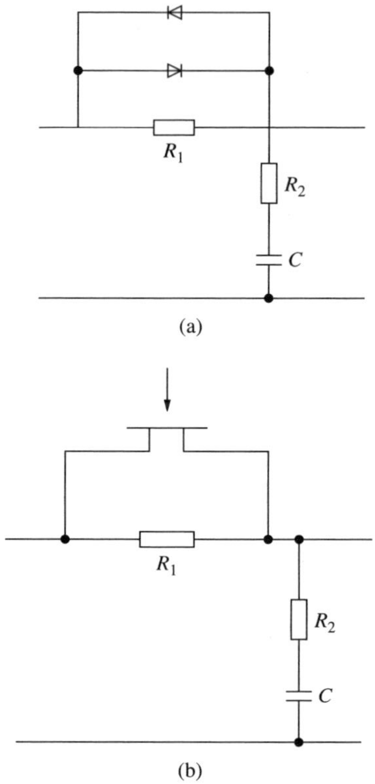

Figure 7.5 Principal arrangement for increasing the natural frequency ω_n by shunting the filter resistor: (a) with diodes and (b) with a switch.

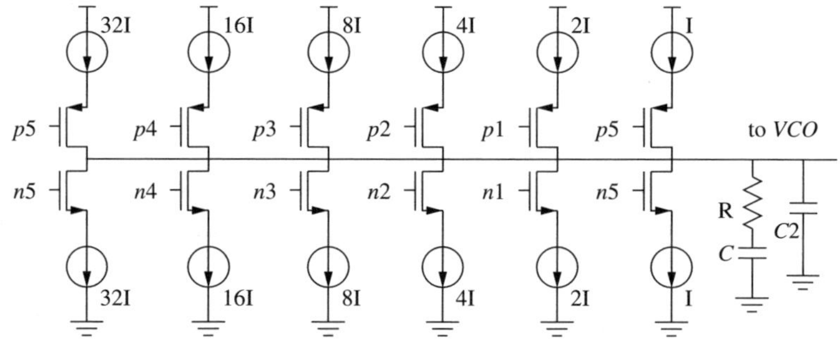

Figure 7.6 Charge pump with switched currents and a loop filter (Reproduced from G.-T. Roh *et al.*, "Optimum phase-acquisition technique for charge-pump PLL", *IEEE Trans. Circuits Syst.*, **44**(9), 729–739, 1997 by permission of IEEE, New York, 2002).

of (2.10) or (2.20) indicates two ways of increasing ω_n, either by reduction of the time constant T_1 or by enlarging the gain K.

In the first case we can reduce, at least partly, the loop filter resistance, as shown schematically in Fig. 7.5 – in the simplest way, this can be done by connecting two diodes in a reverse arrangement (cf. Fig. 7.5(a)). If the frequency difference is large,

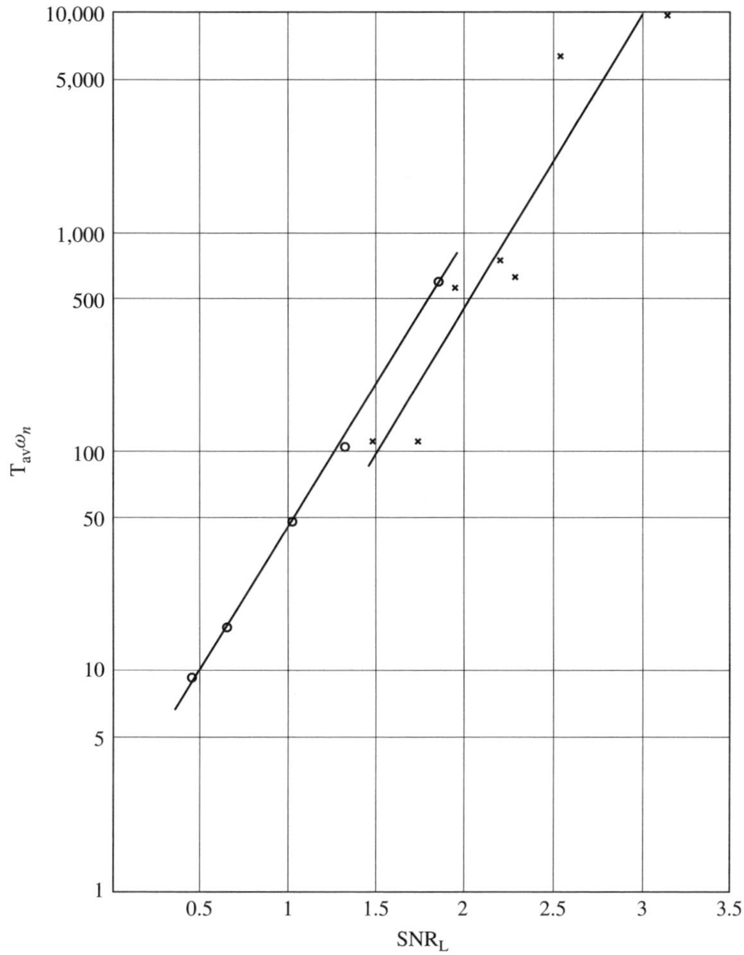

Figure 7.7 Normalized mean time to unlock ($T_{av}\omega_n$) versus (SNR)$_L$ in high-gain second-order loop ($\zeta = 0.7$): (O) theoretical results (Sanneman and Rowbotham [7]) and ($\times$) experimental verification (Keblawi [8]) ([© Kroupa [9]).

the output of the PD $v_d(t)$ is large and the effective resistance of R_1 in Fig. 7.5(a) is equal to the resistance of the diodes in the conducting state. As soon as the PLL lock takes place, the voltage $v_d(t)$ is reduced, the diodes are closed and their resistance increases to an effective infinity.

Another solution presents a complementary metal oxide semiconductor (CMOS) switch controlled by the lock indicator over the resistor R_1. The advantage of this arrangement, in comparison with diodes, is in preventing transient peaks generated in the PD to influence activities of PLLs.

Modern phase frequency detectors make possible the perfection of the above suggested changes of PLL properties, by multiple levels of the current I_p. One arrangement is shown in Fig. 7.6 [6].

7.4 TIME TO UNLOCK

If noise is present, as always happens, there is a very small probability of cycle-slipping in the phase locked loop. Sanneman and Rowbotham [7] investigated the behavior of the high-gain second-order loop ($\zeta = 0.707$) by computer simulation. The result is a simple exponential law between the average time (T_{av}) to unlock and the $(\mathrm{SNR})_\mathrm{L}$

$$T_{av} = \frac{2}{\omega_n} \exp[\pi (\mathrm{SNR})_\mathrm{L}] \tag{7.31}$$

This relation was verified experimentally by Keblawi [8] (see Fig. 7.7). For $(\mathrm{SNR})_\mathrm{L} = 1$, T_{av} is of the order of seconds and the loop performance is certainly unsatisfactory. However, the behavior of the phase lock loop improves rapidly with $(\mathrm{SNR})_\mathrm{L}$. If the extrapolation in Fig. 7.7 to larger values is justified, then for $(\mathrm{SNR})_\mathrm{L} = 10$ to 20 the expected T_{av} will be of the order of 10^8 s or even more, which is compatible with reliability requirements of solid-state devices.

REFERENCES

[1] F.M. Gardner, *Phaselock Techniques*. 2nd ed. New York: John Wiley, 1979.

[2] D. Richman, "Color-carrier reference phase synchronization accuracy in NTSC color television", *Proc. IRE*, **42**(1), 106–133, 1954.

[3] G.A. Korn and T.M. Korn, *Mathematical Handbook for Scientists and Engineers*. New York: McGraw-Hill, 1961.

[4] W. Rosenkranz, "On the pull-in time of second order phase locked loops with noise input signals", *AEÜ*, **38**(1), 9–14, 1984.

[5] J. Eijselendoorn and R.C. den Duik "Improved phase-locked loop performance with adaptive phase comparators", *IEEE Trans. Aerospace Electron. Syst.*, **AES-18**(3), 323–332, 1982.

[6] G.-T. Roh, Yong Hoon Lee and Beomsup Kim "Optimum phase-acquisition technique for charge-pump PLL", *IEEE Trans. Circuits Syst.*, **44**(9), 729–739, 1997.

[7] R.W. Sanneman and J.R. Rowbotham, "Unlock characteristics of the optimum Type II phase locked loop", *IEEE Tr. Aerospace Navigational Electronics*, **ANE-11**, 15–24, 1964.

[8] F.S. Keblawi, "Unlock behavior of the second order phase-locked loop with and without interfering carriers", *RCA Rev.*, **28**, 277–296, 1967.

[9] V.F. Kroupa, *Frequency Synthesis: Theory, Design et Applications*. London: Ch. Griffin, 1973; New York: John Wiley, 1973.

8

Basic Blocks of PLLs

8.1 FILTERS WITH OPERATIONAL AMPLIFIERS

From the discussion of second-order type 2 PLLs, we encountered operational amplifiers with a high gain in the loop filter. Its principle arrangement is illustrated in Fig. 8.1. To find its properties, we apply Kirchhoff's law on the point "a"

$$i_1 + i_2 - i_3 = 0 \tag{8.1}$$

After introduction of the effective voltage ε, we have for the output

$$v_o = A\varepsilon \tag{8.2}$$

and application of all three voltages, in accordance with (8.1), reveals

$$\frac{v_i - \varepsilon}{Z_1} + \frac{v_o - \varepsilon}{Z_2} - \frac{\varepsilon}{R_v} = 0 \tag{8.3}$$

where we have introduced the input resistance of the amplifier R_v. With the assistance of some simple computation, we arrive at the transfer function of the network in Fig. 8.1

$$\frac{v_o}{v_i} = \frac{AZ_2R_v}{-AZ_1R_v + Z_2R_v + Z_1R_v + Z_1Z_2} \tag{8.4a}$$

$$\doteq -\frac{Z_2}{Z_1} \qquad (\text{for } A \to \infty) \tag{8.4b}$$

The input impedance with the use of (8.1) and (8.2) is

$$\frac{v_i}{i_1} = Z_1\frac{v_i}{v_i - v_o/A} \doteq Z_1 \tag{8.5}$$

Phase Lock Loops and Frequency Synthesis V.F. Kroupa
 ISBN: 0-470-84866-9

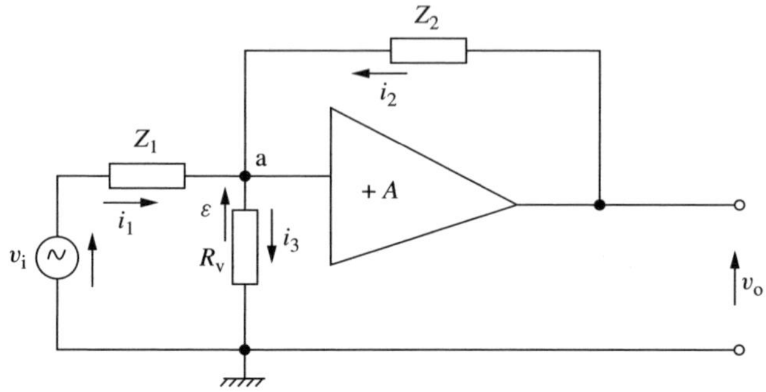

Figure 8.1 Principle filter arrangement of the operational amplifier.

Finally, by replacing impedances Z_1 and Z_2 in accordance with Fig. 2.8(b) we arrive at the desired transfer function (2.25). For a slightly complicated impedance Z_2, as in Fig. 8.2, we find the second-order loop filter with the transfer function $F(\mathrm{j}\omega)$ as suggested for the third-order loop (cf. eqs. (3.45) and (3.46)):

$$F(\mathrm{j}\omega) = \frac{1 + \mathrm{j}\omega R_2(C + C_3)}{\mathrm{j}\omega R_1 C(1 + \mathrm{j}\omega R_2 C_3)} \tag{8.6}$$

A popular arrangement of operational amplifiers (OPAMPs) is a differential connection, illustrated in Fig. 8.3. Similar to the solution applied in the previous simple case, we introduce auxiliary voltages ε_1 and ε_2

$$\varepsilon_1 = \frac{v_{i1} Z_2 + v_o Z_1}{Z_1 + Z_2} \tag{8.7}$$

and

$$\varepsilon_2 = v_{i2} \frac{Z_4}{Z_3 + Z_4} \tag{8.8}$$

By assuming, as earlier,

$$\varepsilon_2 - \varepsilon_1 = v_o / A \approx 0 \tag{8.9}$$

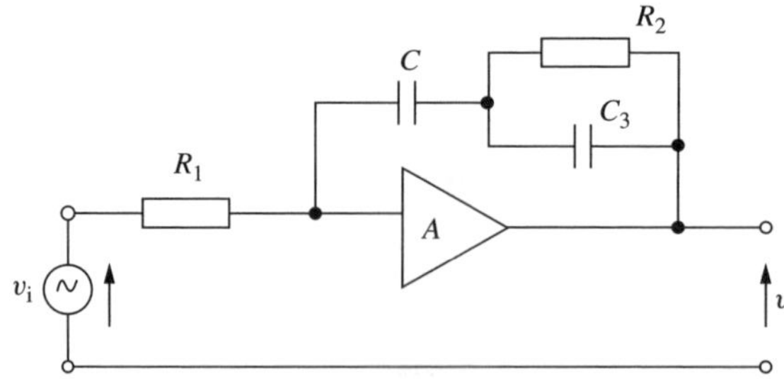

Figure 8.2 Third-order operational amplifier loop filter.

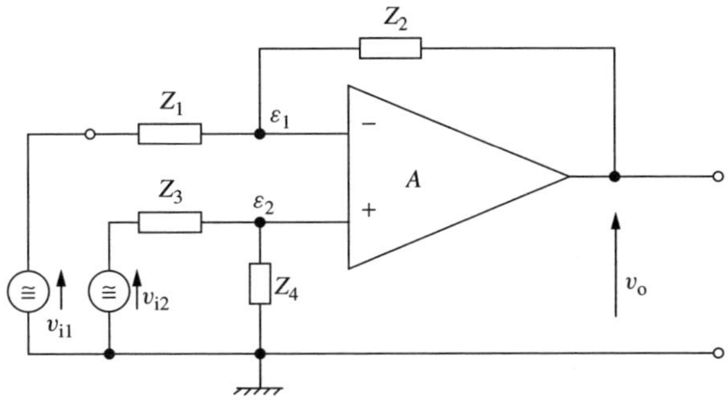

Figure 8.3 Differential connection of the operational amplifier.

we get for the output voltage v_o

$$v_o = \frac{Z_1 + Z_2}{Z_1}\left(v_{i2}\frac{Z_4}{Z_3 + Z_4} - v_{i1}\frac{Z_2}{Z_1 + Z_2}\right) \tag{8.10}$$

A very important case, encountered in practice, is the following equality of the impedances

$$Z_3 = Z_1; \qquad Z_4 = Z_2 \tag{8.11}$$

After introducing the above relations into eq. (8.10), the transfer function of this active filter is

$$\frac{v_o}{v_{i2} - v_{i1}} = \frac{Z_2}{Z_1} \tag{8.12}$$

as required for type 2 PLLs. By arranging the impedance Z_2 as in Fig. 8.2, we again change the second-order PLL into the third-order PLL (cf. (8.6)).

Another type of low-pass active filter, used in fourth- and fifth-order PLLs (in Chapter 3), is shown in Fig. 8.4. Again, application of Kirchhoff's law on point "b" reveals

$$\frac{v_i - \varepsilon}{R} + \frac{v_o - \varepsilon}{Z_1} - \frac{\varepsilon}{R + Z_2} = 0 \tag{8.13}$$

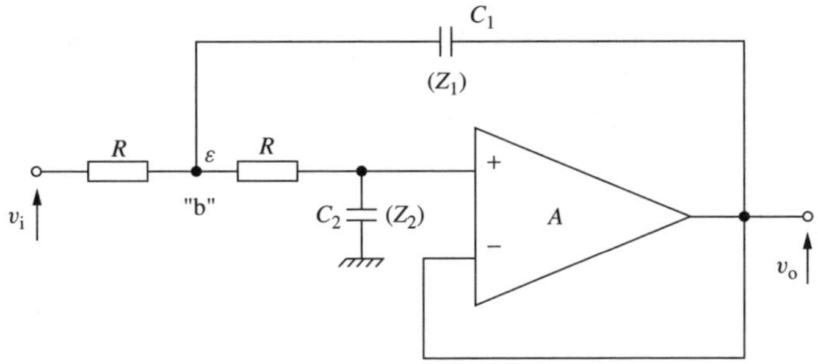

Figure 8.4 Active second-order loop filter with the operational amplifier.

The magnitude of the auxiliary voltage ε is computed from the relation

$$\frac{\varepsilon}{R + Z_2} Z_2 - v_o = v_o/A \approx 0 \tag{8.14}$$

Introduction of the above equation into (8.13) reveals, after some simple calculations, the transfer function of this active low-pass filter

$$F(j\omega) = \frac{1}{1 + j\omega 2RC_2 - \omega^2 R^2 C_1 C_2} \tag{8.15}$$

Its properties were discussed earlier in Section 3.1.3.

8.2 INTEGRATORS

Each PLL contains at least one integrator coupled with the integrating properties of the VCO, as explained in Chapter 1. The input voltage to the varactor changes the VCO frequency; however, the active quantity is the phase, which is the integral of the frequency (cf. eq. (1.8)).

8.2.1 Active Integrators with Operation Amplifiers

Type 2 or 3 PLLs require additional integrators. The problem is solved by the introduction of one or two OPAMPs/integrators discussed in the previous section (cf. relation (8.4) and Figs. 2.8 and 8.2).

8.2.2 Passive Integrators

In cases where the simple RC filter is connected to the source of the infinite internal resistance, the circuit would behave as an integrator. However, we have to replace the transfer function $F(s)$ with the impedance $Z(s)$. The principle arrangement is depicted in Fig. 2.11. In practice, the input current will not be a continuous one, but will be supplied via a three-state switch (controlled, for example, by the popular phase-frequency detector) from two ideal current sources during the time proportional to the phase difference ϕ_e [1] (see Section 2.3.2).

8.3 MIXERS

Mixers are encountered mostly in PLL feedback paths (cf. Figs. 3.7 or 9.22). Since the principle is often common with the multiplicative PDs, we shall first mention these types.

8.3.1 Multiplicative Mixers

Multiplicative mixers use the nonlinearity between voltage and current in many electronic elements (transistors, diodes, etc.). The parabolic dependence is an ideal one

$$i = a_0 + a_1 v + a_2 v^2 \tag{8.16}$$

In cases where the voltage v is formed by two sinusoidal signals, owing to the quadratic term in (8.16), the circuit generates two sidebands with the sum and difference of frequencies (cf. eq. (1.4)). One of them together with both leaking "carriers" is undesired and must be removed by the following filters. However, this ideal case is not encountered in practice because semiconductor elements exhibit exponential dependence between current and voltage, that is,

$$i = I_o e^{\alpha v} \tag{8.17}$$

Since exponential functions are expanded into infinite series, the consequence is that the rhs of (8.17) contains higher-order terms, several quite large [2, p. 119], which generate a set of output signals with frequencies

$$f_i = r f_1 + s f_2 \quad (r, s = 0, \pm 1, \pm 2, \ldots) \tag{8.18}$$

Particular attention needs to be given to those mixing products, the frequency of which lies in the pass band of the output filter. A quick explanation about the intermodulation frequencies f_i leads to the diagram in Fig. 8.5. Its application for solving intermodulation problems will be explained with the following example.

Example 8.1

Let us investigate intermodulation properties of the mixer having the input signals with frequencies

$$f_1 = 260 \text{ to } 300\,\text{MHz} \quad \text{and} \quad f_2 = 300 \text{ to } 310\,\text{MHz}$$

the output low-pass filter with the corner frequency and

$$f_{p,max} = 50\,\text{MHz}$$

The solution is simple. We mark extremes $(f_1/f_2)_{min} = 0.84$ and $(f_1/f_2)_{max} = 1$ on the x-axis and $(f_p/f_2)_{min} = 0$ and $(f_p/f_2)_{max} = 0.167$ on the y-axis.

We draw horizontal and vertical lines through these points that bound a rectangle in Fig. 8.5. Parameters (r,s) of all straight lines intersecting its area indicate the order of the intermodulation signals that are present in the output.

$$f_{p,min} \le f_p \le f_{p,max} \tag{8.19}$$

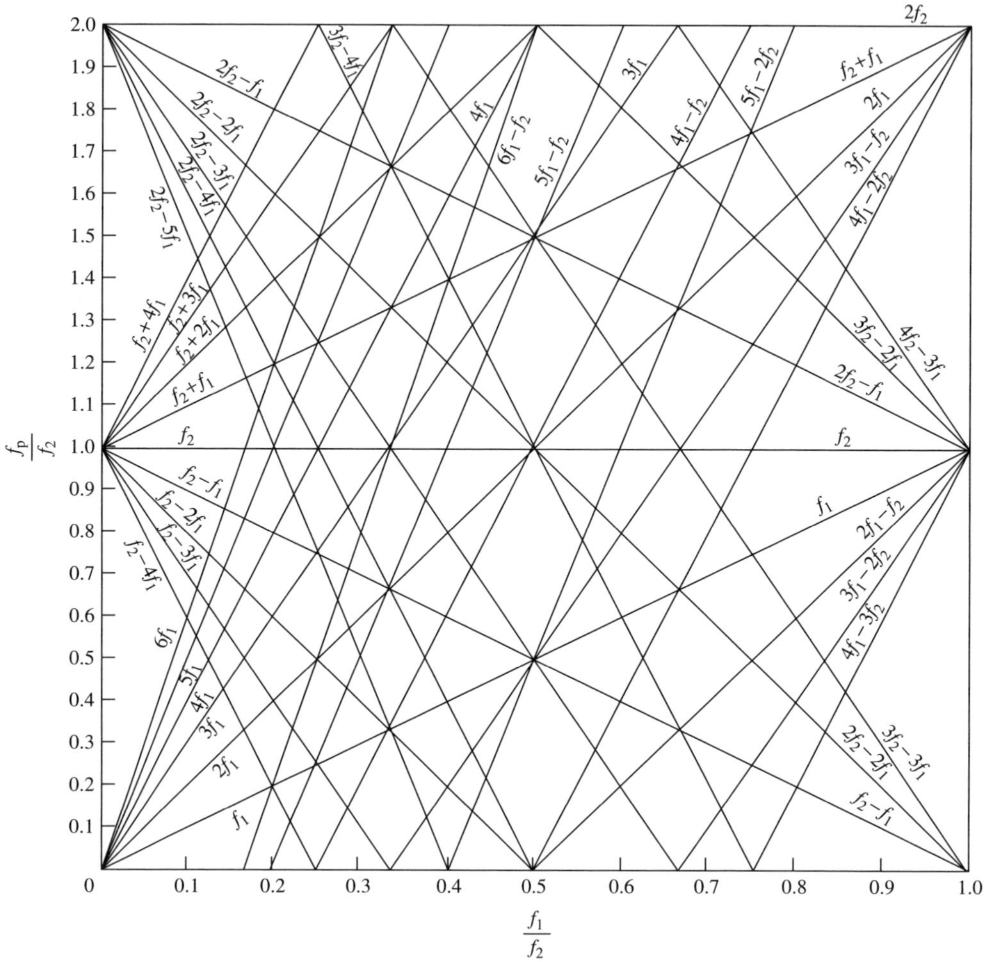

Figure 8.5 Diagram for the investigation of lower order intermodulation signals in the output pass band of a mixer (© [2], adapted from [3]).

With the assistance of Fig. 8.5 we shall find $(f_2 - f_1)$, $(2f_2 - 2f_1)$, $(3f_2 - 3f_1)$, and so on. Note that higher-order terms are not indicated; because of the clarity, we have limited the intermodulation order in Fig. 8.5 to

$$|r| + |s| = 7 \tag{8.20}$$

The other reason is that higher-order terms than those in eq. (8.20) are generally below the level of $-80\,\text{dB}$ and tolerable in most instances (cf. [4] and Tab. 8.1).

Note that the rectangle shown in Fig. 8.5 is in reality a polygon. However, we feel that the application of computers will provide the desired correct information for investigation of intermodulation signals.

Table 8.1 Intermodulation suppression in double-balanced mixers in decibels; r corresponds to the high-level (LO) input and s to the low-level (RF) input; $\Delta P = P_{LO}(\text{dBm}) - P_{RF}(\text{dBm})$

Harmonics s, r	1	2	3
1	0	$\Delta P + 41$	$2\Delta P + 28$
2	35	$\Delta P + 39$	$2\Delta P + 44$
3	10	$\Delta P + 32$	$2\Delta P + 18$
4	32	$\Delta P + 39$	–
5	14	–	$2\Delta P + 14$
6	35	$\Delta P + 39$	–
7	17	–	$2\Delta P + 11$

8.3.2 Switching Mixers

A nonlinear electronic element, often encountered, is a switch, schematically depicted in Fig. 8.6(a). The working mode is such that the input signal either passes to the output or is blocked. The mathematics is very simple:

$$v_2(t) = v_1(t)p(t) \tag{8.21}$$

where $p(t)$ is a periodic rectangular function with two levels, either $+1$ or 0, with the repetition frequency Ω_{LO}. Consequently, the above equation can be rewritten as

$$v_2(t) = V_1 \sin(\omega_s t) \cdot \left[\frac{\Theta}{2\pi} + \sum_{r=1,2,\ldots}^{\infty} \frac{2}{\pi r} \sin\frac{r\Theta}{2} \cos(r\Omega_{LO}t)\right] \tag{8.22}$$

In instances in which the switching has space-to-mark ratio, 1:1, that is, the current flow angle is $\Theta = \pi$, the above relation is simplified to

$$V_2(t) = V_1(\sin\omega_s t)\left[\frac{1}{2} + \sum_{r=1,3,5,\ldots}^{\infty} \frac{2}{\pi r} \sin\frac{r\pi}{2} \cos(r\Omega_{LO}t)\right] \tag{8.23}$$

Inspection reveals that the output signal contains the component with the fundamental frequency

$$\tfrac{1}{2}V_1(\sin\omega_s t) \tag{8.24a}$$

and the sidebands

$$\frac{V_1}{\pi r}\sin(\omega_s \pm r\Omega_{LO})t \quad (r = 1, 3, 5, \ldots) \tag{8.24b}$$

Note that theoretically the switching frequency is missing. The situation being such, we speak about the balanced modulators.

There are two principal forms, each of which uses four diodes in a bridge arrangement (Figs. 8.6(b) and (d)). Assuming that the amplitude of the switching voltage is much larger than the signal amplitude, the operation of rectifiers is controlled by the switching voltage alone – except for negligibly short intervals when this voltage is near zero. Thus, the concept of time-varying resistance may be applied, and the resulting linear theory is sufficient when solving problems of the impedance match between filters and mixers and when determining mixer losses.

8.3.3 Ring Modulators

The ring modulator represented in Fig. 8.7(a) is equivalent during one-half period of the switching signal to the lattice-resistive network in Fig. 8.7(b) and in the following half period to a similar network with r_b and r_f interchanged. However, this is equivalent to a fixed lattice-resistive network followed by the periodic phase inverter (Fig. 8.7(c)). Being an ideal ring modulator ($r_f = 0$, $r_b = \infty$), it reduces to the phase inverter alone. The transformers shown in Fig. 8.7 are only provided for supplying the switching voltage and in the following discussions are assumed to have a ratio of 1:1.

By investigating properties of these mixers, we can start from relation (8.22), however, by subtracting the same component shifted by π; thus

$$v_d(t) = v_2'(t) - v_2'(t + \pi/\Omega_{LO}) \tag{8.25}$$

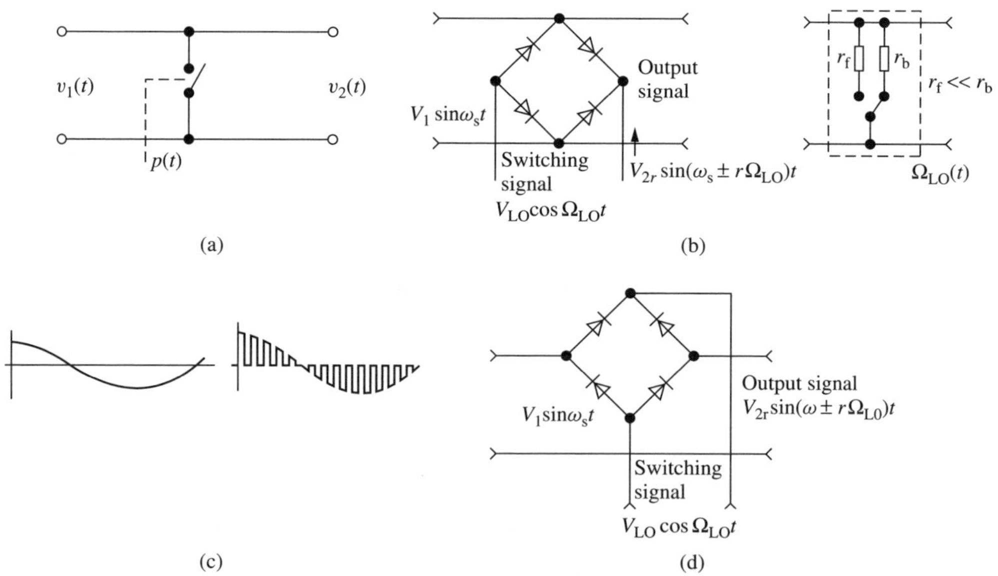

Figure 8.6 Switching mixer: (a) idealized circuit; (b) parallel diode switch; (c) input and output signal of the parallel switch; and (d) series diode modulator (© [2]).

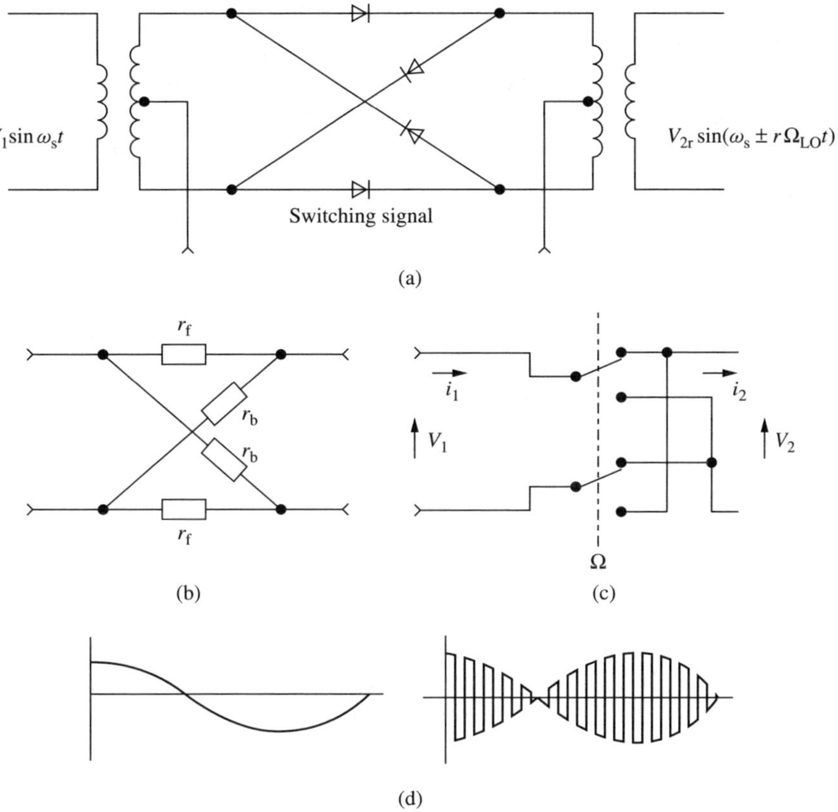

Figure 8.7 Ring modulator: (a) arrangement of diodes and balanced transformers; (b) the equivalent lattice-resistive network during one-half period of the switching signal; in the following half period r_b and r_f are interchanged; (c) the ideal phase inverter; and (d) idealized input and output waveforms. (© [2]).

and after application of the above voltage in eq. (8.22) we get

$$v_d(t) = \frac{2}{\pi} V_1 \sum_{r=1,3,5,\ldots}^{\infty} \frac{1}{r} \sin(\omega_s \pm r\Omega_{LO})t \tag{8.26}$$

The relation between the amplitude of the input signal and that of the desired sideband is called *conversion loss* and for the ring modulator is theoretically

$$\mathrm{L} = 20 \log(\pi/2) \approx 4 \text{ dB} \tag{8.27}$$

The ring modulator is not an ideal switch; consequently, the conversion loss is larger, practically in the range from 5.5 to 7 dB. In addition, in accordance with the relation (8.22) no harmonics of the input signals should be present. However, this rarely happens in reality. The effective diode resistance in the conduction mode is dependent on the effective voltage that is formed by three components: $v_{LO}(t)$, $v_1(t)$, and output $v_2(t)$. The theoretical investigation performed in [2, p. 116] reveals the presence of odd intermodulation signals. The spurious level is in good agreement with eq. (8.23) (see the components (3,1), (5,1), and (7,1) in the intermodulation table *(r;s)* – Tab. 8.1). These formulae for approximating suppression of certain intermodulation products were adopted from [4]. Slightly lower values were published by Rohde in [5] confirming the rule of thumb in (8.20).

8.4 PHASE DETECTORS

Frequency mixers, discussed in the previous section, change into PDs in instances in which the input frequency is equal to the switching frequency. Then from relation (8.24b) or (8.26) it follows that the lower sideband is actually a DC component, the magnitude of which will be maximum in the case in which both voltages $v_s(t)$ and $v_{LO}(t)$ are in the quadrature, that is, shifted by 90°; this is in agreement with the assumptions made in Chapter 1.

8.4.1 A Simple Switch

For introduction into the phase detector properties, we shall apply the switching circuit depicted in Fig. 8.8. In the case that both frequencies in (8.24b) are equal, that is, $\omega_s = \omega_{LO}$, we get for the DC component or a slowly varying component

$$v_d = \frac{V_1}{\pi} \sin(\phi_s - \phi_{LO}) \tag{8.28}$$

Inspection of the above equation reveals that the output voltage v_d is proportional to the sine difference of the input signals. This is why we designate these detectors as sine phase detectors. For the PD gain, K_d, we have, in agreement with relation (1.4) or (1.5),

$$K_d = V_1/\pi \tag{8.29}$$

Typical voltages of these PDs are illustrated in Fig. 8.8(b).

The switch can be realized either with the assistance of a special transistor or with the diode "mixer" shown in Fig. 8.6. The drawback of these simple PDs is the presence of rather large spurious signals with frequencies ω_s, $2\omega_s$, and $3\omega_s$, as depicted in Fig. 8.9.

8.4.2 Ring Modulators

The operation is similar to that of the switching PDs discussed in the previous section. Only the spurious signal with the fundamental frequency is compensated to the level −35 to −40 dB. The expected suppression of other IM signals is summarized in Tab. 8.1. The advantage of the ring modulators for application as PDs is its delivery in the compact form by many manufacturers. In addition, they are useful from very low

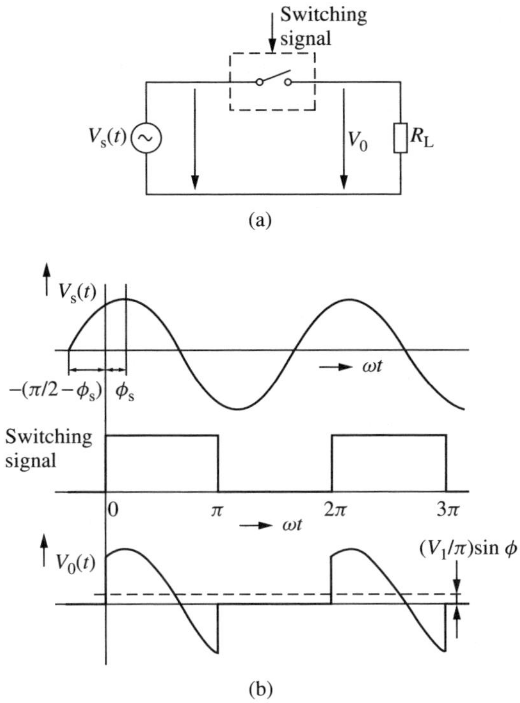

Figure 8.8 Switching phase detector: (a) idealized arrangement and (b) typical voltages.

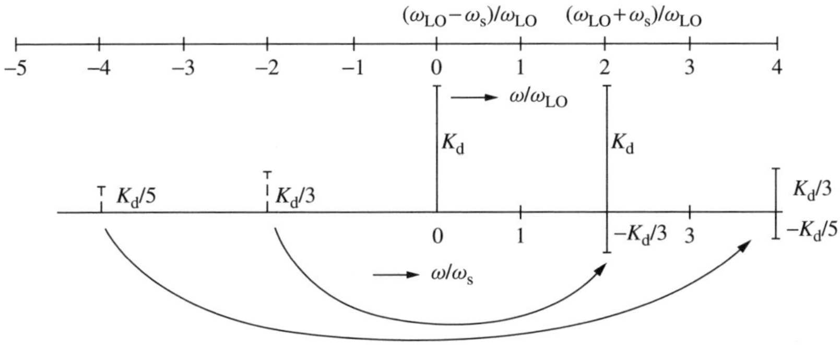

Figure 8.9 Spectral diagram of the switching phase detector.

frequencies up to the gigahertz ranges. Another advantage is very low noise, perhaps the lowest of all PDs. A difficulty might be the moderate gain that is computed from (8.28) as

$$v_d(t) = \frac{2}{\pi} V_1 \tag{8.30}$$

where V_1 is the voltage at the "RF input."

8.4.3 Sampling Phase Detectors

The principle block diagram is shown in Fig. 8.10. During each period, the memory capacitor is connected, for a very short time ΔT, to the reference signal source and keeps information about the instantaneous voltage during ΔT to the arrival of the next sampling impulse when the above process is repeated.

In the case of the sinusoidal reference voltage, the working characteristic of the PD is also sinusoidal. Similarly, if the reference voltage was a sawtooth signal, the PD characteristic would be also of the sawtooth type.

The sampling phase detector process can be expressed with the following relation:

$$v_{\mathrm{d}} \approx V_{\mathrm{s}} \sin(\omega_{\mathrm{s}} k T_{\mathrm{s}} + \phi_k) \quad (k = 0, 1, 2, \ldots) \tag{8.31a}$$

$$v_{\mathrm{d}} \doteq V_{\mathrm{s}} \sin(\phi) \tag{8.31b}$$

The background phase error, at the rate of the reference frequency, is ideally generated by the finite duration of the sampling time ΔT and the inevitable leakage resistance. In accordance with the preceding equation and Fig. 8.11, the voltage on the memory capacitor C_{H} changes during the sampling time ΔT by about

$$\Delta v \doteq V_{\mathrm{s}} \omega_{\mathrm{s}} \Delta T = V_{\mathrm{s}} 2\pi \Delta T / T_{\mathrm{s}} \tag{8.32}$$

By taking into account that the voltage Δv is of the narrow impulse shape, the harmonics of which are constant to a very high order and are equal to

$$A_1 \doteq A_2 \doteq \cdots \approx \frac{\Delta v \Delta T}{T_{\mathrm{s}}} \tag{8.33a}$$

$$\doteq 2\pi V_{\mathrm{s}} \left(\frac{\Delta T}{T_{\mathrm{s}}} \right)^2 \tag{8.33b}$$

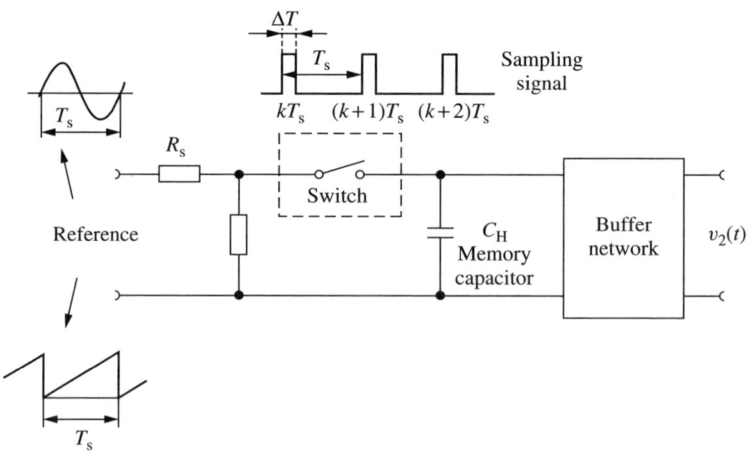

Figure 8.10 Principle block diagram of the sampling phase detector.

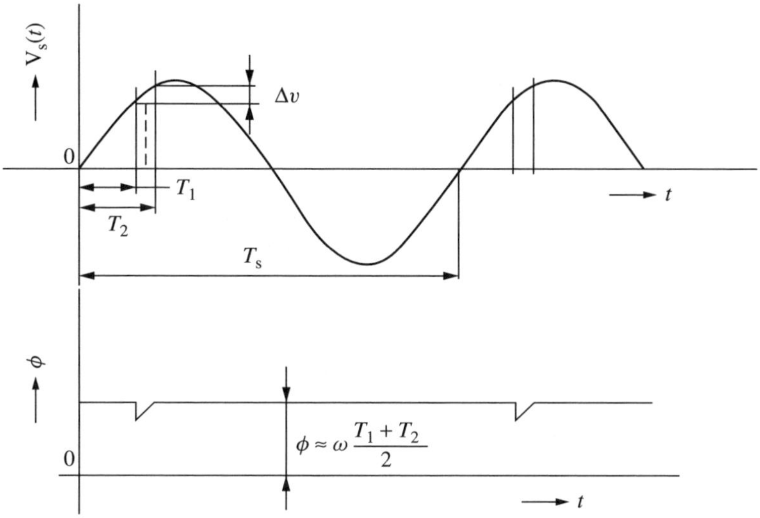

Figure 8.11 Leakage of the reference frequency to output of the sampling phase detector.

we easily compute that the duration of the sampling impulse ΔT is only one-hundredth of T_s, so the level of the harmonic components of the leaking reference signal will be below -70 dB. This is an important advantage in respect to the PDs discussed up to now. However, these achievements are possible only when the time constant formed by the inner resistance of the reference generator and the memory capacity C_H is small compared with the duration of the sampling time ΔT

$$\tau_1 = R_s C_H \ll \Delta T \tag{8.34}$$

The above condition limits the application of the sampling phase detector to the rather low reference frequency f_s.

Example 8.2

Let us estimate maximum input frequency for the sampling phase detector in cases in which the source resistance $R_s = 10\,\Omega$ and the memory capacitor $C_H = 1000$ pF. From the relation (8.34), we compute ΔT

$$\Delta T \approx 10\tau_1 = 10^{-7} \text{ [s]} \tag{8.35}$$

The required level of the background noise, -70 dB, puts the upper bound of the sampling frequency (in accordance with (8.33b)) to

$$f_s \doteq 10^5 \text{ [Hz]} \tag{8.36}$$

The designer must also take into consideration the reduction of the voltage on the memory capacitor due to discharging via inevitable leaking resistance R_{leak}. This process is governed by the time constant

$$\tau_2 = C_{\mathrm{H}} R_{\mathrm{leak}} \ [\mathrm{s}] \tag{8.37}$$

As long as $\tau_2 \gg T_{\mathrm{s}}$, the disturbing voltage will be approximately of the sawtooth shape with a peak voltage $V_{\mathrm{s}} T_{\mathrm{s}}/\tau_2$. Consequently, amplitudes of the respective spurious signals are

$$A_k = \frac{1}{k\pi} V_{\mathrm{s}} \frac{T_{\mathrm{s}}}{\tau_2} \quad (k = 1, 2, 3, \ldots) \tag{8.38}$$

Example 8.3

By assuming $C_{\mathrm{H}} = 1000\,\mathrm{pF}$ and $R = 10^{10}\,\Omega$, the time constant, τ_2, will be 10 [s]. From this we can estimate the minimum input frequency for the spurious level of $-70\,\mathrm{dB}$ with the assistance of (8.37):

$$f_{\mathrm{s}} \doteq 10^3 \ [\mathrm{Hz}] \tag{8.39}$$

The foregoing examples confirm the earlier recommendation by Egan, for this type of PD, to be used in the range from 4 to 60 kHz [6].

However, there are other sources of jamming or interference, namely, leakage of the sampling pulses into the VCO tuning circuit. The compensation might help, but a careful design is a better solution.

Finally, a few practical comments on this type of PD. Note that the gain K_{d} is proportional to the input sine voltage V_{s} (cf. eq. (8.31)). However, this voltage may vary from one application to the other. An easy remedy would be to change it into a sawtooth one by charging an auxiliary capacity, C_{R}, with a constant current (see Fig. 8.12) during the whole sampling period T_{s} with complete discharging at its end.

$$v_{\mathrm{s}}(t) = V_1 \frac{t}{T_{\mathrm{s}}} \quad (0 \le t < T_{\mathrm{s}}) \tag{8.40}$$

The charging process is interrupted during the sampling pulse (see Fig. 8.12(a)) and the respective voltage $v_{\mathrm{s}}(t)$ is read by the memory-holding capacitor C_{H} (see Fig. 8.12(b)). In this way the spurious impulse Δv reported in relation (8.32) is suppressed. Consequently, the limiting condition (8.34) is substantially weakened.

However, there are other difficulties with the application of these PDs. In accordance with Examples 8.2 and 8.3, a time delay is introduced. Consider, for example, in Fig. 8.13, that the information about the phase difference in time $T_{\mathrm{s},k}$ is retained on the holding capacitor till time $T_{\mathrm{s},k+1}$, when new information is provided. Circuits of this type are designated as "zero order data hold" and introduce a time delay T_{s}.

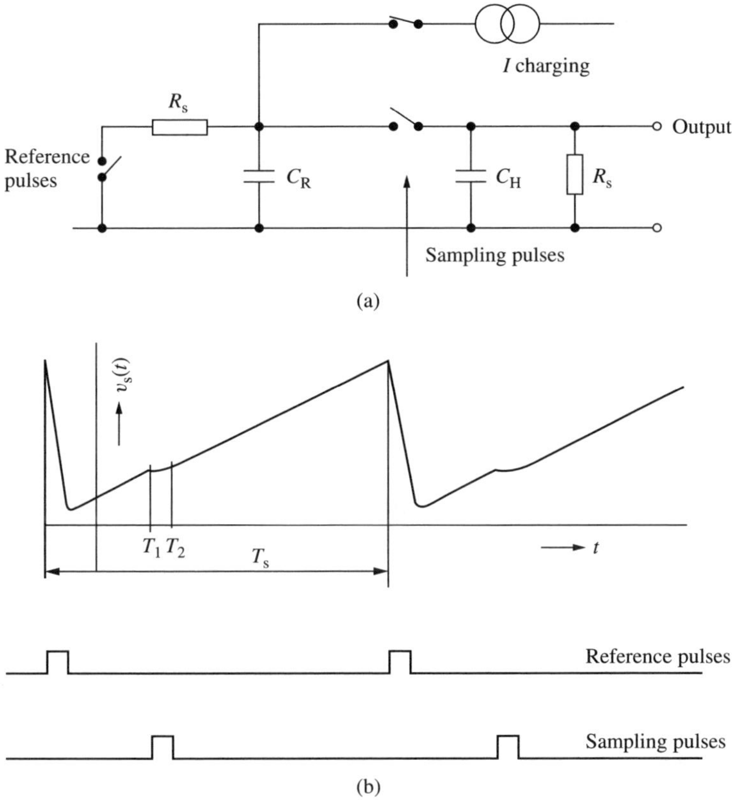

Figure 8.12 Phase detector with double sampling: (a) arrangement for the sawtooth wave performance and (b) input voltage with respect to timing of the reference and sampling pulses.

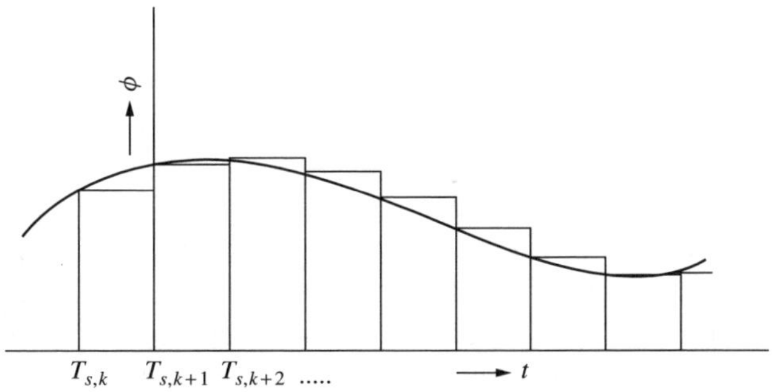

Figure 8.13 Output voltage of the "zero-order data hold" circuit.

Consequently, we must correct the PD transfer function as explained in Section 3.1.6 and expect a reduction in the PLL stability.

Note that the system with two holding capacitors, as in Fig. 8.14, exhibits the function of an additional RC filter. That is, in the case that the second capacitor C_{H2} is much larger than C_{H1}, the charges and consequently the voltages are equalized

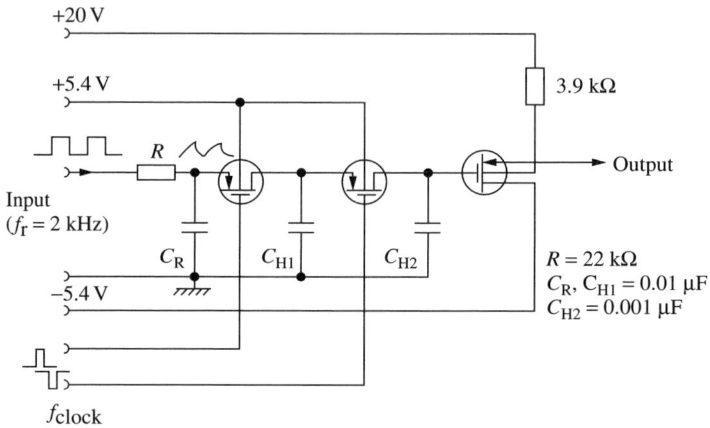

Figure 8.14 An earlier type of double-sampling phase detector: the first sampling pulse activates the second switch after decay of transient effect on C_{H1}; note the filtering properties.

gradually. The result is that the effective memory system exhibits the function of the RC filter with the time constant

$$T = T_s \frac{C_{H1} + C_{H2}}{C_{H1}} \approx T_s \frac{C_{H1}}{C_{H1}} \tag{8.41}$$

8.4.4 Digital Phase Detectors

Digital PDs are based on applications of simple flip-flop gates or more complicated combinations.

8.4.4.1 Flip-flop phase detectors

We can use both R-S and J-K flip-flops as PDs. By changing the input signals in the PD into pulse rates fed to the set (S) input and the other to the reset (R) pin, we get on the Q or $\overline{Q}$ outputs a rectangular waveform with the mean value proportional to the phase difference between the steering signals. The DC output is a sawtooth wave indicating the phase difference range from 0 to 2π or from $-\pi$ to π (see Fig. 8.15). The difficulties with this type of PD are a voltage bias at the VCO varactor (or any other tuning element) and large amplitudes of the reference signal and its harmonics. By choosing the operation in the middle of the sawtooth wave characteristic, we have for the PD gain

$$\begin{aligned} v_d(t) &= \frac{2A}{\pi}\left[\frac{\pi}{4} + \sum_{r=1,3,\ldots}^{\infty} \frac{1}{n}\sin\frac{n\pi}{2}\cos(n\omega_s t)\right] \\ &= \frac{A}{2} + \frac{2A}{\pi}\sum_{r=1,3,\ldots}^{\infty} \frac{1}{n}\sin\frac{n\pi}{2}\cos(n\omega_s t) \end{aligned} \tag{8.42}$$

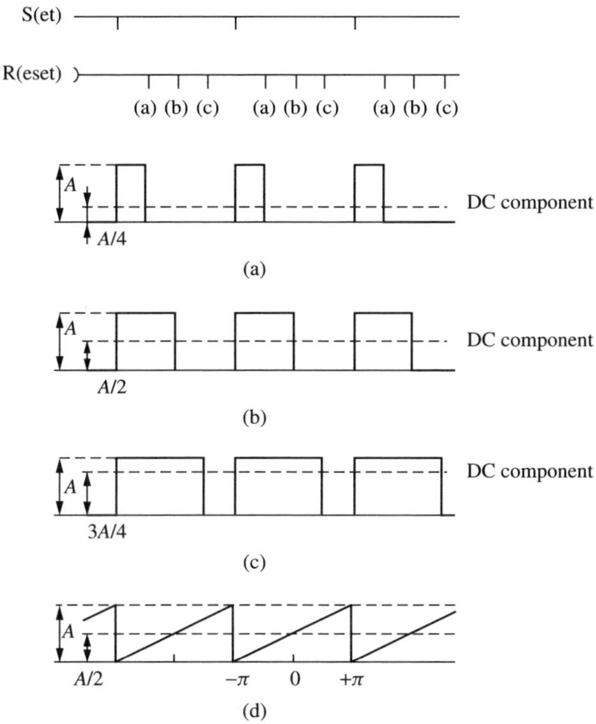

Figure 8.15 Output voltages of the flip-flop (sawtooth wave) PD: (a) for the phase difference $-\pi/2$; (b) for the zero phase difference; (c) for the phase difference $+\pi/2$; and (d) the smoothed output for slowly changing phase difference.

The advantage of these PDs is their simplicity and compatibility with IC design; the difficulty is a lot of spurious reference harmonics (note that the voltage A is approximately equal to V_{cc}).

8.4.4.2 Exclusive-OR gate phase detectors

This set of PDs retains the advantages of those discussed in the previous section, that is, simplicity and IC compatibility. In addition, they present another benefit, that is, in the ideal case they suppress the reference frequency as ring modulators do. Their logic operation can be described as

$$X = A\overline{B} + \overline{A}B \tag{8.43}$$

With its assistance we are able to investigate the detector output voltage. If both input signals have the space-mark ratio 1:1, then for different values of the phase difference we have a triangular output wave as depicted in Fig. 8.16(d). Inspection reveals that the peak amplitude, equal to the logic 1 (H), is for phase difference $\pi/2$

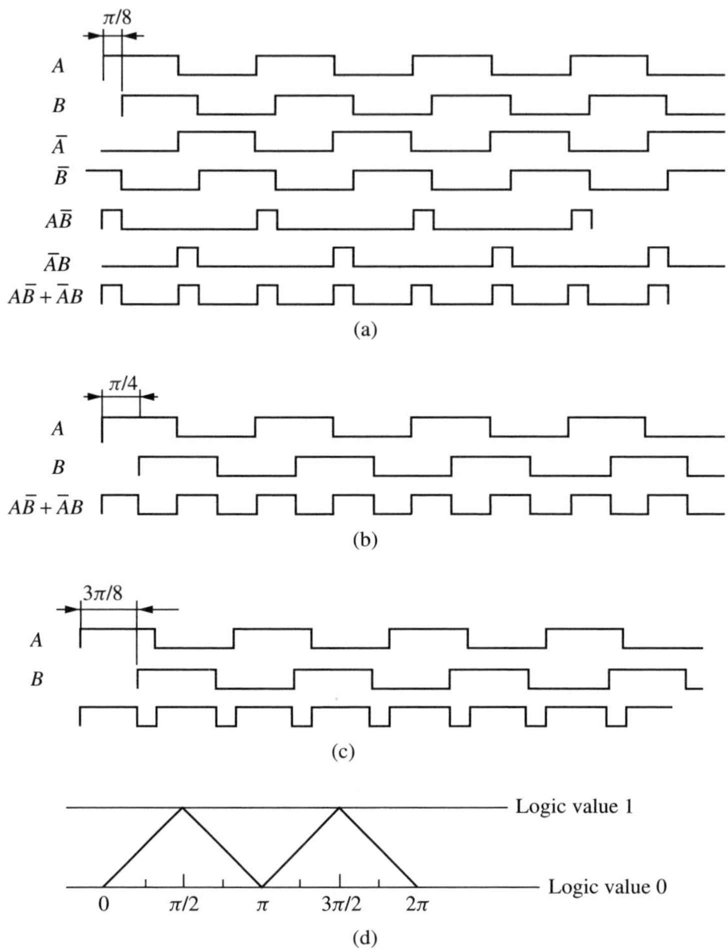

Figure 8.16 Output voltages of the exclusive-OR (a triangular wave) PD: (a) for the phase difference $-\pi/8$; (b) for the phase difference $-\pi/4$; (c) for the phase difference $+3\pi/8$; and (d) the smoothed output for slowly changing phase difference.

and the respective output voltage

$$v_d(t) = V_H\phi(t) \quad \text{for} \quad \left(0 < \phi < \frac{\pi}{2}\right) \tag{8.44}$$

8.4.4.3 Phase frequency detectors

At present, these PDs are most often encountered in practice. One reason is acceleration of the phase lock by combination of the frequency detector properties (for larger detuning) with final phase detection. Another appreciated property is their IC production by many companies, generally under the type number 4046.

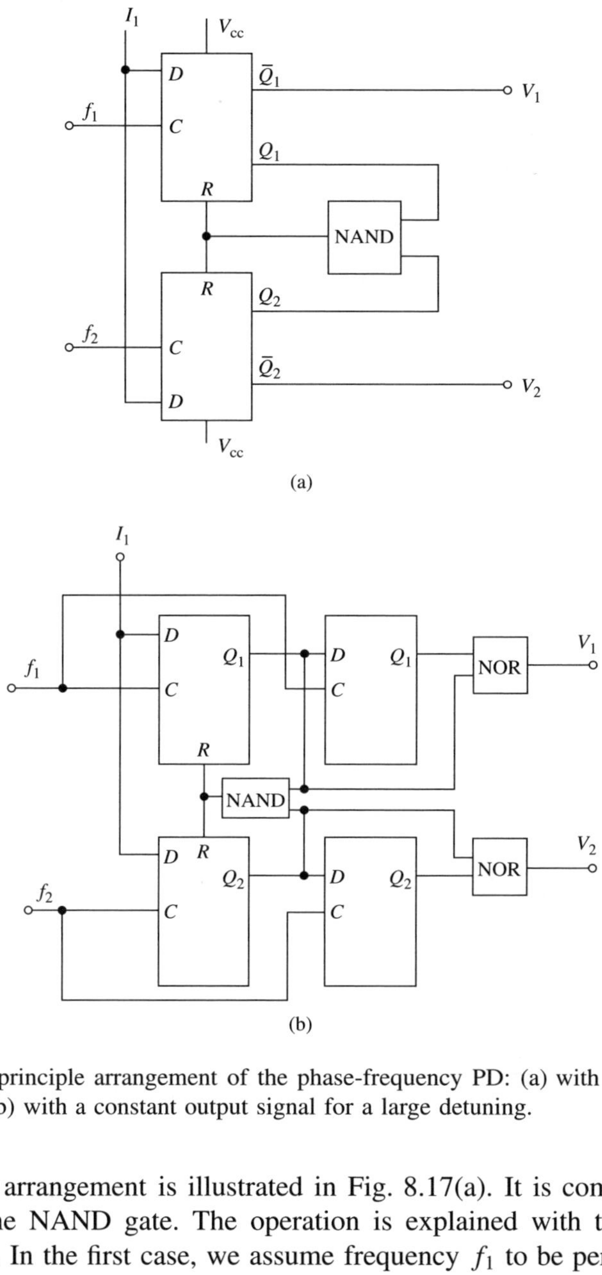

Figure 8.17 The principle arrangement of the phase-frequency PD: (a) with the sawtooth wave output signal and (b) with a constant output signal for a large detuning.

The principle arrangement is illustrated in Fig. 8.17(a). It is composed of two D flip-flops and one NAND gate. The operation is explained with time diagrams in Fig. 8.18(a)–(d). In the first case, we assume frequency f_1 to be permanently higher than f_2. Consequently, only output V_1 in Fig. 8.18(a) is activated. In the opposite circumstances again only the output V_2 will be working. The PD output characteristic is shown in Fig. 8.19. In both cases, we face the sawtooth operation from 0 to $+2\pi$ or -2π and the mean output voltage, either positive or negative, with the magnitude

$$v_{\mathrm{d,DC}} = \frac{V_{\mathrm{cc}}}{4} \tag{8.45}$$

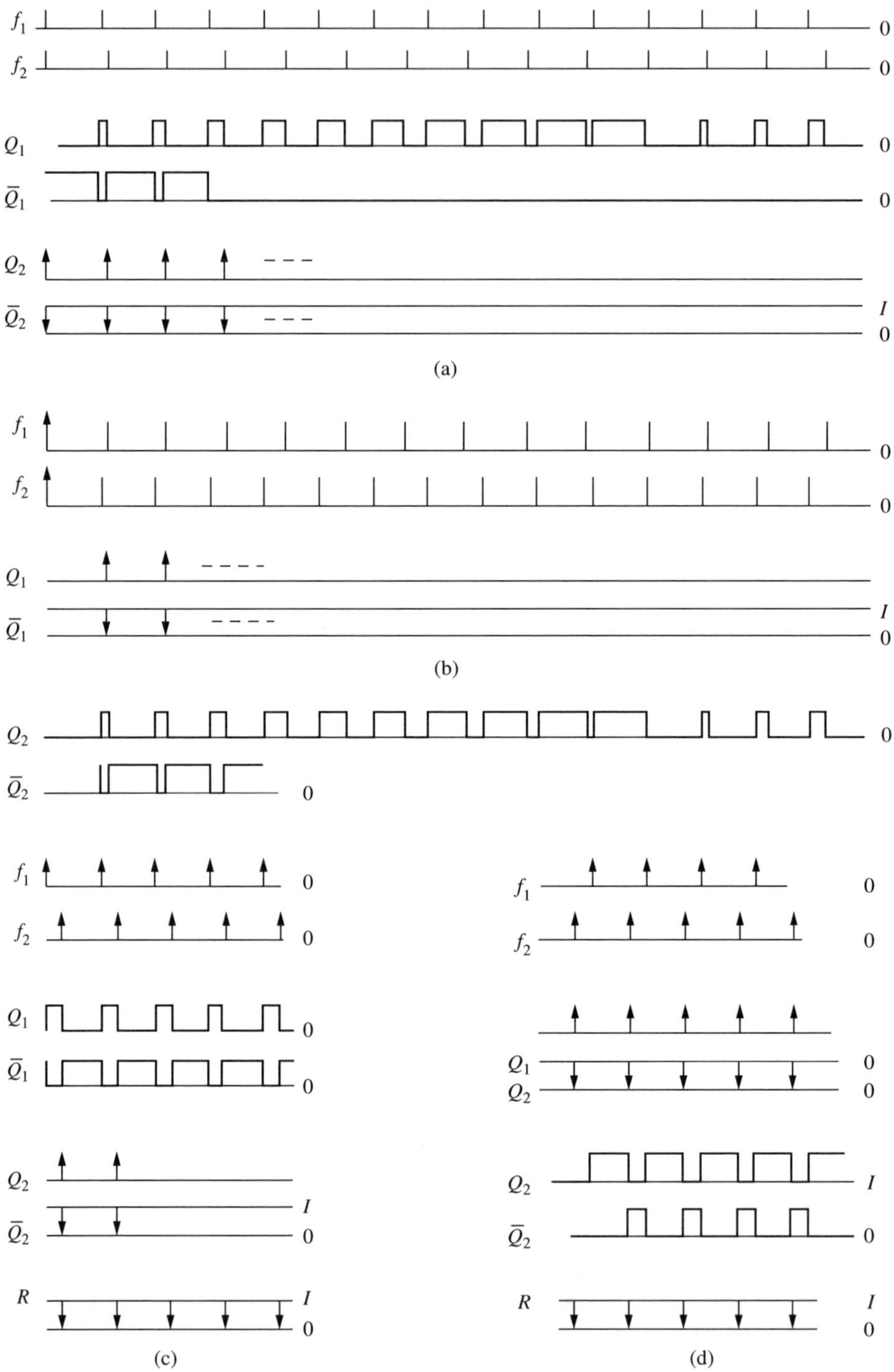

Figure 8.18 Time diagrams of the phase-frequency PD: (a) frequency f_1 is higher than f_2; (b) $f_1 < f_2$; (c) $f_1 = f_2$; however, the leading edge of f_2 does delay f_1; and (d) $f_1 = f_2$; however, the leading edge of f_1 does delay f_2.

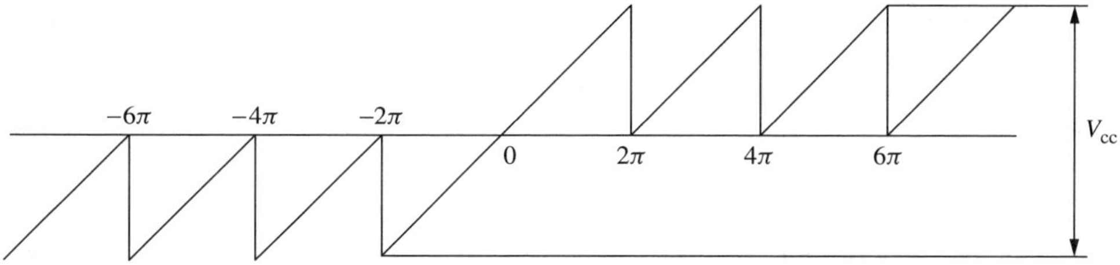

Figure 8.19 Output voltage of the phase-frequency PD.

In the range where phase lock is possible, we can compute the phase detector gain to be constant:

$$K_d = \frac{V_{cc}}{4\pi} \tag{8.46}$$

Some perfection was introduced by several manufacturers by the application of four D flip-flops (the so-called quad-D circuits) depicted in Fig. 8.17(b). In instances in which one of the input frequencies is more than twice as high as the other, these detectors exhibit a constant output voltage. We mention, for example, types #CD 4046 by RCA or #MC 14046 by Motorola.

There is one difficulty connected with these phase frequency detectors. From Figs. 8.18 and 8.19 it follows that after a certain time, designated in Chapter 7 as the *pull-in time*, both input frequencies are equal and the *lock-in time* starts. The consequence is that pulses supplied by one of the outputs, say $\overline{Q}_1$, became narrower

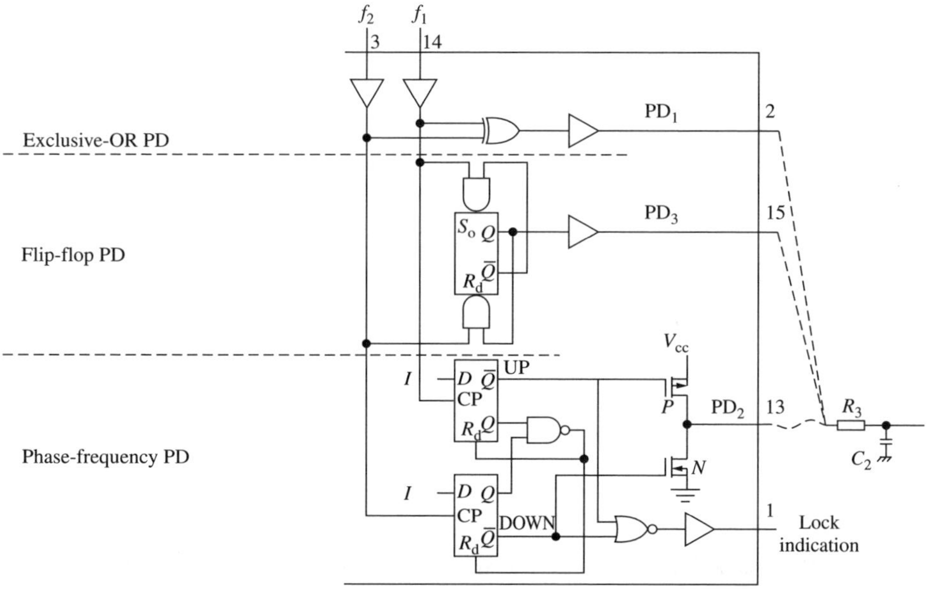

Figure 8.20 Combined IC phase detector: the PD set contains an exclusive-OR PD, a simple flip-flop PD, and the phase-frequency detector © Philips, 2002.

and narrower (cf. Fig. 8.18) and the integration, due to filtration, reduces the phase difference $\phi_e(t)$ to zero. At this instance there is no output from the PD and the VCO is effectively disconnected. This "ideal" state is undesirable since the overall gain of the PLL is reduced to zero and we encounter the so-called dead zone. However, this situation is unrealistic since the switching times generated by the used D flip-flop and by other components are not equal. The situation being such, we witness at the PD output the random pulses that have no relation to the output phase $\phi_e(t) = 0$. The consequence is an additive noise experimentally investigated by Egan [6].

The problem can be solved by connecting a defined leakage resistance (e.g., 1 MΩ) in parallel with the integrating capacitor. The introduction of a time delay may provide another solution.

An effective solution is the combination of several phase detectors such as that of the phase frequency and sampling PD or a commercial arrangement, the principle

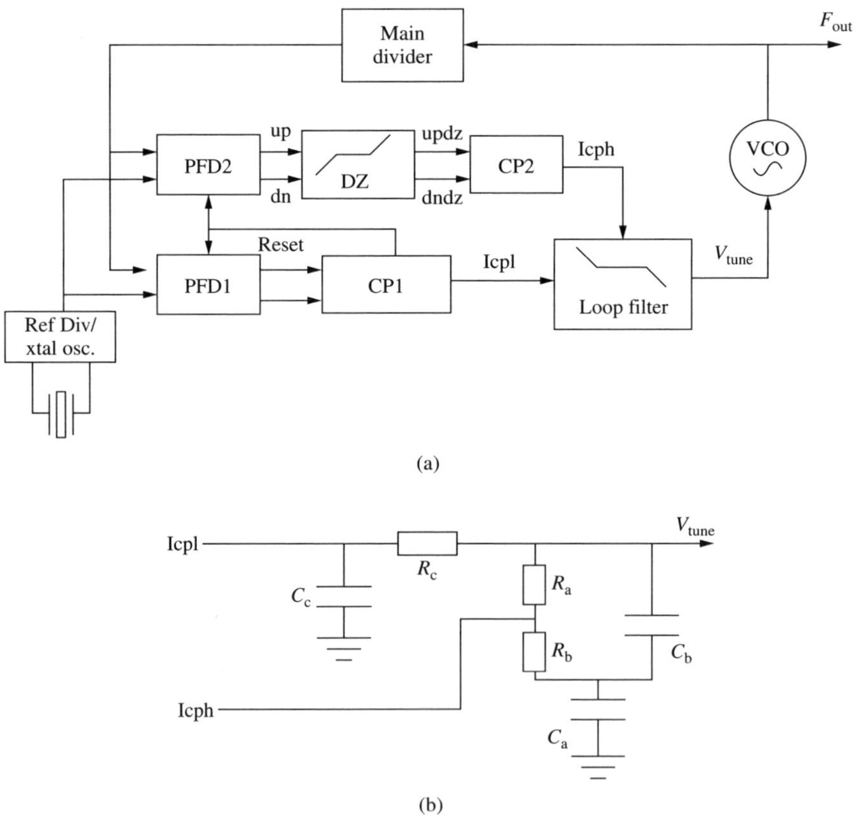

Figure 8.21 Dead zone reduction with a combination of two PDs with different gains in K_{di} (Reproduced from C.S. Vaucher, "An Adaptive PLL System Architecture Combining High Spectral Purity and Fast Settling Time", *IEEE J. Solid-State Circuits*, **35**, 490–502, 2000 by permission of IEEE, 2002).

of which is shown in Fig. 8.20: the PD set combines an exclusive-OR PD, a simple flip-flop PD, and the phase frequency detector [7].

Another possibility is to reduce the PD gain either by several levels of I_p before locking (cf. Fig. 7.6) or by combining, for example, two PDs with different gains of K_d – as suggested in [8] (see Fig. 8.21).

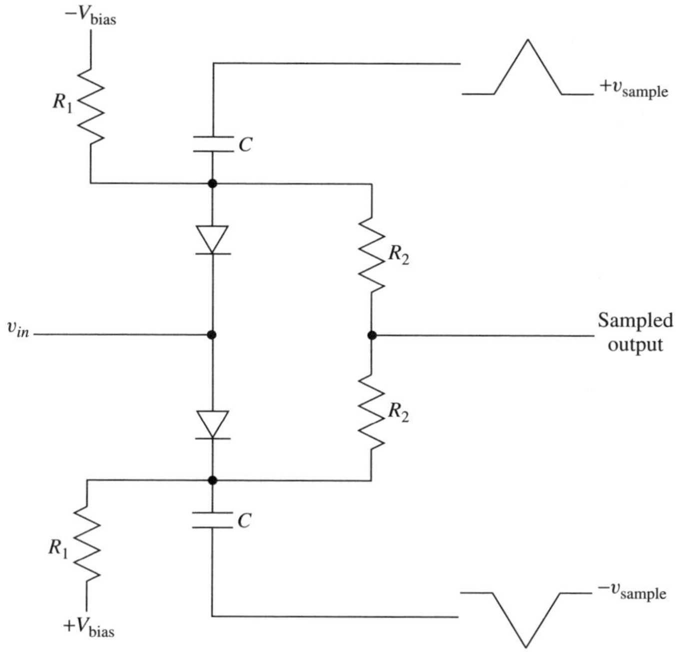

Figure 8.22 Sampling PD suitable for microwave ranges (Reproduced from W.F. Egan, *Frequency Synthesis by Phase Lock*. 2nd ed. New York: Wiley 1981, 2000 by permission of John Wiley & Sons, Inc, 2002).

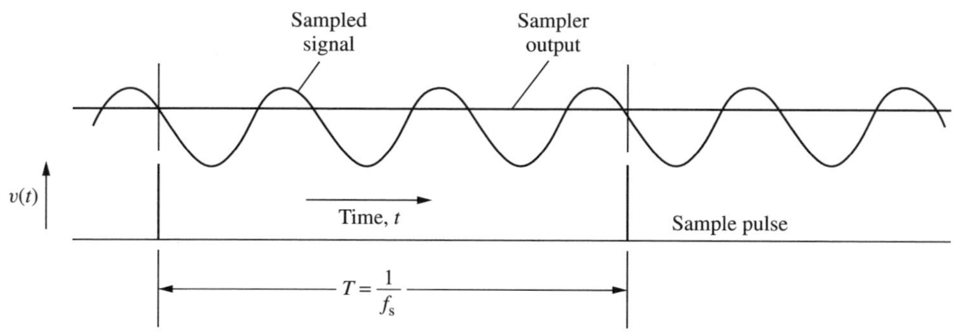

Figure 8.23 Phase detection by sampling on higher harmonics (Reproduced from W.F. Egan, *Frequency Synthesis by Phase Lock*. 2nd ed. New York: Wiley 1981, 2000 by permission of John Wiley & Sons, Inc, 2002).

8.4.4.4 Microwave phase detectors

From the beginning of the twentieth century, the range of communications systems has been pushed to higher and higher frequencies with the result that today general applications exceed tens of gigahertz on one hand and reach optical transmissions on the other hand. No wonder that PLLs follow the trend.

Mainly nonlinear properties of semiconductor diodes are exploited (cf. Section 8.3.1) for PDs . For lower microwave frequencies, PDs based on narrow pulses generated in the step-recover diodes were suggested [6]. The basic block diagram of this sampler PD is shown in Fig. 8.22. Egan suggests its application up to the 10-GHz range. Since the sampling pulses are very narrow, phase detection on harmonic frequencies can be used (see Fig. 8.23). Some difficulties may be expected from locking on neighboring false harmonics.

There is another possibility with ring modulators, the application of which is possible, with present technology, up to 20 to 30 GHz. Also, digital systems are applicable in the low gigahertz ranges.

8.5 FREQUENCY DIVIDERS

PLLs working in microwave ranges are almost without exception equipped with frequency dividers. Today, digital systems are prevailing; however, we still come across regenerative dividers at very high frequencies [9].

8.5.1 Regenerative Frequency Dividers

A block diagram of this type of frequency divider is shown in Fig. 8.24. The operation is as follows. Without the input signal $A_\mathrm{i}\cos(\omega t+\phi_\mathrm{i})$, regenerative action cannot take place. However, there might be difficulties with starting, even at its presence: when the circuit is switched on, a small signal $a_n\cos\{[(n-1)/n]\omega t+\Psi\}$ is generated through the transient phenomenon at the second input of the mixer. After mixing with the input signal already present, the component with the fractional frequency ω/n may be filtered out and amplified. From the output there is a feedback loop in which the multiplication $(n-1)$ times is performed. If the overall gain exceeds “one”, the regeneration takes place; finally, saturation effects in either of the nonlinear circuits (including the amplifier) limit the amplitude. The oscillation requires the zero phase shift around the loop, that is,

$$\omega t+\phi_\mathrm{i}-\left(\frac{n-1}{n}\omega t+\phi_{n-1}\right)+\phi_\mathrm{a}=\frac{\omega t}{n}+\phi_\mathrm{o} \tag{8.47}$$

and

$$\left(\frac{\omega t}{n}+\phi_\mathrm{o}\right)(n-1)+\phi_\mathrm{b}=\phi_{n-1}+\frac{n-1}{n}\omega t \tag{8.48}$$

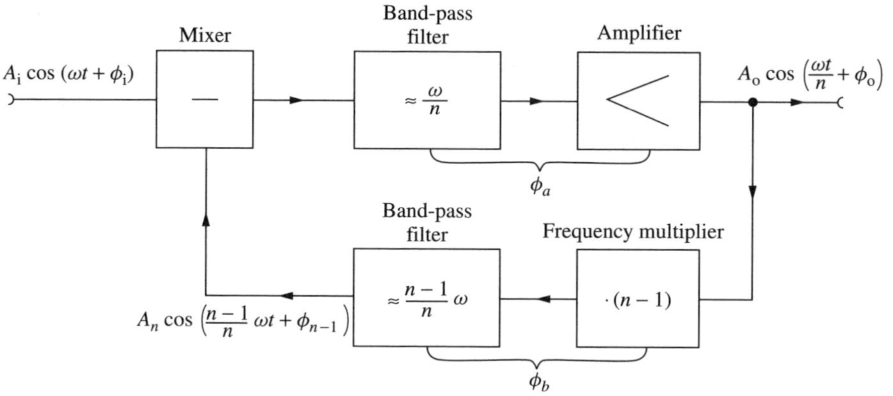

Figure 8.24 Block diagram of a regenerative frequency divider (© [2]).

The properties of phases ϕ_i, ϕ_{n-1}, ϕ_a, ϕ_b, and ϕ_o are evident from Fig. 8.24. After eliminating ϕ_{n-1} from (8.47) and (8.48), we get the relation between the output and the input phases ϕ_o and ϕ_i, respectively,

$$\phi_o = \frac{\phi_i + \phi_a - \phi_b}{n} \tag{8.49}$$

The maximum division ratio depends predominantly on the selectivity of the band-pass filter ($\omega_m \approx \omega/n$) following the mixer. The loop gain for the next possible working modes, that is, $\omega/(n \pm 1)$ (and naturally for all others, $\omega/(n \pm k)$ as well) must be safely below "one."

Dividing circuits of this type may be very simple; however, the resonant circuits present some difficulty.

8.5.2 Digital Frequency Dividers

Digital frequency dividers used in PLL are based on application of the IC gates forming flip-flop memory systems. In practical applications we encounter three basic IC families:

TTL (transistor–transistor logic – not recommended for new applications)

CMOS (complementary metal oxide semiconductor)

ECL (emitter coupled logic)

More information about IC families is summarized in Section 8.6.

8.5.2.1 Binary-coded dividers

The basic building blocks are bistable flip-flops, either of the D or JK type, generally four in one package. The output pins are arranged so that divisions by 2, 4, 8, or

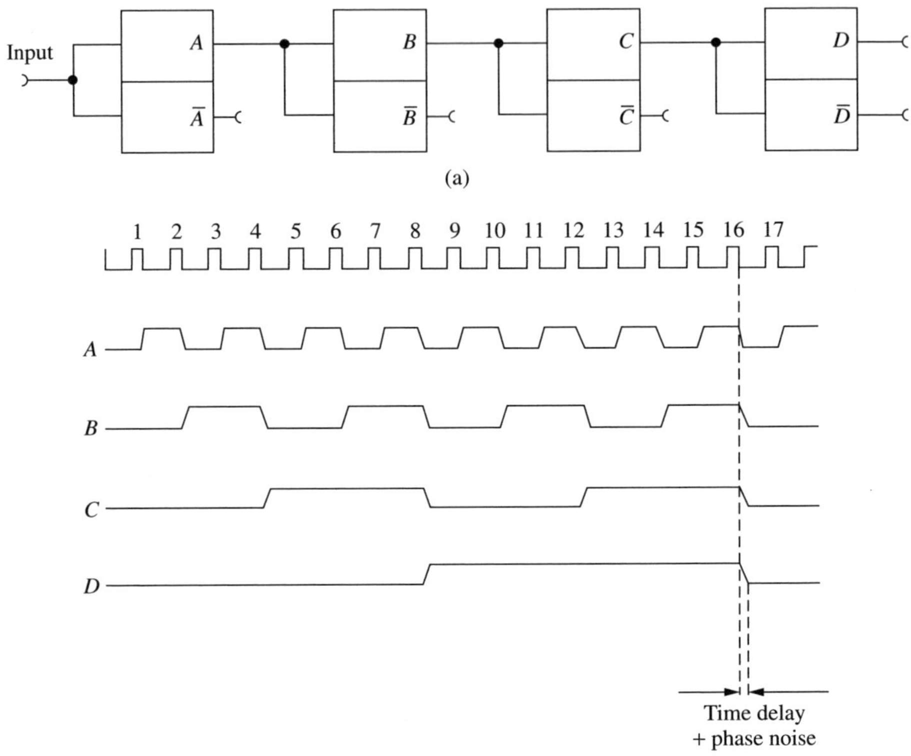

Figure 8.25 The four-stage binary counter-divider: (a) schematic diagram and (b) the respective waveforms with the time jitter (© [2]).

16 are possible. A schematic arrangement is illustrated in Fig. 8.25(a). Note that in each bistable stage, there is some delay between the arrival of the input pulse and the instant when the stage settles into its new state. Owing to noise, temperature variations, voltage variations, and so on, there is a small time jitter of the leading and trailing edges of the output rectangular waveform as well (cf. Fig. 8.25(b)). This difficulty is removed with synchronous clocking, as shown in a more detailed circuit in Fig. 8.26. For division by any arbitrary number N, we need more bistable elements, namely,

$$2^{n-1} \leq N \leq 2^n \tag{8.50}$$

Example 8.4

Design digital divider by 6

Combination of three flip-flops is necessary in accordance with (8.50). Figure 8.27 shows that the two first bistables divide by 3 and the last by 2. In this way we arrive at a space-mark ratio of 1:1.

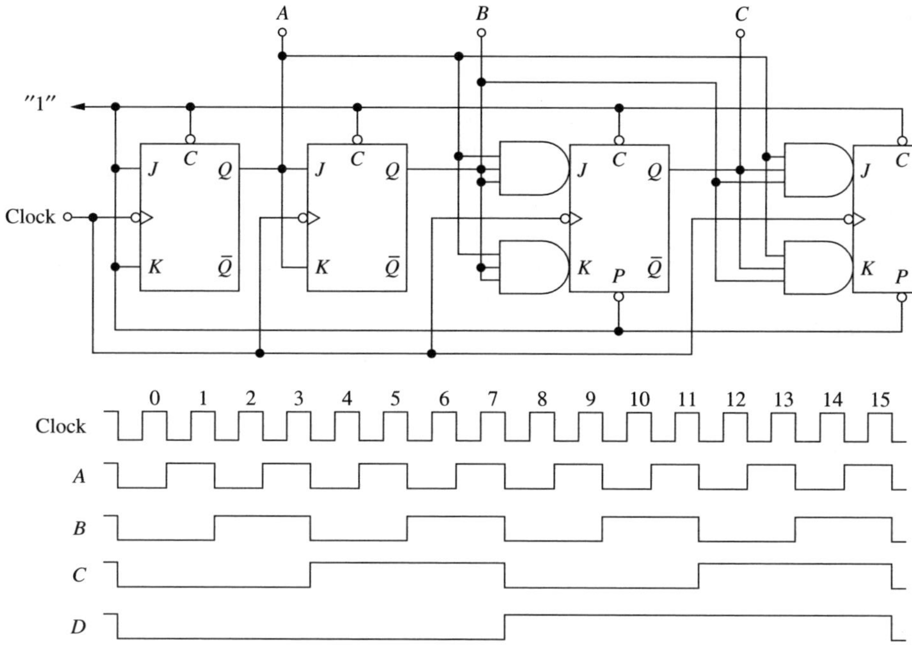

Figure 8.26 Synchronous binary divider (Reproduced from T.S. Aytur and B. Razavi, "A 2-GHz, 6-mW BiCMOS Frequency Synthesizer", *IEEE J. Solid-State Circuits*, 1457–1462, 1995 by permission of IEEE, 2002).

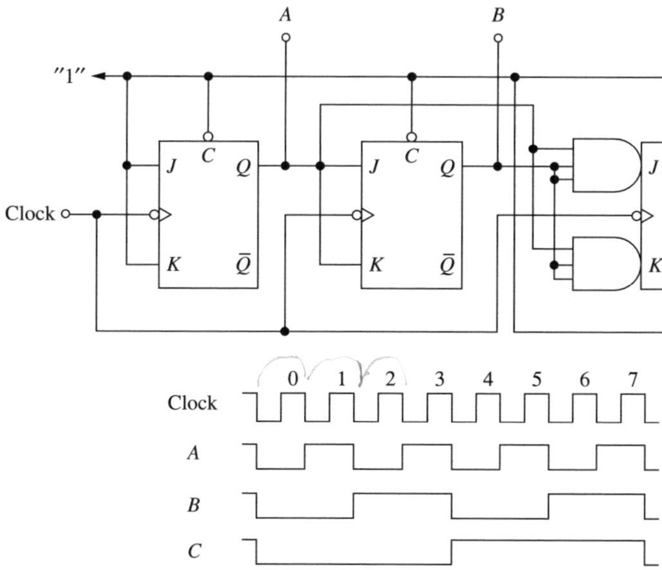

Figure 8.27 Binary divided by 6 (Reproduced from U.L. Rohde, *Microwave and Wireless Synthesizers*. New York: John Wiley, 1997 by permission of John Wiley & Sons, Inc, 2002).

Example 8.5

Design a very important binary-coded decimal counter (BCD).
According to (8.50), a four-stage counter is necessary. If the binary-decimal code – 1248 – is desired, the flip-flop state configurations corresponding to the numbers 10 to 15 must be inhibited. From Tab. 8.2 it may be easily deduced that after the ninth pulse both the first and the fourth stages must be "reset," whereas the activation of the second binary circuit must be prevented.

A logical diagram of the synchronous BCD counter using IC J-K bistable elements is shown in Fig. 8.28. The divider is effectively composed of a divide-by-2 and a divide-by-5 circuit. Note that the output wave is an asymmetrical rectangular wave. This difficulty is easily resolved as in the previous example by interchanging the division sequence, the output of which then results in a 1:1 space-mark ratio.

8.5.2.2 *Variable-ratio digital divider*

In most practical applications, in the PLL feedback path we need dividers with larger division ratios, very often programmable over several decimal orders. The actual dividing process may be performed in different ways:

(i) *Upcounter with digit recognition logic* The principle is shown in Fig. 8.29. The input pulses accumulate in the counter until the programmed number N is reached, whereupon an output pulse is generated in the recognition logic circuits and the counter resets to the "0" state. It is evident that all these operations must be performed before the $(N+1)$th (the first pulse of the next series) input pulse

Table 8.2 Logical states in the four-stage binary counter-divider during one counting cycle

Number of the input pulse P_k	Bistable circuit output			
	D	C	B	A
0	0	0	0	0
1	0	0	0	1
2	0	0	1	0
3	0	0	1	1
4	0	1	0	0
5	0	1	0	1
6	0	1	1	0
7	0	1	1	1
8	1	0	0	0
9	1	0	0	1
10	1	0	1	0
11	1	0	1	1
12	1	1	0	0
13	1	1	0	1
14	1	1	1	0
15	1	1	1	1
16	0	0	0	0

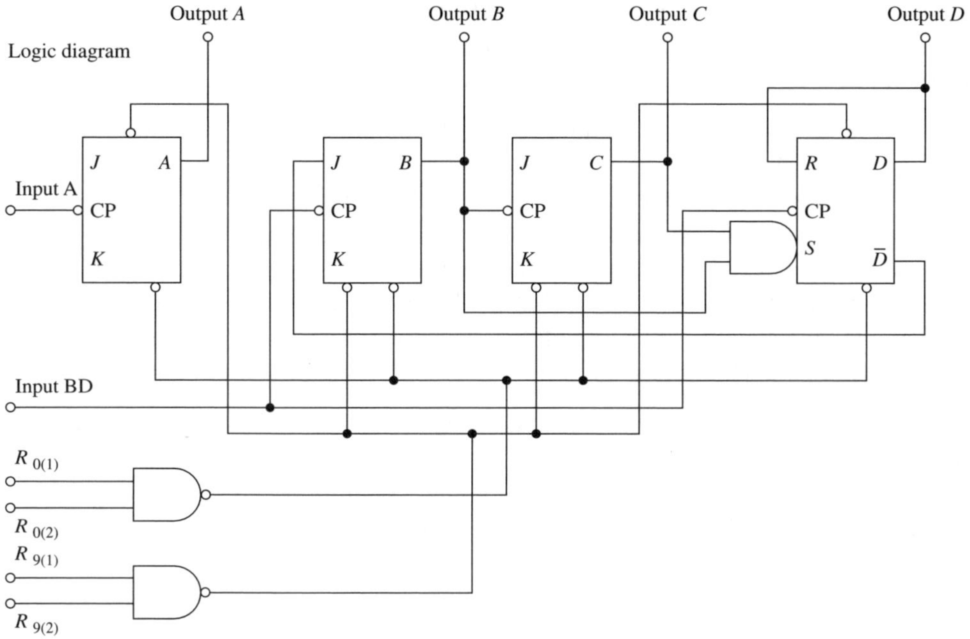

Figure 8.28 Block diagram of the BCD (1248) decade counter-divider using JK flip-flops.

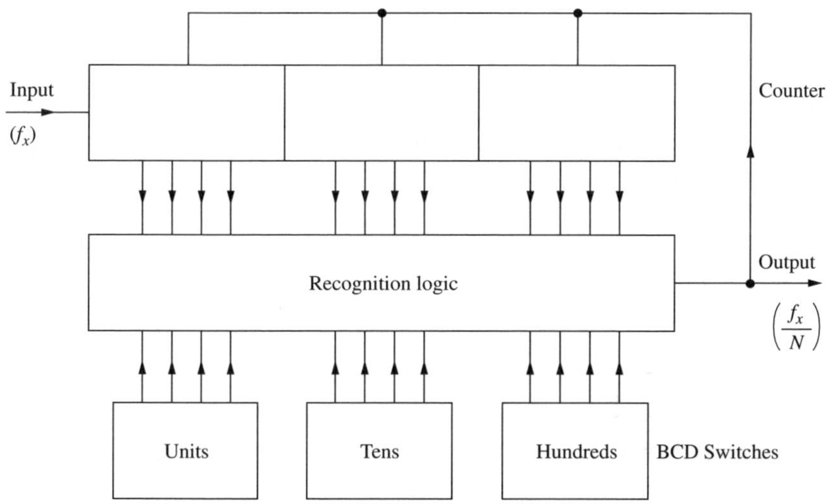

Figure 8.29 Variable divider employing an upcounter and digit recognition (© [2]).

appears. Propagation delays of individual counting decades form a large recognition delay which, if not fought against, will cause the maximum allowable input frequency (f_x) to be reduced, eventually, far below the resolution capabilities of individual flip-flops. This disadvantage can be overcome by preventing a known number of input pulses from entering the counter, while at the same time retaining their information value (e.g., by resetting the counter to 0100 instead of 0000). The penalty one must pay for this compromise is that the minimum division factor will be, in practice, much larger than 2.

(ii) *Preset downcounter* The idea is that the complement of N to the full counter capacity ($10^n - N$, when n is the number of counting decades) is inserted into the counter by the presetting logic circuits. After N input pulses, the zero state is reached and an output pulse is generated. On its command, the programmed number is transferred into the counter, and only then can the new dividing cycle start (Fig. 8.30). The main disadvantage of this system is that the last (that is, the Nth) pulse must "propagate" through the counter, changing all the decades into the zero state before the output pulse is generated. This difficulty can be overcome by combining both methods.

(iii) *Counter with digit presetting and recognition logic* In the simplest case the recognition state is moved from the "0s" to the "9s." The advantage of this is evident. At the arrival of the last Nth pulse, only the first counting decade is operated because all higher-digit decades are already waiting in the "9" state. The counter must be preset to ($10^n - 1 - N$) or to the 9s complement in each decade.

When general-purpose IC counting circuits of the type 7419xxx are used as building blocks, the variable divider may be realized with the minimum of additional logic circuits.

8.5.2.3 Dual-modulus dividers

The programmable dividers discussed above have limited frequency ranges. The difficulty may be alleviated with the so-called dual-modulus frequency dividers. The principle arrangement is shown in Fig. 8.31. The operation is as follows: First the

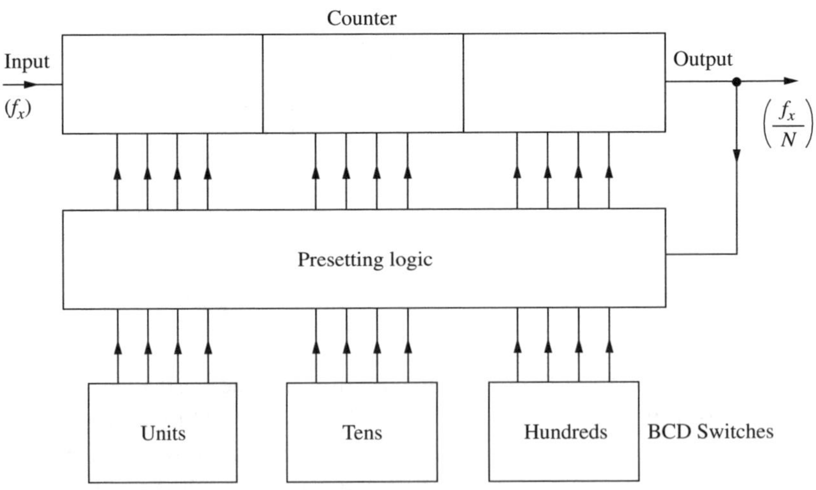

Figure 8.30 Variable divider employing a downcounter and digit-presetting logic (© [2]).

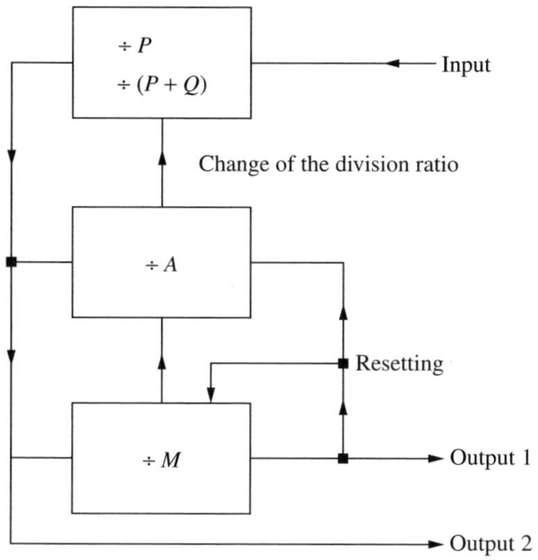

Figure 8.31 Dual-modulus divider (Reproduced from V.F. Kroupa, (ed.) *Direct Digital Frequency Synthesizers*. New York: IEEE Press, 1999 [12], by permission of IEEE, 2002).

input divider operates with the division factor $P + Q$. Its output is fed simultaneously into both auxiliary counters with the division factors A and M. As soon as the divider A overflows, the output signal changes the division factor $P + Q$ to P and blocks the input to itself. This state remains unchanged until the divider M is full. Its output signal resets the main divider to the $P + Q$ state again and the auxiliary divider A to zero, and the cycle starts to repeat. The time available for the change is ideally P periods. The final division ratio from the input to the output (1) is computed from the following relation:

$$N = (P + Q)A + P(M - A) = PM + AQ \tag{8.51}$$

The number of pulses stored in the counter M is at least

$$M_{\min} \geq A_{\min} + 1 \tag{8.52}$$

In the case in which only one pulse should be removed, that is,

$$Q = 1 \tag{8.53}$$

the division ratio N would comprise the whole set of integers from $PM_{\min}$ to $(PM_{\max} + A_{\max})$ if the division ratio A is in the range

$$0 \leq A \leq P - 1 \tag{8.54}$$

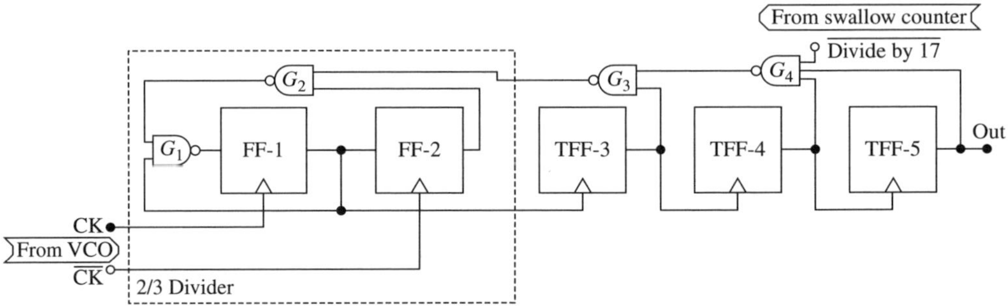

Figure 8.32 Dual-modulus 16/17 divider prescaler (Reproduced from T.S. Aytur and B. Razavi, "A 2-GHz, 6-mW BiCMOS Frequency Synthesizer", *IEEE J. Solid-State Circuits*, 1457–1462, 1995 by permission of IEEE, 2002).

From the above discussion we find the minimum division factor N

$$N_{\min} = PM_{\min} + A_{\min} = P^2 \tag{8.55}$$

Evidently, the smallest possible dual-modulus divider is $P/Q = 2/3$. These dual-modulus frequency dividers make it possible to extend the range of the fractional-N frequency synthesizer into gigahertz ranges [10]; particularly those switching along binary numbers to a neighboring number by one higher range (see Fig. 8.32).

8.5.2.4 *Prescalers*

Digital dividers with short switching times are power-hungry, which is hardly compatible with the power-sparing mobile applications such as GSM (Global System for Mobile Communications) and other emerging communications systems in gigahertz ranges [11, 13, 14]. One solution is the division of the PLL feedback divider into the so-called fast prescaler dividing by small binary numbers, such as 2, 4, 8, and the main divider by N. The penalty would be with the choice of the eventual channel spacing. The remedy is in the application of fractional-N frequency synthesis and its modifications, for example, via $\Sigma-\Delta$ (sigma–delta) modulations. The problem will be discussed in detail in Chapter 12.

8.6 DIGITAL CIRCUITS

Modern integrated circuits (ICs) with high or very high integration can be manufactured with an intensive application of digital circuits. In PLL systems they make it possible to produce small and highly sophisticated PDs, frequency dividers, and eventually whole PLL devices. Here, we shall mention the properties of the basic building blocks that are simple memory circuits as gates and flip-flops. They allow two binary operations denoted as logical addition (+) and multiplication ($\cdot$).

8.6.1 Gates – the Logic Levels and Symbols

In a binary system there are only two states, referred to as H(igh) or L(ow). H is relatively the more positive level than L.

Positive-logic means that the voltage level assigned to the binary "1" (H) state is more positive than that assigned to the "0" (L) state. In negative-logic, "1" is less positive than "0".

If the output of a gate is high only if all its inputs are high, then the gate performs the AND operation (logical multiplication) – the logical symbol is shown in Fig. 8.33(a). If the output of a gate is high if one of its inputs is high, the gate performs the OR operation (logical addition) – the logical symbol is shown in Fig. 8.33(b). A small circle at the input line of a logic symbol indicates that a low (L) level activates the function (Fig. 8.33(c)); if it is at the output line, then the low (L) level corresponds to the activated gate (Fig. 8.33(d)). Thus, the small circle indicates inversion – the logical symbol of the inverter is shown in Fig. 8.33(e). The equivalence between the arrangements shown in Fig. 8.33(a) to (d) and (a′) to (d′) is easily proved when using DeMorgan's theorems. Finally, if the output of a gate is high only if one of the inputs is high and the other is low (Fig. 8.33(f)), then the gate, or rather the system of gates, is called an *EXCLUSIVE-OR* gate.

When positive-logic symbology is used to represent the negative-logic functions, the dual of the functions is produced. For example, the positive-logic AND (NAND) gate becomes a negative-logic OR (NOR) gate.

The typical gate arrangements are formed by the original TTL family illustrated in Fig. 8.34, which is replaced with the power-saving CMOS circuits (see Fig. 8.35). Their advantage is that in either logic state (L or H), they do not require current. Only transition from one state to another is connected with loading and unloading of the spurious capacitances – refer to Tab. 8.3. Another advantage is their resistance to the spurious signals and larger voltage ranges. The difficulty with this is the chance

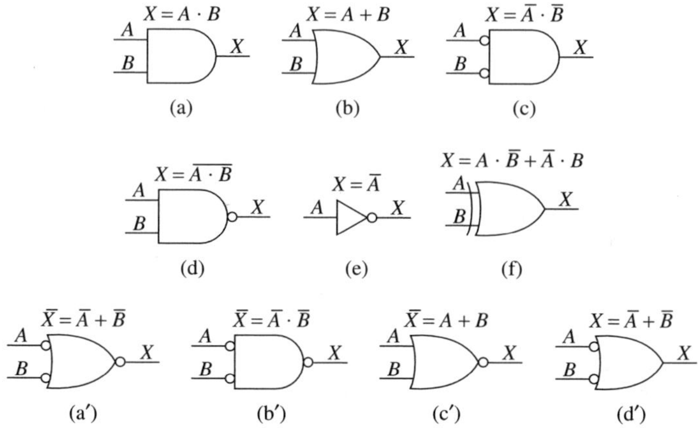

Figure 8.33 Logical symbols: (a a′) AND gate; (b b′) OR gate; (c c′) NOR gate; (d d′) NAND gate; (e) INVERTER; and (f) EXCLUSIVE-OR gate (© [2]).

Table 8.3 Comparison of CMOS and TTL technologies: supply voltage $V_{cc} = 5\,\text{V}$, ambient temperature $T_{amb} = 25°\text{C}$, load capacitance $C_L = 15\,\text{pF}$ (Reproduced from Philipps: Data handbook IC06 – High-Speed CMOS Family by permission of IEEE, 2002)

Parameters	Technology	HCMOS	Metal gate CMOS	Standard TTL	Low-power Schottky TTL	Schottky TTL	Advanced low-power Schottky TTL	Advanced Schottky abbTTL	Fairchild advanced Schottky TTL
	Family	74HC	4000 CD HE	74	74LS	74S	74ALS	74AS	74F
Power dissipation, type. (mW)									
Gate	Static Dynamic @ 100 kHz	0.0000025 0.075	0.001 0.1	1010	22	1919	1.2 1.2	8.5 8.5	5.5 5.5
Counter	Static Dynamic @ 100 kHz	0.000005 0.125	0.001 0.120	300300	100100	500500	6060	– –	190190
Propagation delay (ns)									
Gate	Typical Maximum	814	94 40 190 80	1020	9.515	35	47	1.5 2.5	34
Delay/power product (pJ)									
Gate	At 100 kHz	0.52	9 4	100	19	57	4.8	13	16.5
Maximum clock frequency (MHz)									
	Typical	55	4 12	25	33	100	60	160	125
D-type flip-flop									
	Minimum	30	2 6	15	25	75	40	–	100
Counter	Typical Minimum	4525	2 6 1 3	3225	3225	7040	45 –	– –	125100
Output drive (mA)									
	Standard outputs Bus outputs	46	0.51 0.8 1.6	1648	824	2064	824	2048	2064
Fan-out (LS-loads)									
	Standard outputs Bus outputs	1015	1 2 4	40120	2060	50160	2060	50120	50160

Note: HCMOS: High speed CMOS

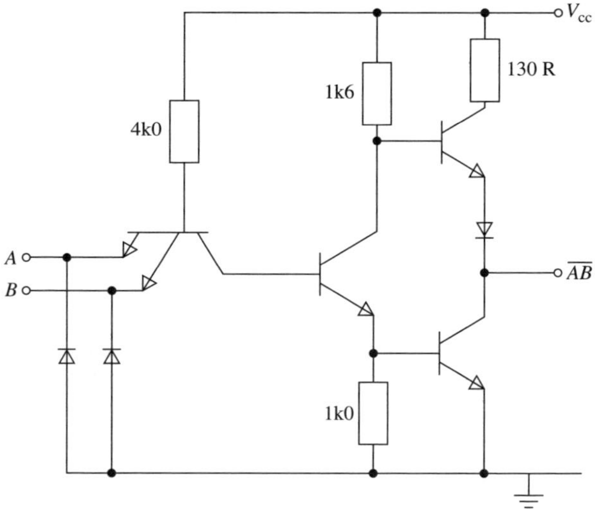

Figure 8.34 TTL technology gate.

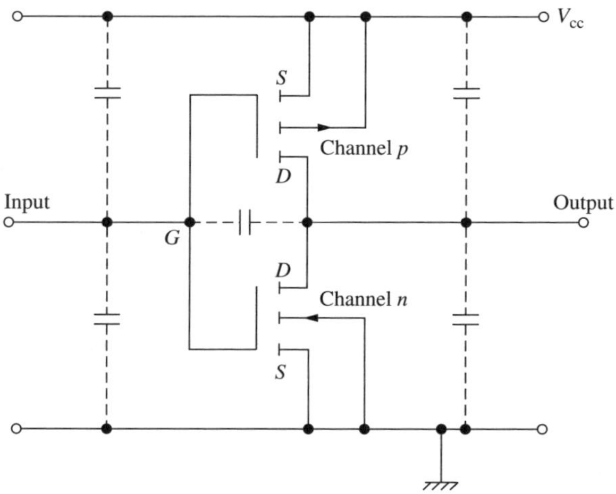

Figure 8.35 CMOS logic technology gate.

of being destroyed by electrostatic changes, despite the diode protection. Maximum working frequency of these IC gates hardly exceeds 100 MHz. However, this is not sufficient for PLLs in microwave frequency ranges. The solution is the application of ECL circuits with much larger power consumption – power consumption of different logic families as a function of the frequency is depicted in Fig. 8.36.

8.6.2 Flip-flops

Among memory circuits with several electrical states, the most important are the combinations of gates into flip-flops manufactured by IC technology.

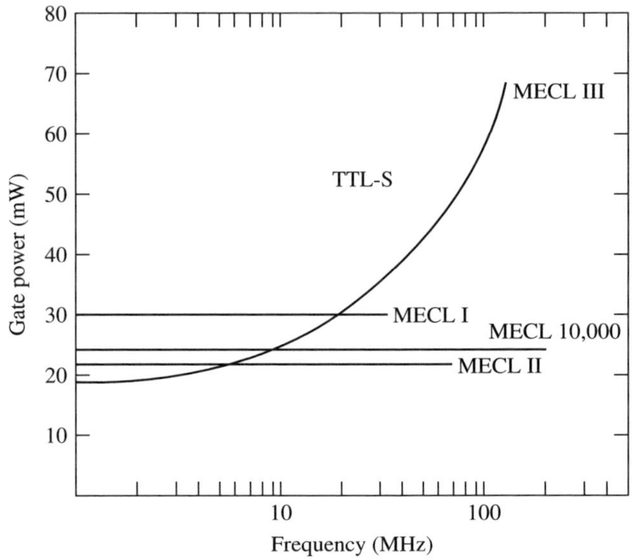

Figure 8.36 Power consumption of different logic families (Reproduced from Philipps: Data handbook IC06–High-Speed CMOS Family by permission of IEEE, 2002).

8.6.2.1 Set-reset flip-flops

A set-reset (R-S) flip-flop is schematically shown in Fig. 8.37. As long as both inputs are at the high state (H), the output does not change. Putting one signal (A or B) to 0 (low state L) causes an (H) state at either S or R. Or more exactly, a 1 (H) at S input "sets" the flip-flop such that it causes Q output to be at (H). A 1 at R "resets" the circuit to the state where Q is 0 (L). When that input returns to 1(H), the state of the flip-flop does not change. Consequently, the R-S circuit is a memory element.

8.6.2.2 D flip-flops

This bistable element, depicted in Fig. 8.38, is basically a shift-register. On the effective edge of the clock signal (say, positive slope), the logic state of the output Q becomes equal to the logic state on D (see the logic table, Tab. 8.4).

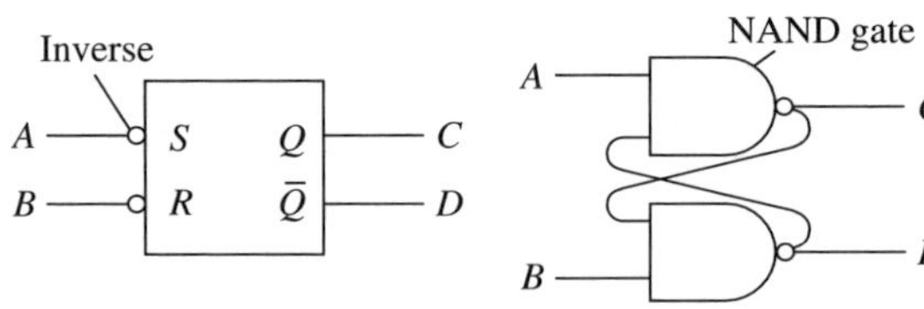

Figure 8.37 R-S flip-flop (Reproduced from W.F. Egan, *Frequency Synthesis by Phase Lock.* 2nd ed. New York: Wiley 1981, 2000 by permission of John Wiley & Sons, Inc, 2002).

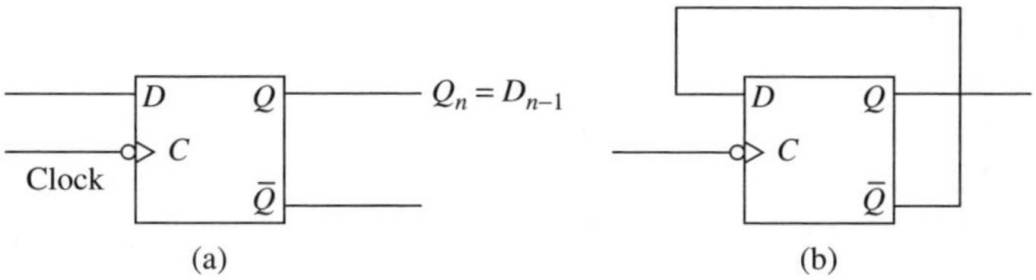

Figure 8.38 D-flip-flop: (a) schematic arrangement and (b) external connection for binary division (Reproduced from W.F. Egan, *Frequency Synthesis by Phase Lock*. 2nd ed. New York: Wiley 1981, 2000 by permission of John Wiley & Sons, Inc, 2002).

Table 8.4 Truth table of the D-flip-flop

T_n	T_{n+1}	
D	Q	$\overline{Q}$
L	L	H
H	H	L

Table 8.5(a) Set-reset truth table for J-K flip-flop; asynchronous entry, independent of "clock" and "synchronous input"

S	R	Q_{n+1}	$\overline{Q}_{n+1}$
0	0	1	1
0	1	1	0
1	0	0	1
1	1	Q_n	$\overline{Q}_n$

Table 8.5(b) Truth table of the JK flip-flop; synchronous entry, JK mode operation

J	K	Q_{n+1}	$\overline{Q}_{n+1}$
0	0	Q_n	$\overline{Q}_n$
0	1	0	1
1	0	1	0
1	1	$\overline{Q}_n$	Q_n

By connecting D with $\overline{Q}$ via an external connection, the Q output changes state at each effective signal on the C (clock) input. (Note $\overline{Q}$ is the opposite state of Q, that is, if Q is H, $\overline{Q}$ is L). D flip-flops find application in divider circuits.

8.6.2.3 J-K flip-flops

J-K flip-flops are versatile bistable elements, performing simultaneously additional logical functions. Multiport AND gates are provided on the J and K inputs, and the direct "Set" and "Reset" (also called "Preset" and "Clear") operations of the bistable element independent of the clock condition are possible. A functional diagram of J-K

master–slave flip-flop circuit is shown in Fig. 8.39. (This is not the only possible arrangement; nevertheless, its properties as described below are representative.) The complexity of the resultant discrete element circuit is evident. For reasons of economy, therefore, only the IC technique versions are practical.

The *J-K* bistable element is a clocked device, the state of which depends on the logical states of *J* and *K* inputs just before the arrival of the clock pulse according to the truth table – Tab. 8.5.

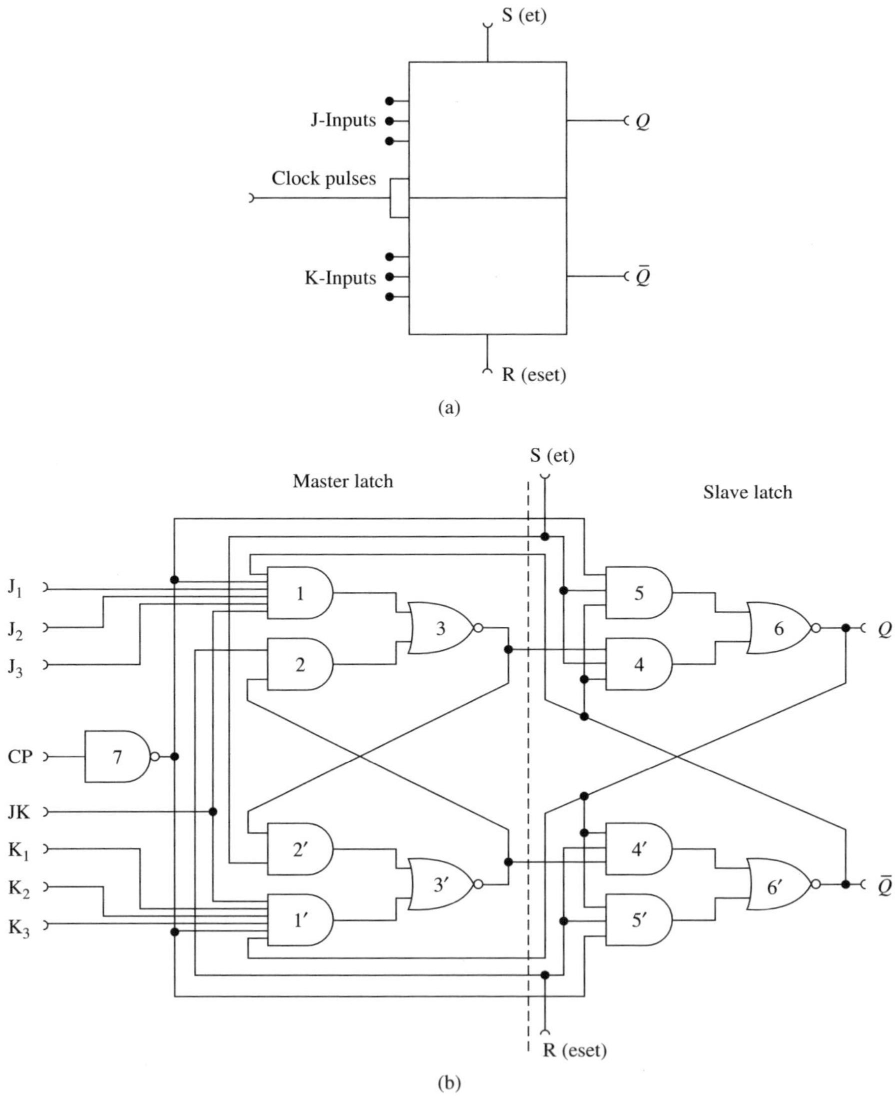

Figure 8.39 *J-K* flip-flop: (a) schematic arrangement and (b) functional logic diagram in the "master–slave" configuration (© [2]).

REFERENCES

[1] F.M. Gardner, "Charge-pump phase-lock loops", *IEEE Trans. Commun.*, **COM-28**, 1849–1858, 1980; Reprinted by W.C. Lindsey and C.M. Chie, *Phase-Locked Loops*. New York: IEEE Press, 321–330, 1986.

[2] V.F. Kroupa, *Frequency Synthesis: Theory, Design et Applications*. London: Ch. Griffin, New York: John Wiley, 1973.

[3] T.T. Brown, "Mixer harmonic chart", *Electronics*, **24**, 132, 134, 1951.

[4] B.C. Henderson, "Predicting intermodulation suppression in double-balanced mixers," © 1983, *Watkins-Johnson Company*, vol. 10, No. 4 July/August; revised and reprinted © 2001 WJ Communications, Inc.

[5] U.L. Rohde, *Digital PLL Frequency Synthesizers, Theory and Design*. Englewood Cliffs: Prentice Hall, 1983.

[6] W.F. Egan, *Frequency Synthesis by Phase Lock*. 2nd ed. New York: Wiley 1981, 2000.

[7] Philipps: Semiconductors Data handbook IC06 – High-Speed CMOS Family, Sunnyvale, CA: Phillips Electronics N.V., 1995.

[8] C.S. Vaucher, "An Adaptive PLL System Architecture Combining High Spectral Purity and Fast Settling Time", *IEEE J. Solid-State Circuits*, **35**, 490–502, 2000.

[9] M. Mossammaparast, C. McNeilage, P. Stockwell and J.H. Searls "Phase noise of X-band regenerative frequency dividers", *2000 IEEE International Frequency Control Symposium, Proc.* June 2000, Kansas City Missouri, pp. 531–535.

[10] T.S. Aytur and B. Razavi, "A 2-GHz, 6-mW BiCMOS Frequency Synthesizer", *IEEE J. Solid-State Circuits*, 1457–1462, 1995.

[11] U.L. Rohde, *Microwave and Wireless Synthesizers*. New York: John Wiley, 1997.

[12] V.F. Kroupa, ed., *Direct Digital Frequency Synthesizers*. New York: IEEE Press, 1999.

[13] Ch. Wuensch, Fractional-N Synthesizers Meet HSCSD and GPRS Demands", Mixcrowaves & RF, 105, August 1999.

[14] T.-H. Lin *et al.*, "A900-MHz 2.5 mA CMOS Frequency synthesizer with an Automatic SC tuning loop", *IEEE J. Solid-State Circuits*, **36**, 424–430, 2001.

9

Noise and Time Jitter

9.1 INTRODUCTION

Random fluctuations of phase and amplitudes (generally designated as noise) of frequency generators are often limiting factors for many applications even in PLLs. Consequently, we must assume, in general, the output voltage (or current) to be

$$v(t) = V_o[1 + a(t)]\cos[\omega_o + \varphi(t)] \tag{9.1}$$

Owing to the limiting processes in many frequency generators, amplitude modulation $a(t)$ can be neglected. Therefore, in the following equation, we shall consider only phase or frequency fluctuations

$$v(t) = V_o\cos[\omega_o t + \varphi(t)] = V_o\cos[\omega_o t + \int \dot{\varphi}(t)\,\mathrm{d}t] \tag{9.2}$$

where

$$\dot{\varphi}(t) \approx \Delta v(t) + \text{discrete sp. signals} + \text{secular terms} \tag{9.3}$$

The pertinent frequency stability can be specified either in the frequency domain using spectral densities or in the time domain in terms of autocorrelation functions, which are also called *autocovariance*.

9.2 TYPES OF NOISE

All physical processes are subjected to some sort of uncertainties due to fluctuations of individual internal or external parameters. If the respective changes are small and random, we speak about noise. In practice, we encounter three fundamental types of noises that differ by the power in the time or frequency unit $S(f)$ (generally 1 Hz), the latter being called the *Power Spectral Density* (PSD) – see Fig. 9.1.

Phase Lock Loops and Frequency Synthesis V.F. Kroupa
 ISBN: 0-470-84866-9

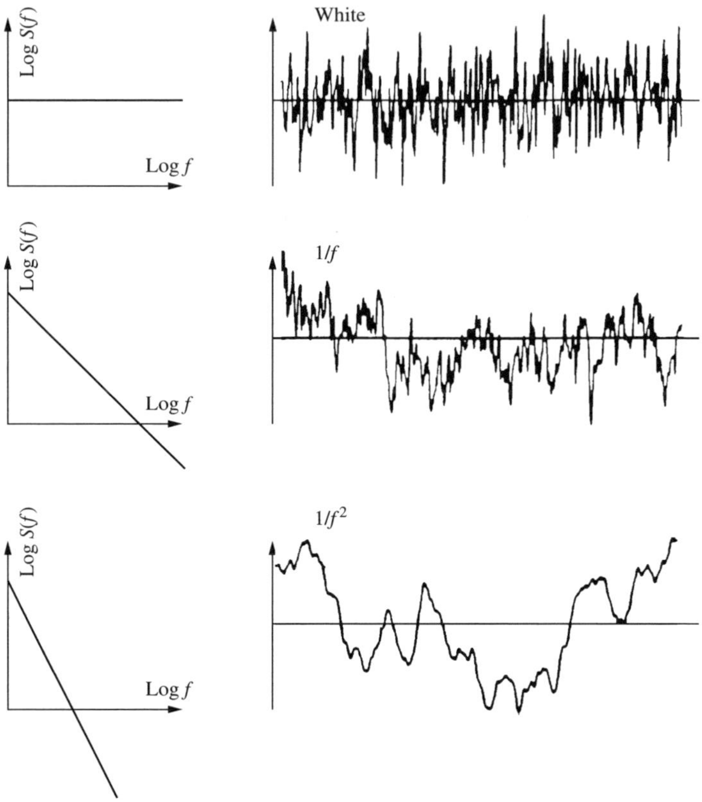

Figure 9.1 Fundamental types of noises (Reproduced from R.F. Voss, "$1/f$ (flicker) noise: a brief review", *Annual Frequency Control Symposium*, 1979, Proceedings, pp. 40–46 by permission of IEEE, 2002 [1]).

The three types of noises are as follows:

1. *white noise* with a constant power spectral density (PSD): $S(f) = \text{constant}$;
2. *flicker noise* or $1/f$ noise with the PSD $S(f) \sim 1/f^{\alpha}$, where the power α is very close to 1;
3. *random walk* or Brownian motion (often designated as the Wiener–Levy process) with $S(f) \sim f^2$.

9.2.1 White Noise

Typical representative of white noise is the *black body radiation* on one hand and thermal noise of resistors or shot noise in electronic devices on the other.

9.2.1.1 Thermal noise

In 1928, Johnson [2] and Nyquist [3] published their theory explaining the existence of thermal noise in conductors. It is caused by short current pulses generated by collisions of a large number of electrons. The result is that a noiseless conductor is connected in series with a generator with rms noise voltage, e_n (see Fig. 9.2),

$$\langle e_n^2 \rangle = 4kTR\Delta f \tag{9.4}$$

where k is the Boltzmann constant and T the absolute temperature (see Tab. 9.1), R is the resistance of the conductor in (Ω) and Δf is the frequency bandwidth in Hz used for the appreciation of the noise action. After dividing the relations (9.4) and (9.6) by the frequency bandwidth Δf, we arrive at the PSD in 1 Hz, that is,

$$S_{e,n}(f) = 4kTR \ [V^2/Hz] \tag{9.5}$$

Similarly, with the assistance of the Thevenin theorem we get the noise current, i_n, flowing into the resistance R or the conductivity $G = 1/R$ (see Fig. 9.2),

$$\langle i_n^2 \rangle = 4kTG\Delta f \ [A^2] \tag{9.6}$$

With the PSD

$$S_{i,n}(f) = 4kTG \ [A^2/Hz] \tag{9.7}$$

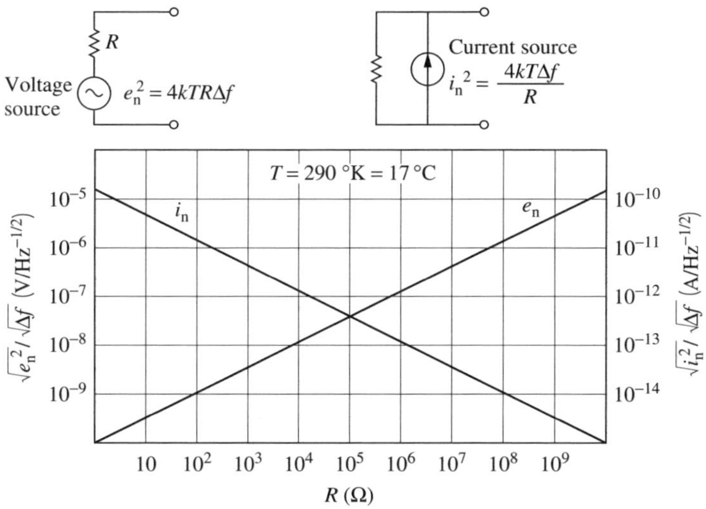

Figure 9.2 Thermal noises of conductors (Reproduced from S. Lezter and N. Webster, "Noise in amplifiers", *IEEE Spectrum*, **August**, 67–75, 1970 by permission of IEEE, 2002 [4]).

Table 9.1 Several physical constants

Planck constant	h	6.625×10^{-35} (Js)
Boltzmann constant	k	1.380×10^{-23} (J/K)
Electron charge	q	1.6×10^{-19} (C)
Noise voltage	$4kTR$; $T = 296\,K$, $R = 1\,\Omega$	$10^{-19.8}$ [V^2 (rms)]
Noise voltage	$4kTR$; $T = 296\,K$, $R = 50\,\Omega$	$10^{-18.1}$ [V^2 (rms)]

In instances with a general impedance or admittance, we introduce only the real parts into the above equations. Further, since both PSDs (9.5) and (9.7) are constant in a very large frequency band (with no filter at the output), we call this type of noise *white* (in accordance with optical physics).

By considering the noise power in a frequency range $\Delta f = f_{\mathrm{high}} - f_{\mathrm{low}}$, we get the noise power

$$P_{\mathrm{n}} = \int_{f_{\mathrm{l}}}^{f_{\mathrm{h}}} \frac{S_{\mathrm{e,n}}(f)}{R}\,\mathrm{d}f = 4kT(f_{\mathrm{h(igh)}} - f_{\mathrm{l(ow)}}) \approx 4kTf_{\mathrm{h}}\ [\mathrm{W}] \tag{9.8}$$

However, by increasing the upper bound, f_{h}, above all limits, the noise power, P_{n}, would also increase above all limits. But this is not possible and the correct solution is provided by quantum mechanics, which changes noise PSD for extremely high frequencies into

$$P_{\mathrm{n}} = \frac{4hf}{e^{hf/kT} - 1}\ [\mathrm{WsHz}^{1}] \tag{9.9}$$

where h is the Planck constant, $h = 6.625 \times 10^{-35}$ [Js].

Example 9.1

Compute the thermal noise generated in the 1-Hz bandwidth in the case where $R = 50\,\Omega$

$$\begin{aligned} k &= 1.38 \times 10^{-23}\ [\mathrm{J/K}] \\ T &\approx 296 \\ R &= 50\,\Omega \\ \langle e_{\mathrm{n}}^2 \rangle &= 4 \times 1.38 \times 10^{-23} \times 296 \times 50 = 8.17 \times 10^{-19} = 10^{-18.1}\ [\mathrm{V}^2] \end{aligned}$$

The noise power P_{n} is

$$P_{\mathrm{n}} = 1.6 \times 10^{-20}\ [\mathrm{W}]$$

or

$$P_{\mathrm{n}} = -168\ [\mathrm{dBm}]$$

9.2.1.2 Shot noise

In all cases in which the output current is composed of random arrivals of a large number of particles, we again witness fluctuations of the white noise type [5].

By considering an idealized transition in Fig. 9.3 where electrons flow randomly from A to B and holes from B to A, in a negligible transit time, each particle arrival

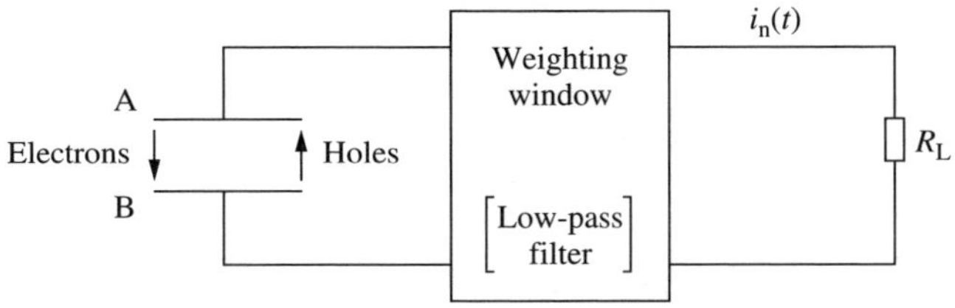

Figure 9.3 Circuit model for shot noise.

is connected with transport of a current pulse; consequently, in a time unit τ[s], the number of n charges generates the current

$$i = \frac{q}{\tau} n \tag{9.10}$$

where q is the electron or hole charge, $q = 1.6 \times 10^{-19}$ [C].

It was shown earlier [5] that the probability of the transition of the charge carriers was subjected to the Poisson distribution

$$p(n) = \frac{\langle n \rangle^n}{n!} \mathrm{e}^{-\langle n \rangle} \tag{9.11}$$

where $\langle n \rangle$ is the mean value of the number of carriers in the time unit. In such a case, the variance is equal to

$$\sigma^2(n) = \langle n \rangle \tag{9.12}$$

By reverting to the relation (9.10), we get for the mean current

$$I = \frac{q}{\tau} \langle n \rangle \tag{9.13}$$

and for its variance value

$$\sigma^2(I) = \frac{q^2}{\tau^2} \langle n \rangle = \frac{q}{\tau} I \tag{9.14}$$

To arrive at the PSD, we use a slightly heuristic approach with the assistance of autocorrelation.

$$\begin{aligned} S_{\mathrm{i,n}}(f) &= 2 \int_0^\infty \sigma^2(I) \cos(\omega t)\, \mathrm{d}t \approx 2 \int_0^\tau \sigma^2(I) \cos(\omega t)\, \mathrm{d}t \\ &= 2\sigma^2(I) \left. \frac{\sin \omega t}{\omega} \right|_0^\tau \approx 2 \frac{q}{\tau} I \frac{\omega \tau}{\omega} = 2qI \end{aligned} \tag{9.15}$$

Example 9.2

Find the PSD $S_{i,n}(f)$ for the shot noise of the transistor current

$$I = 1\,\mathrm{mA} \quad S_{i,n} = 2 \times 1.6 \times 10^{-19} \times 10^{-3} = 3.2 \times 10^{-22}$$

For PSD of other currents, see Fig. 9.4.

9.2.2 Flicker or 1/*f* Noise

In the mid-1920s, it was found by Johnson [6] that at very low frequencies the shot noise did not follow the law in accordance with (9.13), but the magnitude of the excess spectral density varied as the current squared, and in addition it was frequency-dependent. Schottky [7] introduced the term "*flicker noise*," which is still used today.

The same type of noise was observed with India ink resistors, carbon microphones, and many other electronic devices. Some years later, Bernamont [8] suggested for PSD the law

$$S_n(f) \sim \frac{1}{f^{\alpha}} \tag{9.16}$$

where the power of α was in the vicinity of 1. Subsequent observations proved the 1/*f* law for a much larger set of physical phenomena on one hand and its validity at very low frequencies on the other hand, even at such low frequencies that the experimental set failed, or the patience of the experimenter failed.

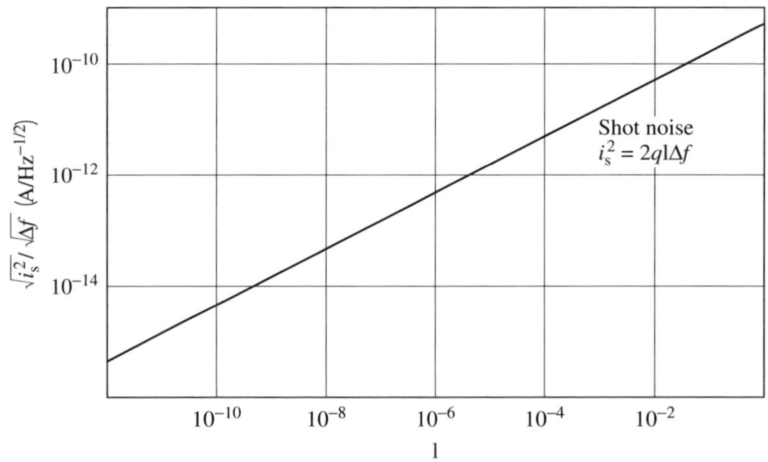

Figure 9.4 Noise current through semiconductor junction (Reproduced from S. Lezter and N. Webster, "Noise in amplifiers", *IEEE Spectrum*, **August**, 67–75, 1970 by permission of IEEE, 2002 [4]).

Example 9.3

One example is shown in Fig. 9.5 presenting the PSD $S_R(f)/R^2$ for an India ink resistor in accordance with (9.16); the validity extends for more than 10 decades with $\alpha \approx 1.21$ [1].

Other measurements provided for the exponent values from $\alpha = 0.8$ to $\alpha = 1.4$.

9.2.3 Noise $1/f^2$

The PSD of this type of noise is

$$S(f) \sim \frac{1}{f^2} \tag{9.17}$$

and it is closely connected with the Brownian motion. The first observations (1827) were of mechanical particles. However, later studies proved that many thermal, electric, and electronic processes might be solved with the same mathematical approach.

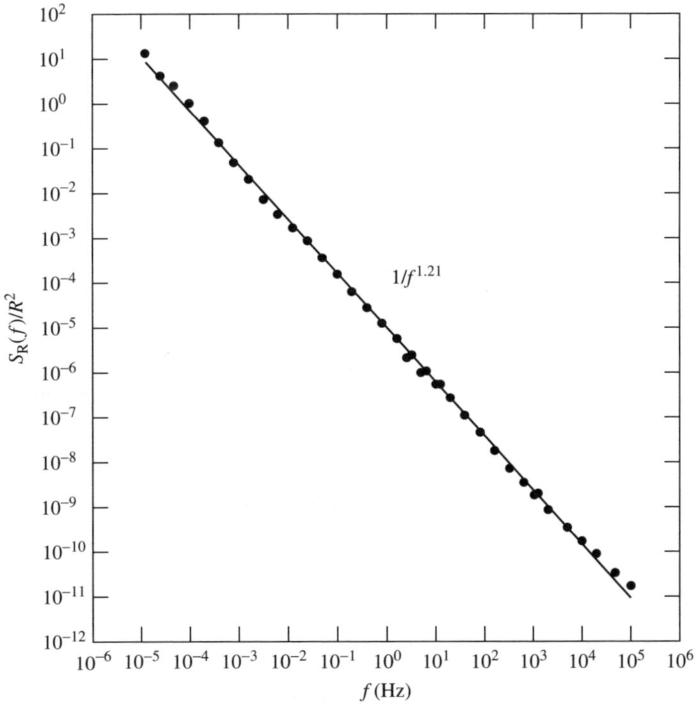

Figure 9.5 Relative resistance fluctuation spectrum, $S_R(f)/R^2$, for India ink resistor (Reproduced from R.F. Voss, "$1/f$ (flicker) noise: a brief review", *Annual Frequency Control Symposium*, 1979, Proceedings, pp. 40–46 by permission of IEEE, 2002 [1]).

Let us start with the velocity of a free particle in a viscous medium. With the assistance of the laws of motion, we arrive at the following differential equation:

$$m\frac{\mathrm{d}v(t)}{\mathrm{d}t} + bv(t) = B(t) \equiv m \cdot n(t) \tag{9.18}$$

where m is its mass, b the friction force proportional to the velocity, $v(t)$, and $B(t)$ represents the collision force [9]. Furthermore, $v(t)$, $B(t)$, and $n(t)$ are stochastic processes and $n(t)$ is normal white noise with zero mean and spectrum, $S_{\mathrm{n}}(f) = a$. In this case, in the long term

$$t \gg \frac{m}{b} = \frac{1}{\beta} \tag{9.19}$$

one may consider $v(t)$ as a stationary process and eq. (9.18) as stochastic. With the rules of derivation of spectra of stochastic processes, we get

$$\omega^2 S_{\mathrm{v}}(\omega) + \beta^2 S_{\mathrm{v}}(\omega) = S_{\mathrm{n}}(\omega) \tag{9.20}$$

Consequently, we arrive at the Lorentzian spectrum

$$S_{\mathrm{v}}(f) = \frac{a}{f^2 + (\beta/2\pi)^2} \doteq \frac{a}{f^2} \quad (\omega \gg \beta) \tag{9.21}$$

Since random fluctuations of phase and amplitudes in frequency generators are often limiting factors for many applications, even in PLLs, we shall discuss the problem in some depth.

9.2.4 Piecewise Approximations of Noise Characteristics

In previous sections we have discussed three different types of noises with their PSD proportional to f^0, f^{-1}, and f^{-2}. In the instance that these three noises are related to frequencies, we arrive at phase PSD after integration, that is, after division by f^2. Consequently, we have five types of phase noises encountered with oscillators, phase lock loop, and so on (see Tab. 9.2). A piecewise linearized phase noise characteristic of an oscillator is shown in Fig. 9.6 and recalled in Tab. 9.2.

Table 9.2 Types of noises

Random walk of frequency	$1/f^4$
Flicker noise of frequency	$1/f^3$
White noise of frequency	$1/f^2$
Flicker noise of phase	$1/f$
White phase noise	f^0

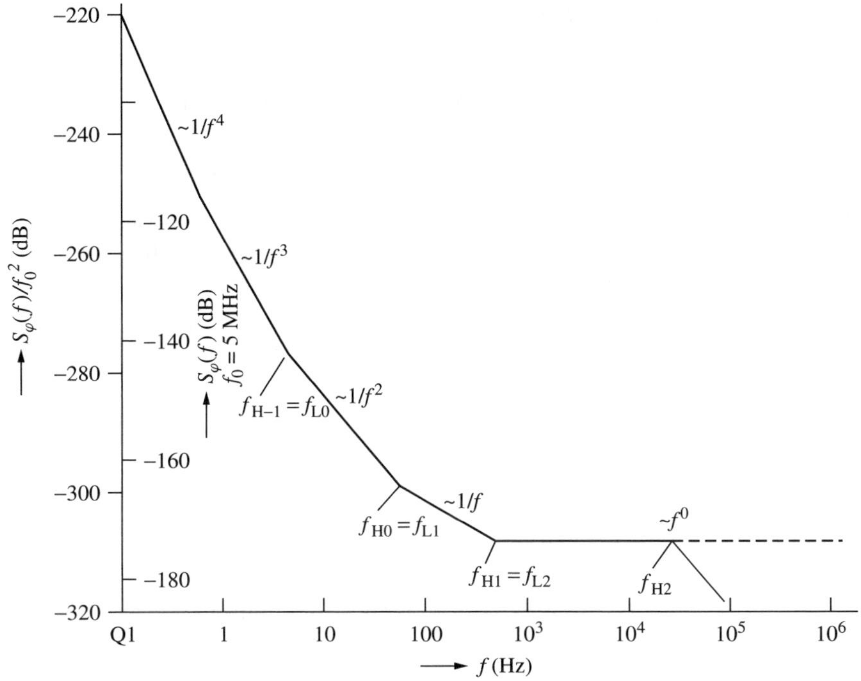

Figure 9.6 A piecewise linearized phase noise characteristic.

9.3 MATHEMATICAL THEORY OF NOISE

Characterization of the frequency stability of all types of generators, inclusive of PLLs, is important for applications, in the first place for designers and also for users. In the mid-1960s, theoretical principles of the phase noise theory in frequency generators were established [10] and later a number of practical papers were published [11, 12]. Here, we shall recall the theory briefly.

9.3.1 Frequency Domain

In the frequency domain we evaluate either the phase noise or the frequency noise of generators, more exactly their PSD.

9.3.1.1 Phase measures

The autocorrelation of the random phase departures, $\phi(t)$, is defined with the assistance of the following relation:

$$R_\phi(\tau) = \langle \phi(t+\tau/2)\phi(t-\tau/2) \rangle = \left[\lim \frac{1}{T} \int_{-T/2}^{T/2} \phi(t+\tau/2)\phi(t-\tau/2)\, \mathrm{d}\tau \right]_{T\to\infty} \tag{9.22}$$

and the respective *power spectral density* (PSD) $S'_\phi(\omega)$[1] is

$$S'_\phi(\omega) = \int_{-\infty}^{\infty} R_\phi(\tau)\mathrm{e}^{-\mathrm{j}\omega\tau}\,\mathrm{d}\tau \ [\text{radian}^2/\text{Hz}] \tag{9.23}$$

We often encounter another definition, that is, $\mathscr{L}(f)$, defining the ratio of the phase power at frequencies $f_o \pm f$ in the 1-Hz bandwidth (where f_o is the carrier frequency and f is the so-called Fourier frequency) in respect of the whole power of the investigated signal

$$\mathscr{L}(f) \doteq S'_\phi(f) \doteq \tfrac{1}{2}S_\phi(f) \tag{9.24}$$

where $S_\phi(f)$ is the so-called *one-sided PSD*.

9.3.1.2 Frequency measures

The difficulty with phase measurements is the definition of the mean value because of the increasing phase with time. Unlike the uncertainty about the first moment of phase fluctuations, the first moment of frequency fluctuations can be made zero in nearly all frequency generators, that is,

$$m_1 = \langle\Delta\omega(t)\rangle \approx 0 \tag{9.25}$$

However, this is not the case with the second moment, which can be defined with the PSD as

$$m_2 = \langle[\Delta\omega(t)]^2\rangle = \int_0^{\infty} S_{\Delta\omega}(f)\,\mathrm{d}f \tag{9.26}$$

A further simplification will be achieved by normalizing frequency fluctuations in respect to the carrier frequency

$$y(t) = \frac{\Delta\omega(t)}{\omega_o} \tag{9.27}$$

having the power spectral density (PSD), $S_y(f)$. The advantage of the fractional PSD, $S_y(f)$, is its invariance in respect to frequency multiplication and division. In addition, we can plot in one diagram the noise characteristics of sources with different carriers and investigate individual influences on the final noise output. After taking into account that phase is the time integral of the frequency, we get, with the assistance of the Laplace transform, the following relation between $S_y(f)$ and the above-introduced $S_\phi(f)$

$$S_\phi(f) = S_y(f)\left(\frac{f_0}{f}\right)^2 \tag{9.28}$$

[1] Prime indicates two-sided spectra.

9.3.2 Time Domain

At very low frequencies, direct evaluation of phase PSD is difficult. The problem is solved with sample variances that provide other, very effective frequency stability measures. Nevertheless, in practice we encounter the Allan variance (two sample variances), which is defined with the assistance of the fractional frequency deviations

$$\sigma_y^2(\tau) = \tfrac{1}{2}\langle(\overline{y}_{k+1} - \overline{y}_k)^2\rangle \tag{9.29}$$

where the fractional frequency deviations are defined or computed as

$$\overline{y}_k = \frac{1}{\tau}\int_{t_k}^{t_k+\tau} y(t)\,\mathrm{d}t; \quad t_k + \tau = t_{k+1} \tag{9.30}$$

Often, we encounter the so-called modified Allan variance that has been introduced to distinguish between the white and the flicker phase noise parts of the time domain characteristics.

$$\mathrm{Mod}\,\sigma_y^2 = \frac{1}{2n^2}\left\langle\left[\sum_{i=1}^{n}(\overline{y}_{i+n} - \overline{y}_i)\right]^2\right\rangle \tag{9.31}$$

9.3.3 Conversion Between Frequency and Time Domain Measures

Both frequency stability measures, the PSD and the Allan variance, are related with the assistance of a transfer function [11]

$$\sigma_y^2(\tau) \approx \int_0^\infty S_y(f)|H_A(\mathrm{j}f)|^2\,\mathrm{d}f = \int_0^\infty S_y(f)\frac{2\sin^4(\pi\tau f)}{(\pi\tau f)^2}\,\mathrm{d}f \tag{9.32}$$

The difficulty is that we can evaluate the integral in (9.32), in the closed form, only for a very particular form of $S_y(f)$, that is, a piecewise linearized form (see Fig. 9.6 and Tab. 9.3).

$$S_\phi(f) = f_o^2 S_y(f) \approx \frac{f_o^2}{f^2}\left(\frac{h_{-2}}{f^2} + \frac{h_{-1}}{f} + h_0 + h_1 f + h_2 f^2\right); \quad h_i = f^{-i}S_y(1) \tag{9.33}$$

We shall revert to the problem later when discussing oscillator noise.

Table 9.3 PSD noise constants of fractional frequency noise

The random walk of frequency with the noise constant	h_{-2}
The flicker frequency noise with the noise constant	h_{-1}
The white frequency noise with the noise constant	h_0
The flicker phase noise with the noise constant	h_1
The white phase noise with the noise constant	h_2

Table 9.4 Conversion between spectral densities, $S_y(f)$, two sample variances, $\sigma_y^2(\tau)$, and "Modified Allan variances" Mod $\sigma_y^2(\tau)$ for power law spectral density model

$S_y(f)$	$\sigma_y^2(\tau)$	Mod $\sigma_y^2(\tau)$
h_{-2}/f^2	$(2\pi)^2\tau h_{-2}/6$	$\approx 5.4n\tau_0 h_{-2}$
h_{-1}/f	$2h_{-1}\ln 2$	$\approx 0.94h_{-1}$
h_0	$h_0/2\tau$	$\approx h_0/4n\tau_0$
$h_1 f$	$h_1(2\pi\tau)^{-2}[1.38 + 3\ln(\omega_H\tau)]$	$\approx 0.084h_1/(n\tau_0)^2$
$h_2 f^2$	$3h_2 f_H(2\pi\tau)^{-2}$	$3h_2 f_H/n(2\pi n\tau_0)^2 \approx 0.076\tau_0 h_2 f_H/(n\tau_0)^3$

Inspection of Fig. 9.6 reveals that from the knowledge of the h_i coefficients in relation (9.33), we can estimate both phase noise PSD and Allan variances (see Tab. 9.4).

9.3.4 Phase and Time Jitter

Phase and time jitter is another measure of the instabilities of oscillators and frequency synthesizers. They find applications in communications for appreciation of errors encountered with modern digital systems.

For estimation, we may apply the measured phase noise characteristic or the Allan variance. For direct measurement, we apply either the period-jitter evaluation with a digital storage oscilloscope (DSO) or a time interval analyzer (TIA).

9.3.4.1 Phase noise jitter

Generally, we refer to the rms of the phase disturbances, which may be computed from the known phase PSD as

$$\langle\Delta(\phi)\rangle \approx \sqrt{\langle\phi(t)^2\rangle} \approx \sqrt{\int_{f_{\text{low}}}^{f_h} S_\phi(f)\,\mathrm{d}f} \tag{9.34}$$

The problems with the above relation are the integration limits. By taking into account the nonlinearity of the PSD characteristic, the integration might proceed only after applying the piecewise linearization (see Fig. 9.6). Starting from the very low limit of integration, the part proportional to $1/f^3$ would supply the largest contributions of the rms of $\langle\Delta(\phi)\rangle$ which is not important for error evaluation in communications. We shall see later that the lower-bound integration may be close to several Hz (it depends on the type of the service, e.g., 20 Hz). As to the upper limit, the actual output frequency f_{out} might be a realistic choice. Practically, the white noise part of output PSD would be decisive for the rms of the phase noise jitter. Note that reliable technical data should mention f_{low} when indicating phase or time jitter.

Example 9.4

As an example, we consider here the quadrature phase shift keying (QPSK) (see Fig. 9.7). Note that between individual messages the phase shift is 90°. Evidently, the phase noise may introduce the error $\pm\Delta\Psi$. In the rms we have

$$\Delta\Psi_{\mathrm{rms}} \approx \frac{180}{\pi}\sqrt{\Delta\langle(\phi)^2\rangle} = \frac{180}{\pi}\sqrt{\int_{f_{\mathrm{low}}}^{f_{\mathrm{h}}} S_\phi(f)\,\mathrm{d}f} \tag{9.35}$$

By taking into account that the peak phase excursion might be 6 to 14 times the rms, we arrive at the value for $\Delta\Psi$ of several degrees only, that is, in the range from 3° to 6°. Now let us consider some numerical values, namely, a PLL output with the VCO having two different PSDs in the white noise range:

$$f_{\mathrm{out}} = 1\,\mathrm{GHz}, \quad S_\phi = -170\,\mathrm{dB}$$
$$\Delta\Psi_{\mathrm{rms}} = \mathrm{SQR}(10^{-17} \times 10^{9}) = 10^{-4}$$

in degrees

$$\Delta\Psi_{\mathrm{rms}} = 180 \times 10^{-4}/\pi \approx 6 \times 10^{-3}\ [^\circ]$$

for

$$f_{\mathrm{out}} = 1\,\mathrm{GHz}, \quad S_\phi = -150\,\mathrm{dB}$$
$$\Delta\Psi_{\mathrm{rms}} = 180 \times 10^{-4}/\pi \approx 0.6\ [^\circ]$$

9.3.4.2 Time jitter

Since the time jitter is closely related to the phase noise, we may write for its mean value

$$\langle\Delta t\rangle = \frac{\langle\Delta\phi\rangle}{2\pi f_{\mathrm{o}}} \tag{9.36}$$

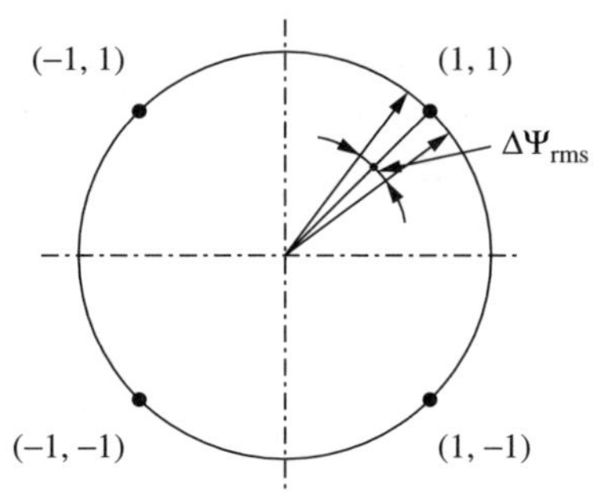

Figure 9.7 Phase distribution in quadrature phase shift keying (QPSK).

We have seen earlier that phase is a slowly varying function of time in a frequency or time interval. Consequently, the time jitter might be computed as follows:

$$\langle \Delta t \rangle = \frac{\sqrt{\langle \phi^2 \rangle}}{2\pi f_o} = \frac{\sqrt{\int_{f_{low}}^{f_{high}} S_\phi(f)\,\mathrm{d}f}}{2\pi f_o} \tag{9.37}$$

Example 9.5

Compute the time jitter for the phase error from the previous example: for

$$f_{out} = 1\,\mathrm{Ghz}, \quad S_\phi = -170\,[\mathrm{dB}]$$

$$\Delta\Psi_{rms} = \mathrm{SQR}(10^{-17} \times 10^9) = 10^{-4}$$

and

$$\Delta t_{rms} = 10^{-4}/10^9 = 10^{-13}\,[\mathrm{s}] \quad \text{or} \quad 0.1\,[\mathrm{ps}]$$

for

$$f_{out} = 1\,\mathrm{Ghz}, \quad S_\phi = -150\,[\mathrm{dB}]$$

$$\Delta\Psi_{rms} = \mathrm{SQR}(10^{-15} \times 10^9) = 10^{-3}$$

we get

$$\Delta t_{rms} = 10^{-3}/10^9 = 10^{-12}\,[\mathrm{s}] \quad \text{or} \quad 1.0\,[\mathrm{ps}]$$

9.3.4.3 Time jitter measurements

1. Frequency domain measurement process:
   (a) demodulate to obtain the modulation versus time or as a voltage versus time;
   (b) use filter and measure the jitter with a true rms voltmeter;
   (c) measure the PSD $S_v(f)$ or $S_\phi(f)$ and calculate the rms time jitter.
2. Time domain measurement process.

Some authors state that the Allan variance is a measure of average frequency over a time interval that is the same as the phase change over the interval; consequently, Allan variance multiplied by the time interval equals the time jitter over that interval:

$$\langle \Delta t(\tau) \rangle = \tau \cdot \sigma_y(\tau, f_h) \tag{9.38}$$

However, the above relation is valid for the white and flicker phase parts of the PSD characteristics (cf. relation (9.37) with the conversion table – Tab. 9.4). Nevertheless, eq. (9.38) is a good measure when the Allan variance for short τ is known.

9.3.5 Small and Band-limited Perturbations of Sinusoidal Signals

Up to now we have considered single frequency generators, that is, oscillators with rather small and continuous amplitude and phase perturbations. However, in frequency synthesizers, particularly in direct digital synthesizers (DDFS), we encounter a lot of spurious signals. In the following sections we shall investigate their properties [13].

9.3.5.1 Superposition of one large and one set of small signals

In actual frequency synthesizers, not excluding PLL systems, we always encounter many, generally small, spurious signals accompanying the carrier. The composite signal may be written as

$$v(t) = \sum_{n=1}^{N} V_n \cos(\omega_n t + \phi_n) \tag{9.39}$$

After introducing

$$\omega_n t + \phi_n = \omega_1 + (\omega_n - \omega_1)t + \phi_n = \omega_1 t + \Phi_n(t) \tag{9.40}$$

and putting

$$\alpha_n = \frac{V_n}{V_1} \tag{9.41}$$

we get

$$\begin{aligned} v(t) &= V_1 \sum_{n=1}^{N} \alpha_n \cos[\omega_1 t + \Phi_n(t)] \\ &= V_1 \cos(\omega_1 t) \sum_{n=1}^{N} \alpha_n \cos[\Phi_n(t)] - V_1 \sin(\omega_1 t) \sum_{n=1}^{N} \alpha_n \sin[\Phi_n(t)] \end{aligned} \tag{9.42}$$

or

$$v(t) = V_1 \alpha(t) \cos[\omega_1 t + \Phi(t)] \tag{9.43}$$

where $\alpha(t)$ is the normalized instantaneous amplitude and $\Phi(t)$ is the instantaneous phase departure

$$\alpha^2(t) = \left[\sum_{n=1} \alpha_n \cos \Phi_n(t)\right]^2 + \left[\sum_{n=1} \alpha_n \sin \Phi_n(t)\right]^2 \tag{9.44}$$

and

$$\Phi(t) = \arctan \frac{\sum_{n=1}^{N} \alpha_{\rm n} \sin \Phi_{\rm n}(t)}{\sum_{n=1}^{N} \alpha_{\rm n} \cos \Phi_{\rm n}(t)} \tag{9.45}$$

Without any loss of generality, it is possible to choose the timescale in such a way that

$$\phi_1 = \Phi_1(t) = 0 \tag{9.46}$$

If only small perturbations are assumed, one may put, further,

$$\alpha_1 \gg \alpha_{\rm n}; \quad (n = 2, 3, \ldots N) \quad (\text{note: } \alpha_1 = 1) \tag{9.47}$$

The situation being such,

$$\alpha(t) \approx 1 + \sum_{n=1}^{N} \alpha_{\rm n} \cos \Phi_{\rm n}(t) \tag{9.48}$$

and

$$\Phi(t) = \arctan \sum_{n=1}^{N} \alpha_{\rm n} \sin \Phi_{\rm n}(t) \approx \sum_{n=1}^{N} \alpha_{\rm n} \sin \Phi_{\rm n}(t) \tag{9.49}$$

Example 9.6

For the superposition of one strong signal $V_1 \cos(\omega_1 t)$ and one weak signal $V_2 \cos(\omega_2 t + \phi_2)$, that is, $V_1 \gg V_2$, we get

$$v(t) \approx V_1 \left[1 + \frac{A_2}{A_1} \cos(\Omega t + \phi_2)\right] \cos\left[\omega_1 t + \frac{A_2}{A_1} \sin(\Omega t + \phi_2)\right] \tag{9.50}$$

Evidently, in the first approximation, we face a simultaneous amplitude and phase modulation of the stronger signal at the rate of difference frequency, $\Omega = |\omega_2 - \omega_1|$, with the modulation indexes, A_2/A_1. For its larger value, higher-order terms must be added.

9.3.5.2 Narrow bandwidth noise

In instances where the noise power is concentrated in a relatively narrowband around the frequency ω_1, the noise voltage can be expressed as

$$e(t) = e_{\rm c}(t) \cos \omega_1 t - e_{\rm s}(t) \sin \omega_1 t \tag{9.51}$$

where the slowly varying time functions, $e_c(t)$ and $e_s(t)$, are statistically independent. Note that the above relation resembles (9.42). Consequently, the product of the mean values is zero if $\langle e(t)\rangle$ is zero, that is,

$$\langle e_c(t)e_s(t)\rangle = \langle e_c(t)\langle e_s(t)\rangle = 0 \tag{9.52}$$

and also

$$\langle e_c(t)\rangle = \langle e_s(t)\rangle = 0; \quad \langle e^2(t)\rangle = \langle e_c^2(t)\rangle = \langle e^2(t)\rangle \tag{9.53}$$

9.4 COMPONENT NOISES

All major parts in PLLs may be generators of frequency or phase fluctuations. Because of the limiting actions, we may neglect, in the first approximation, the amplitude noise contributions. In the following sections we shall investigate the noise properties of individual parts.

9.4.1 Amplifiers

We encounter amplifiers in PLLs in instances in which we need to increase the PLL gain K by a DC factor K_A. However, much more important is the application of operation amplifiers in type 2 loops. Since amplifiers are generators of noise, we shall, in the following section, investigate their noise properties.

9.4.1.1 Noise sources in amplifiers

In this connection, we encounter thermal noises generated in the associated network, noises provided by the active semiconductor layers, and current shot noises. Noise generators in an operation amplifier are shown in Fig. 9.8. The expected values of the noise currents and voltages are illustrated in Fig. 9.2.

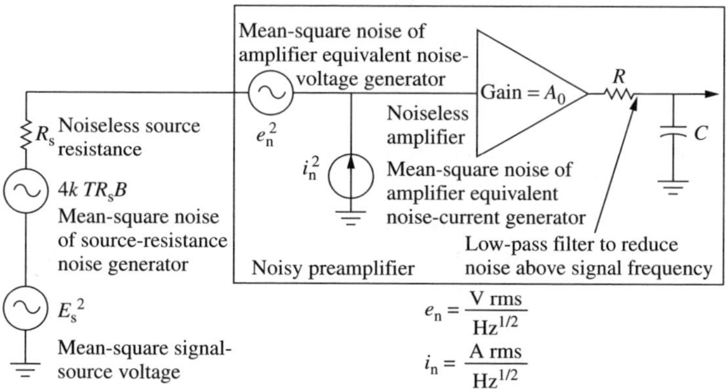

Figure 9.8 Noise generators in an operation amplifier (© [4]).

9.4.1.2 *Total noise*

Let us group together the thermal and shot noises with the fictitious semiconductor voltage and current fluctuations connected to the input terminals, as shown in Fig. 9.8. Further, to simplify investigation we shall assume both e_n and i_n are white with zero correlation. In this case we get for the total noise output voltage

$$E_{n,out} = [4kTR_s + e_n^2 + (i_n R_s)^2]^{1/2} A_0 \sqrt{F_n} \tag{9.54}$$

where A_0 is the midband gain and $F_n = 1/4RC$ the pass band of the low-pass filter in Fig. 9.8. Note that $E_{n,out}$ can be measured with a true rms voltmeter.

9.4.1.3 *Total signal-to-noise ratio*

By considering the useful signal of the output terminal to be

$$E_{out} = E_s \cdot A_0 \tag{9.55}$$

the SNR can be calculated as

$$\mathrm{SNR} = \frac{E_s}{E_{n,out}} = \frac{E_s}{[4kTR_s + e_n^2 + (i_n R_s)^2]^{1/2} \sqrt{F_n}} \tag{9.56}$$

9.4.1.4 *Noise figure*

A popular figure of merit used to describe an amplifier quality, insofar as noise is concerned, is the noise figure (NF) of the amplifier. It is expressed in decibels. From the previous relations we get for NF

$$\mathrm{NF} = 10 \log \left[\frac{E_s^2/4kTR_s F_n}{E_s^2/[4kTR_s + e_n^2 + (i_n R_s)^2] F_n} \right] \tag{9.57}$$

9.4.2 Frequency Dividers

In the last decade, we have witnessed a widespread use of direct digital frequency synthesizers (DDFS) [14], particularly in high megahertz and low gigahertz communications bands. For such high frequencies, a combination of DDFS with phase locked loops (PLLs) and frequency dividers (FD) in the feedback path is inevitable. However, dividers are sources of the noise. Study of this problem started more than 30 years ago [15] and it was shown that changes of the triggering voltage level, due to superimposed spurious signals (e.g., Johnson and flicker noises generated in semiconductors, resistors, etc.) or ambient effects (variation of the temperature, humidity, etc.), were origins of the time jitter and consequently of the divider noise.

Earlier experimental findings revealed for the PSD of the divider noise

$$S_{\phi,\mathrm{D}}(f) \approx \frac{10^{-14.7}}{f} + 10^{-16.5} \tag{9.58}$$

Authors of [15] only had a limited number of experimental measurements at their disposal and, in addition, for rather low output frequencies only. However, from that time the situation changed and we witness a tremendous progress in digital technology toward higher frequencies (ECL and GaAs logic family). After summarizing a wealth of available noise measurements, performed by later authors [16–19], we got the plot in Fig. 9.9 (rectangles), and after applying asymptotic approximations (full lines) we arrived at the following equations for the PSD of the noise generated

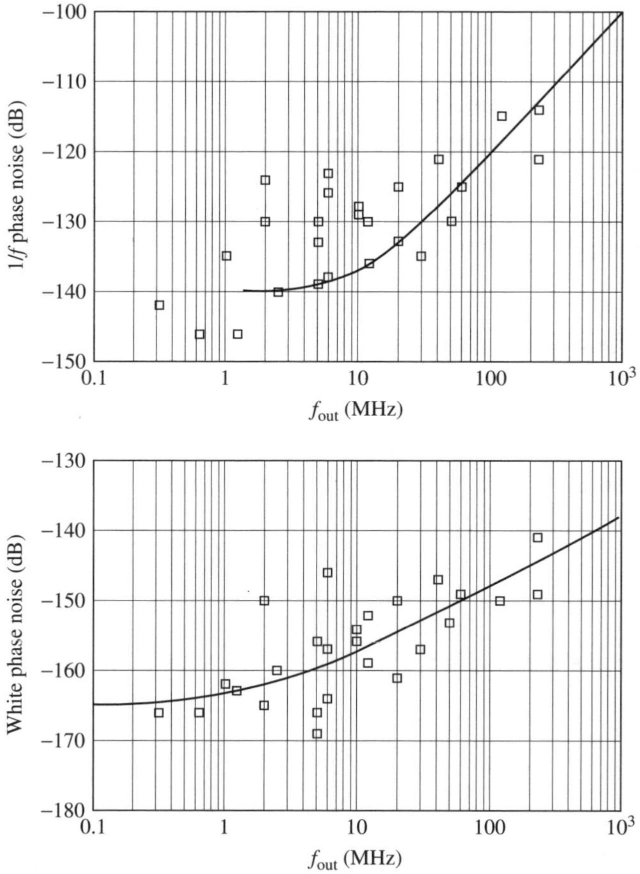

Figure 9.9 (a) Output power spectral density $S_{\phi,fo}(f = 1)$ [dB/Hz] of the flicker phase noise of TTL and ECL digital dividers (Reproduced from V.F. Kroupa, Jitter and phase noise in frequency dividers IEEE Trans. IM – October 2001 by permission of IEEE, 2002 [20]). (b) Output power spectral density $S_{\phi,fo}(f)$ [dB/Hz] of the white phase noise of TTL and ECL digital dividers (© IEEE, 2002).

in the digital frequency dividers [20]:

$$S_{\phi,\mathrm{D}}(f) \approx \frac{10^{-14\pm1} + 10^{-27\pm1} f_o^2}{f} + 10^{-16\pm1} + 10^{-22\pm1} f_o \tag{9.59}$$

Note that the PSD in the $1/f$ noise region increases with the square of the divider output frequency, f_o^2 (Fig. 9.9a), whereas in the white noise region it increases with the first power of f_o (Fig. 9.9b). In addition, the first term on the rhs is nearly the same as in the relation (9.58) for lower output frequencies but proportional to f_o^2 for the higher ones, whereas the second term exhibits proportionality to f_o only.

9.4.3 Phase Detectors

With respect to the noise, the double balanced mixers with Schottky barrier diodes in a ring configuration (see Fig. 8.7) exhibit the lowest phase noise [17]. For a preliminary estimation of the introduced noise we can use relation (9.58), found earlier for dividers. However, generally, we encounter digital phase detectors built either of ECL or CMOS logic families. They exhibit in the flicker range a slightly larger phase noise PSD. Here again we can use for the noise estimation the results found for dividers, particularly relation (9.59).

9.4.4 Noises Associated with Loop Filters

The loop filters with RC sections are generators of the noise. Here, we shall present only the expected output noises of the thermal origin. The theoretical PSD $S_{\phi,\mathrm{L}}(f)$ of this white noise level will be computed from the relation

$$S_{\phi,\mathrm{L}} = \frac{e_n^2}{K_d^2} \approx \frac{4kTR_{\mathrm{eff}}}{K_d^2} \tag{9.60}$$

where k is the Boltzmann's constant and T the room temperature (summarized in Tab. 9.1).

9.4.4.1 RC filter

Figure 9.10 reveals that the effective resistance, R_{eff}, is just the filter resistance R. To get more insight into the PLL noise problem, we shall express the resistance in the ω_n and ζ parameters. With the assistance of (2.10), we get

$$RC = \frac{K_d K_o}{\omega_n^2} \quad \text{and} \quad K_d K_o = \frac{\omega_n}{2\zeta} \tag{9.61}$$

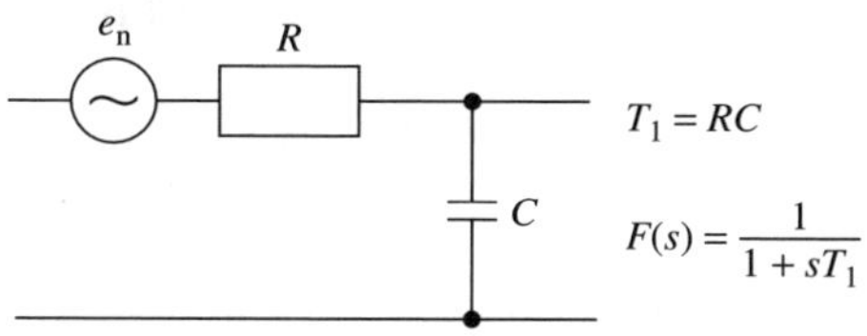

Figure 9.10 Thermal noise generation in one RC section filter.

After introduction into (9.60), we have

$$S_{\phi,L} = \frac{4kT}{2\zeta\omega_n C k_d^2} \tag{9.62}$$

9.4.4.2 *RRC filter*

We easily conclude, with the assistance of Fig. 9.11, that the noise voltage generated in the filter is

$$e_n^2 = e_{n1}^2 + e_{n2}^2 = 4kT(R_1 + R_2) \tag{9.63}$$

The time constant T_1 from Fig. 2.6 is

$$T_1 = C(R_1 + R_2) = \frac{k_d k_o}{\omega_n^2} \tag{9.64}$$

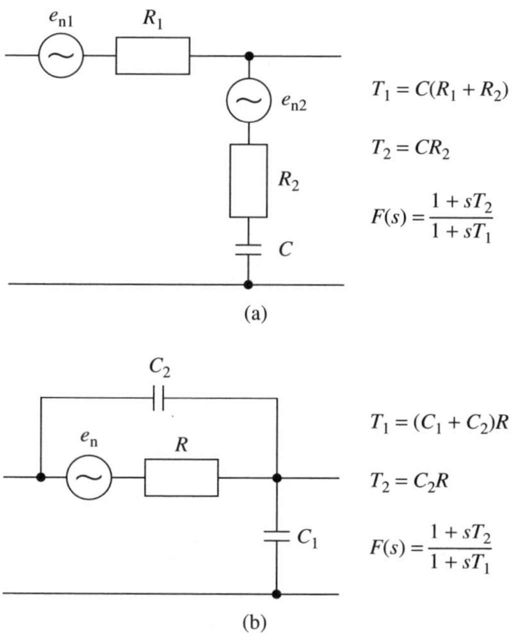

Figure 9.11 Thermal noise generation in the RRC filter configuration.

With the assistance of (2.20), we can express ω_n as follows:

$$\omega_n = \frac{K_d K_o}{2\zeta} \cdot \frac{T_2}{T_1} \tag{9.65}$$

After combination with (9.64) and (9.61), we obtain

$$S_{\phi,L} = 4kT \frac{2\zeta}{\omega_n C k_d^2} \cdot \frac{T_1}{T_2} \tag{9.66}$$

9.4.4.3 RRC filter type 2

With the assistance of Fig. 9.12, we again arrive at the relation (9.62); however, owing to a slightly different definition of the time constant (cf. Fig. 2.8(b)), the noise PSD, $S_{\phi,L}$, is

$$S_{\phi,L} = 4kT \frac{2\zeta}{\omega_n C} \cdot \left(1 + \frac{T_1}{T_2}\right) \tag{9.67}$$

In all three relations (9.62), (9.66), and (9.67), we have introduced the filter capacity C into the denominator. Evidently, the larger the C, the smaller the filter noise generated.

> ***Example 9.7***
>
> Compute the expected phase noise PSD generated in the high-gain PLLs of the second order (Fig. 9.11) by assuming
>
> $$K_d = 0.3 \ \text{[V/rad]}$$
> $$\zeta = 0.7$$
> $$T_1/T_2 \approx 10$$
> $$C = 10^{-6} \ \text{[F]}$$
> $$S_{\phi,L} \approx \frac{10^{-13.5}}{f_n} \tag{9.68}$$
>
> See also Fig. 9.13.

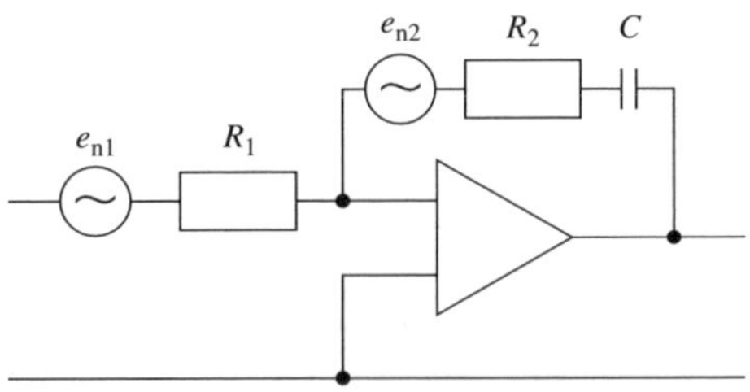

Figure 9.12 Thermal noise generation in the type 2 RRC filter configuration.

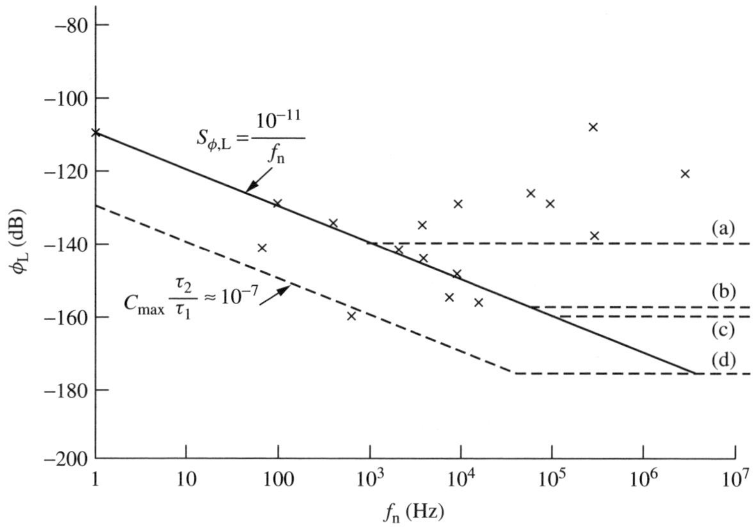

Figure 9.13 PSD $S_{\phi,L}$ of the additive phase noise of different PLL systems (x) together with practical and theoretical limits (−−): (a) ECL level; (b) 10-nV level; (c) TTL level; and (d) ring modulator level. V.F. Kroupa, "Low-noise microwave-frequency synthesizers: design principles", *IEE Proceedings-H*, **130**, 483–488, 1983 (Reprinted in [14]).

9.4.5 Oscillators

Oscillators are inevitable parts of PLLs. In accordance with Fig. 1.1, we encounter a reference oscillator (generally a crystal oscillator, 10 MHz, 100 MHz, etc.) and a voltage-controlled oscillator (VCO) (mostly LC or even digital oscillators).

In the past, many authors have investigated the problem of noise in oscillators [e.g., 21, 22] and in resonators as well [e.g., 23, 24]. The author of this volume summarized all the available experimental data [25] and proceeded with the solution of noise behavior of the basic oscillator circuit shown in Fig. 9.14 by investigating the influence of the additive phase shift introduced via maintaining electronic circuits (the well-known Leeson solution [26]).

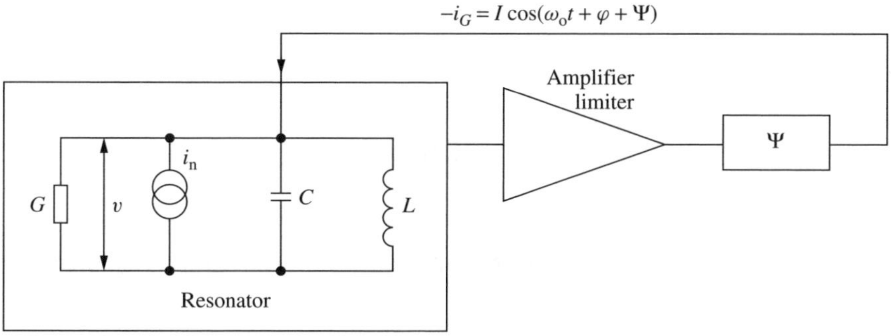

Figure 9.14 Simplified block diagram of the oscillator (© IEEE, 2002).

9.4.5.1 Simplified theory of oscillators

Our investigations will start with the simplified block diagram – Fig. 9.14. The basic differential equation of the oscillator is

$$\dot{v}C + \dot{v}G + \frac{1}{L}\int v\,\mathrm{d}t = i_{-G} + i_\mathrm{n} \tag{9.69}$$

where the oscillator voltage is

$$v(t) = V(t)\cos\Phi(t) = V(t)\cos(\omega_\mathrm{o}t + \phi(t)) \tag{9.70}$$

and the conductivity of the resonant circuit is

$$G = \frac{RC}{L} \tag{9.71}$$

By assuming that the noise is generated by slowly varying parameters of the resonator, we also encounter conductivity fluctuations

$$\Delta G = \frac{\Delta R}{R} + \frac{\Delta C}{C} - \frac{\Delta L}{L} \tag{9.72}$$

that is, ΔG being a function of ΔL, ΔC, and ΔR. Evidently, the steady state term $-i_G/v$ just compensates the conductivity G but leaves a slowly varying component ΔG. Consequently, the solution of the differential eq. (9.69) simplifies to

$$s(C + \Delta C) + \frac{1}{s(L + \Delta L)} + \Delta G = 0 \tag{9.73}$$

Its roots are

$$s_{1,2} = -\frac{\Delta G}{2(C + \Delta C)} \pm \mathrm{j}\sqrt{\frac{1}{(C + \Delta C)(L + \Delta L)} - \frac{1}{4}\left[\frac{\Delta G}{(C + \Delta C)}\right]^2} = -\delta \pm \mathrm{j}\omega \tag{9.74}$$

Finally, for the voltage $v(t)$ we get

$$v(t) = \mathrm{e}^{-\delta t}V_\mathrm{o}\sin(\omega t + \Psi_\mathrm{o}) \tag{9.75}$$

where V_o and Ψ_o are integration constants. Note that in addition both δ and ω are slowly varying time functions. However, the steady state solution with constant output voltage V_o requires that

$$\langle \delta t\rangle = 0 \quad \text{and} \quad \langle v(t)^2\rangle = \langle \mathrm{e}^{-\delta t}V_\mathrm{o}^2\sin^2(\omega t + \Psi_\mathrm{o})\rangle = \tfrac{1}{2}V_\mathrm{o}^2 \tag{9.76}$$

After taking into account the above steady state conditions, we get for fractional frequency fluctuations

$$\left\langle \frac{\Delta\omega}{\omega_o} \right\rangle \approx -\frac{1}{2}\left\langle \frac{\Delta L}{L} + \frac{\Delta C}{C} \right\rangle - \frac{1}{8Q^2}\left\langle \left[\frac{\Delta C}{C} + \frac{\Delta R}{R} - \frac{\Delta L}{L} \right]^2 \right\rangle \tag{9.77}$$

Note that the first moment of the slowly varying fluctuations of $\langle \Delta G(t) = \delta t \rangle$ is zero (at least in the first approximation); this need not be the case with its second moment. Nevertheless, its influence on the fractional frequency fluctuations is of the second order because of the factor $1/8Q^2$.

Another approach to the investigation of the oscillator noise is based on the condition of zero phase around the oscillating loop, that is,

$$\phi = 2\pi N \quad \text{with} \quad N = 0,\ 1,\ 2,\ 3\ldots \tag{9.78}$$

By assuming the gain, G_{loop}, around the loop is equal to 1, the small phase shift, $\mathrm{d}\phi$, anywhere in the oscillation loop causes the frequency change, $\mathrm{d}\nu$, given by

$$\frac{\mathrm{d}\nu}{\nu_o} = \frac{1}{2Q_L}\mathrm{d}\phi \tag{9.79}$$

9.4.5.2 Crystal oscillator output phase noise

After taking into account the noise generated in the resonant circuit itself together with that added by the maintaining feedback system, we get for the output phase of the oscillator

$$\varphi_{\text{out}}(s) = [\varphi(s) + \Psi(s)]\frac{\omega_o}{s \cdot 2Q} + \Psi(s) \tag{9.80}$$

The resonator circuit exhibits the flicker and white noise (the Johnson noise); consequently, the PSD of the fractional frequency fluctuations, $S_y(f)$, is

$$S_y(f) \approx \frac{a_r}{f} + \frac{2kT}{P_r} = \frac{a_r}{f} + \frac{2kTR}{V_o^2/2} \tag{9.81}$$

where $P_r = V_o^2/2R$ is the dissipated power and a_r is the flicker noise constant. Similarly, the PSD of the noise in the maintaining circuit is

$$S_{\phi,\text{el}} = \frac{a_e}{f} + a_o = \frac{a_e}{f} + \frac{2kTFG}{P_r} \tag{9.82}$$

where F is the noise figure and G is the gain of the maintaining electronic circuit. With the assistance of (9.80), we arrive at the PSD of the oscillator phase noise

$$S_{\varphi,\text{out}}(f) \approx \left(\frac{a_r}{f} + \frac{2kT}{P_r} + \frac{a_e}{f} + a_o \right) \frac{f_o^2}{f^2 Q_U^2} + \frac{a_e}{f} + a_o \tag{9.83}$$

where we have introduced the unloaded quality factor of the crystal resonator, Q_U, by putting $2Q_L \approx Q_U$ [27, 28].

9.4.5.3 Oscillator gain K_o

Oscillator gain K_o is a very important factor in the design of PLLs. We can estimate its value with the assistance of relation (9.81). After replacing the noise voltage with a spurious voltage, that is, e_{sp}, and the series resonance voltage with the parallel resonance voltage, we get

$$\frac{\Delta\omega}{\omega_0} = \frac{e_{sp}}{V_0 \cdot Q_U} \tag{9.84}$$

from which

$$K_o = \frac{\Delta f}{e_{sp}} = \frac{f_0}{V_0 \cdot Q_U} \tag{9.85}$$

This result was first introduced, to our knowledge, by Kurokawa [22].

Example 9.8

Compute the gain K_o for a 10-MHz crystal oscillator for the oscillation voltage 1 V.

$$K_o = \frac{\Delta f}{1} = \frac{f_0}{V \cdot Q} = \frac{10^7}{1 \cdot 10^6} = 10 \tag{9.86}$$

Evidently, the estimated gain K_o is too large. In actual crystal oscillators, only a fraction of the voltage v_2 is used for the tuning. Consequently, we must introduce a reduction factor of about $K_{red} \approx 10^{-2}$ or even smaller. In Tab. 9.5, we have plotted K_o for several crystal oscillators.

9.4.5.4 Polynomial interpolations of oscillator noise characteristics

Application of the published noise characteristics of the oscillators is not an easy task. Often, only few discrete noise values are given, and in addition computation of the noise constant h_{-1} from frequency and time domain data often disagree. However, this is not sufficient for the estimation of the noise coefficients in the piecewise linearized

Table 9.5 Crystal oscillator gain K_o for a few 10-MHz OCXO

Crystal oscillator	K_o (Hz/V)
Hewlett-Packard 10811	0.27
Stanford System 10	0.64
Krystaly Hr. Kralove #63	2.27
Krystaly Hr. Kralove #133	1.03

approach. From the frequency stability theory [10, 12], we know that the PSD of the phase noise is given by a polynomial (cf. also (9.83)):

$$S_{\phi}(f) = \frac{a_4}{f^4} + \frac{a_3}{f^3} + \frac{a^2}{f^2} + \frac{a_1}{f^1} + a_o \qquad (9.87)$$

where f is the so-called Fourier or base-band frequency. By knowing only a little data from the PSD characteristic, we shall plot the known points as in Fig. 9.15, then estimate coefficient a_i and try to approximate the characteristic $[S_{\varphi}(f), f]$ with the respective slopes of −40, −30, −20, −10, and 0 dB/decade and plot, with the use of a computer and the sought noise characteristic together with the original points. In case the fit is not a good one, we can get a better fit by slightly changing powers of a_i in (9.83). The procedure will be illustrated with the assistance of the following example.

Example 9.9

As an example, we shall provide asymptotic approximation of the phase noise PSD of the HP crystal oscillator 10 MHz, type 10811 [29]. In Tab. 9.6(a), we recall phase noise data published by the manufacturer and in Tab. 9.6(b), the time domain data.

The asymptotic approximation of the phase noise characteristic reproduced in Fig. 9.15 reveals for the PSD the following equation:

$$S_{\phi} \approx \frac{10^{-10.5}}{f^{-4}} + \frac{10^{-10.5}}{f^{-3}} + \frac{10^{-11.4}}{f^{-2}} + \frac{10^{-13.8}}{f} + 10^{-16.2} \qquad (9.88)$$

9.4.6 Oscillator Properties

There are two major groups of oscillators encountered in PLLs: *reference oscillators* (generally crystal oscillators) and *voltage-controlled oscillators* (LC and digital oscillators).

Table 9.6(a) Phase-noise PSD of the HP crystal oscillator 10 MHz, type 10811 [29] and MTI-Milliren Tech. crystal oscillator 5 MHz [30]

Frequency offset (Hz)	PSD (dB/Hz)	
1	−102	−98
10	−132	−133
100	−150	−148
1,000	−160	−153
10,000	−160	−158
100,000	−160	−160

Table 9.6(b) Time domain data of the HP crystal oscillator 10 MHz, type 10811 [29] and MTI-Milliren Tech. crystal oscillator 5 MHz [30]

$\sigma(\tau)$	Allan variance $\sigma(\tau)$	$\sigma(\tau)$
0.001	1.5×10^{-10}	–
0.01	1.5×10^{-11}	–
0.1	5×10^{-12}	–
1	5×10^{-12}	3.9×10^{-12}
10	5×10^{-12}	–
100	1×10^{-11}	–

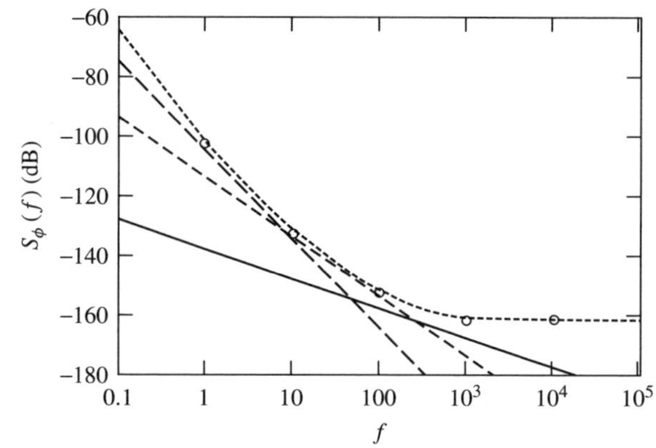

Figure 9.15 Asymptotic approximation of the oscillator noise characteristics.

9.4.6.1 Crystal oscillators

Crystal oscillators are used in PLL systems as reference generators and are often responsible for close-to-the-carrier noise properties. Reverting to (9.83), we ask about the magnitude of the constants, a_r, a_e, and a_o. Their values were appreciated from the noise measurements performed on quartz resonators [23, 24] and on crystal oscillators [25, 27, 28]. For evaluation of the close-to-the-carrier noise in PLL outputs, the most interesting is the value of a_r, which was found to be approximately

$$a_r \approx 10^{-12.75} \tag{9.89}$$

After introducing this value, together with the quartz material constant $f_o Q_U \approx 1.3 \times 10^{13}$, into (9.83), we get for the fractional frequency PSD in the flicker frequency range

$$S_y(1) = h_{-1} \approx \frac{10^{-12.75} + a_e}{Q_U^2} \tag{9.90}$$

See the plot of h_{-1} for crystal oscillators from 5 to 1,000 MHz in Fig. 9.16. Parker [27] arrived at similar results and concluded that the time-independent plateau in the Allan variance is approximately, for all quartz crystal oscillators (since generally $a_r > a_e$),

$$\sigma(\tau) \approx 10^{-6.4} Q_U^{-1} \approx 10^{-19\pm1} f_o \tag{9.91}$$

In Fig. 9.17, we have summarized phase noise characteristics adopted from technical data of present-day crystal oscillators.

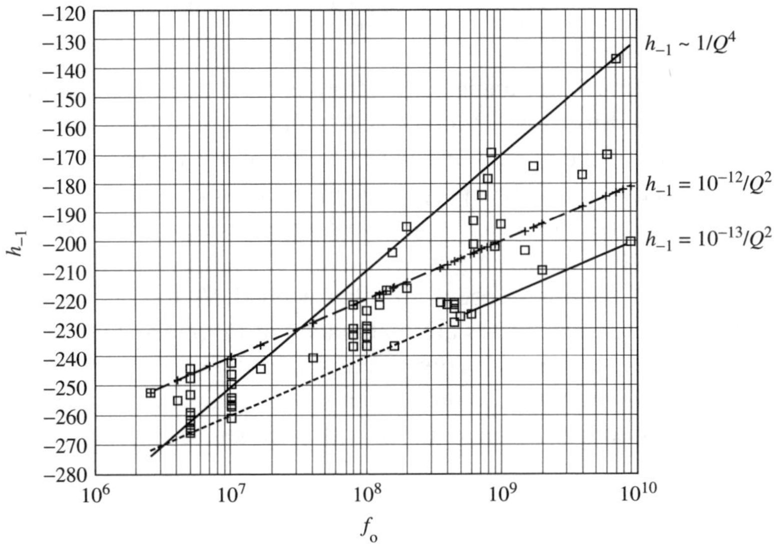

Figure 9.16 Plot of the coefficients h_{-1} for crystal oscillators from 5 to 1000 MHz (© IEEE, 2002).

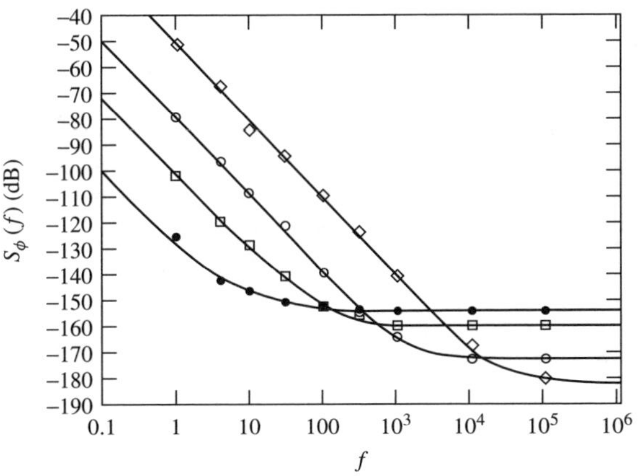

Figure 9.17 Phase noise characteristic adopted from technical data of present-day crystal oscillators, (•) Oscilloquartz 5 MHz type 8607, (□) HP 10 MHz, type 10811, (○) Frequency Electronics 100 MHz, and (◇) SAV500 MHz.

9.4.6.2 LC oscillators

LC oscillators are used as VCOs in the whole frequency range covered by PLLs. The major feature is the low value of the quality factor Q compared with crystal oscillators, from about 100 in RF ranges to only 10 or even smaller in microwave ranges. The fractional frequency noise coefficients h_{-1} and h_0 are plotted in Fig. 9.18, and with their assistance we can write a fairly general oscillator noise equation.

$$\frac{S_\phi(f)}{f_o^2} \approx \frac{1}{f^3}\frac{10^{-11.6}}{Q_L^2} + \frac{1}{f^2}\frac{10^{-15.6}}{Q_L^2} + \frac{1}{f}\frac{10^{-15}}{f_o^2} + 10^{-15} \tag{9.92}$$

Note that the coefficients h_i are mean values from experimental measurements. Actual noise coefficients can differ by -2 to $+1$ order. For a preliminary estimation of the oscillator noise, both of the crystal and the LC oscillators, we can use Fig. 9.19.

Up to now we have discussed properties of oscillators manufactured from discrete transistors and varactors – fairly large and expensive pieces. However, with the growing number of RF services in gigahertz ranges, such as mobile telephone communications, PLLs in the integrated form also needed VCOs in the monolithic IC technology. Nowadays, these VCOs are manufactured in high volumes and marketed inexpensively. Typical monolithic core circuits are shown in Fig. 9.20. However, research and development activities are still in progress and some recently published achievements and properties (as in summer 2002) are summarized in Tab. 9.7.

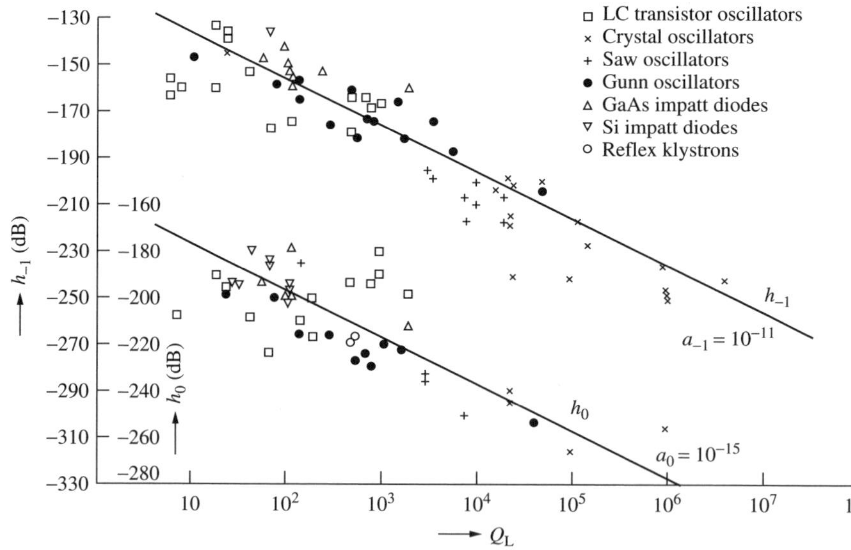

Figure 9.18 Plot of the normalized phase-noise coefficients h_{-1} and h_o with respect to the loaded Q_L of different oscillators (Reproduced from V.F. Kroupa, "Flicker Frequency Noise in BAW and SAW Quartz Resonators", *IEEE Trans.*, **UFFC-35**, 406–420, 1988 by permission of IEEE, 2002 [25]).

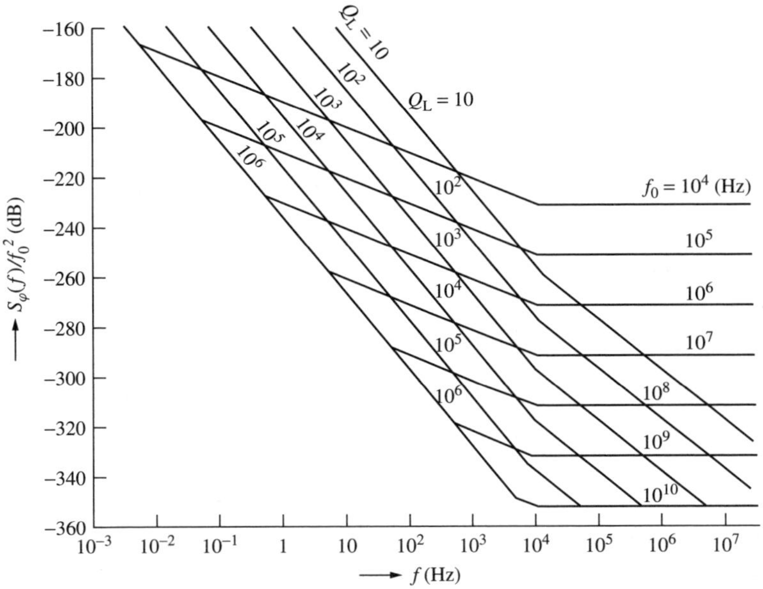

Figure 9.19 Normalized phase noise characteristics of oscillators: parameters are the loaded Q_L of the resonator and the output frequency f_o (© IEEE, 2002).

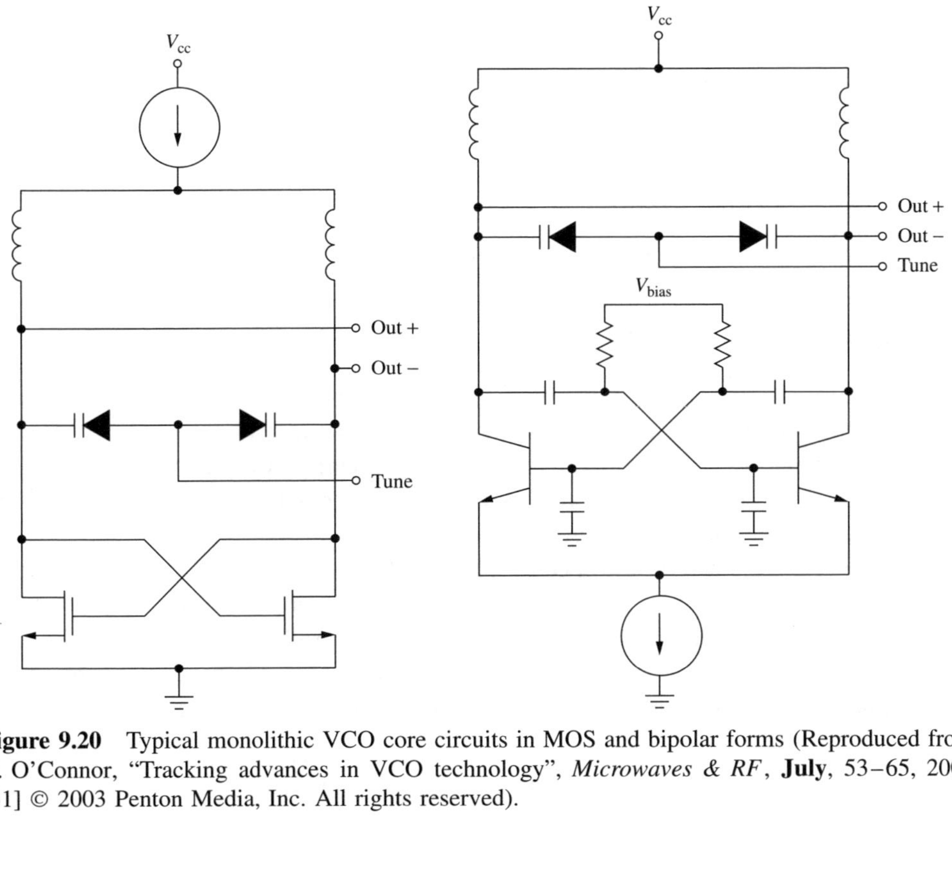

Figure 9.20 Typical monolithic VCO core circuits in MOS and bipolar forms (Reproduced from C. O'Connor, "Tracking advances in VCO technology", *Microwaves & RF*, **July**, 53–65, 2002 [31] © 2003 Penton Media, Inc. All rights reserved).

Table 9.7 Properties of several monolithic LC-VCOs

f_o [GHz]	h_{-1}	h_o	Q_L	K_o [MHz/V]	References
1.1	$10^{-13.7}$	10^{-18}	13	–	[33]
1.5	$10^{-15.2}$	$10^{-20.2}$	3	330	[34]
1.7	$10^{-14.6}$	$10^{-18.6}$	4	170	[35]
2	$10^{-12.3}$	$10^{-17.3}$	20	220	[36]
2.2	$10^{-12.8}$	$10^{-17.8}$	8	18%	[37]
5	$10^{-13.6}$	$10^{-18.1}$	4	150	[38]

9.4.6.3 Ring and other digital oscillators

As stated above, the trend of modern communication systems in gigahertz ranges is a massive application of PLLs in the IC form. Since VCOs constitute an inevitable part, we notice efforts for their inclusion on the same IC chip. One possibility is to constitute LC oscillators; nevertheless, digital circuits present a much more desirable option since their application is very easy. Consequently, multivibrators, ring oscillators, and similar frequency generators are predisposed for this task (cf. Fig. 9.21). The difficulty is the absence of the higher quality factor, Q, which would entail higher noise levels. The advantage is a large oscillator gain, K_o, and an easy way to generate quadrature outputs. PSDs of ring oscillators are discussed in Appendix 9.10.

9.4.7 Phase Noise in PLLs

There are three important sources of noise in PLL systems:

1. noise introduced by the reference generator;
2. noise introduced by the voltage-controlled oscillator (VCO);
3. noise generated in the PLL system blocks (phase detector amplifiers, loop filters, frequency dividers (FD), and others).

For solution of the output noise of the PLL devices, we can use the fact that the noise power generated in individual stages is small compared with the effective power of the corresponding "carrier frequencies." Consequently, we can apply the rule of superposition by adding noise "generators" to inputs or outputs of individual blocks. In addition, application of the Laplace transform may be used for solving the problem and, eventually, we arrive at the spectral density, $S_{\phi,\mathrm{out}}(f)$, of the output phase noise.

By assuming the loop to be locked and by considering the rather general block diagram in Fig. 9.22, we may represent the forward path of the loop as [14, 39] (note that n indicates noise components that are in the Laplace transform notation; however, for simplicity, we leave out the Laplace transform symbol s, throughout)

$$\Phi_{\mathrm{o,n}} = [(\Phi_{\mathrm{i,n}} - \Phi'_{\mathrm{o,n}})K_{\mathrm{d}} + V_{\mathrm{PD,n}} + V_{\mathrm{F,n}}]F_{\mathrm{L}}(s)\frac{K_{\mathrm{o}}}{\mathrm{s}} + \Phi_{\mathrm{osc,n}} \tag{9.93}$$

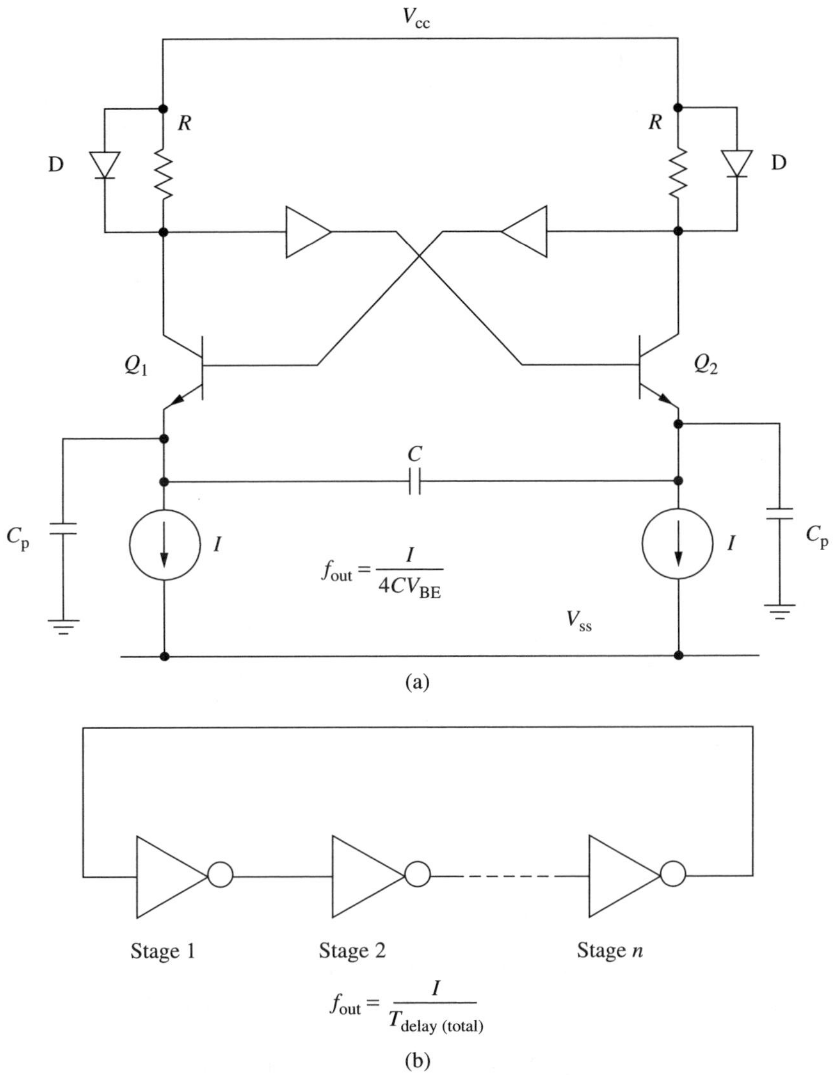

Figure 9.21 "Digital oscillators": (a) emitter-coupled multivibrator and (b) ring oscillator (Reproduced from M. Thamsirianunt and T.A. Kwasniewski, "CMOS VCO's for PLL Frequency Synthesis in GHz Digital Mobile Radio Communications", *IEEE J. Solid-State Circuits*, **32**, 1511–1524, 1997 by permission of IEEE, 2002 [38]).

or the feedback path

$$\Phi'_{o,n} = (\Phi_{o,n} - \Phi_{m,n} + \Phi_{MI,n})\frac{F_M(s)}{N} + \Phi_{DN,n} \tag{9.94}$$

where

$$\Phi_{i,n} = \frac{\Phi_{r,n}}{Q} + \Phi_{DQ,n}; \quad \Phi_{m,n} = M\Phi_{r,n} + \Phi_{MU,n} \tag{9.95}$$

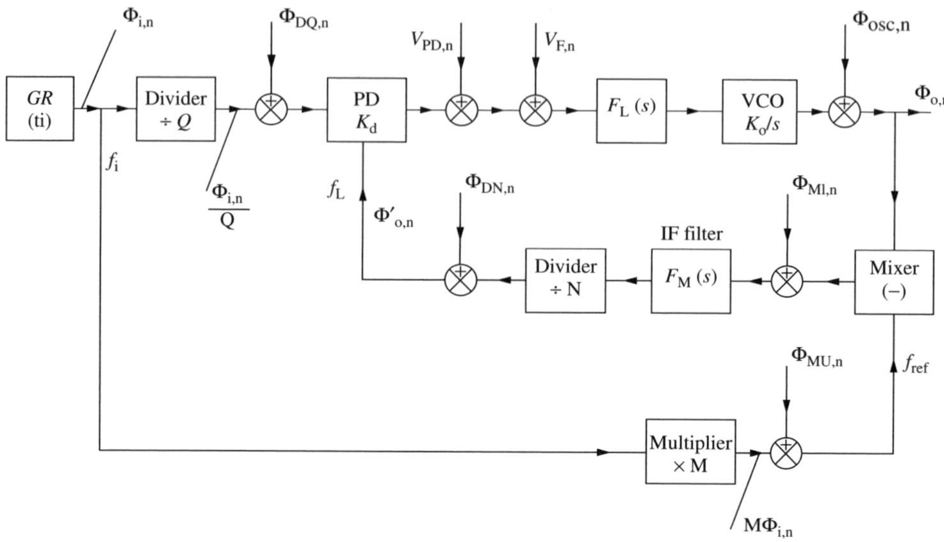

Figure 9.22 Block diagram of the PLL with the generalized feedback path network and additive noise source (© IEEE, 2002).

Since all the noise components are random by nature and uncorrelated, we may sum up the respective power spectral densities $S_{\phi,\dots}(\omega)$. By considering a simpler PLL with only a divider in the feedback path, we get for the output phase noise

$$S_{\phi,o}(\omega) = \left[S_{\phi,i}(\omega) \left(\frac{N}{Q} \right)^2 + S_{\phi,L}(\omega) N^2 \right] |H'(j\omega)|^2 + S_{\phi,osc}(\omega)|1 - H'(j\omega)|^2 \tag{9.96}$$

where we have introduced the effective loop gain $H'(j\omega)$ with the reduced gain in the inverse proportion to the division factor N, that is, $K' = K/N$

$$H'(s) = \frac{KF_L(s)F_M(s)/Ns}{1 + KF_L(s)F_M(s)/Ns} = \frac{K'F_L(s)F_M(s)/s}{1 + K'F_L(s)F_M(s)/s} \tag{9.97}$$

and summed up all additive noises due to the divider, the phase detector, and the circuits associated with the loop filter into

$$S_{\phi,L} = S_{\phi,DQ} + S_{\phi,DN} + \frac{S_{VPD} + S_{VF}}{K_d^2} \tag{9.98}$$

Relations (9.93) to (9.95) and Fig. 9.22 reveal that all individual noises are connected with a frequency as

1. input frequency f_i
2. reference or loop frequency $f_L = f_i/Q$

3. natural frequency f_n
4. output frequency $f_{out} = f_o$.

9.4.8 Application of PLLs for Noise Measurement

For noise measurements of frequency generators, particularly crystal oscillators, applications of PLLs are inevitable.

9.4.8.1 Phase noise measurements of precision oscillators

The block diagram of the measurement set is illustrated with the assistance of Fig. 9.23. The system is the so-called *two-oscillator set* – it is used in instances in which the highest precision is the desired goal. The output voltages of the same oscillators are fed to the phase detector (ring mixer), of which one output is applied to the analyzer and the other closes the PLL via the loop filter and the frequency steering element (varactor) of the second oscillator. For the PSD of the analyzer input we have

$$S_v(f) = S_{\phi,e}(f) \cdot K_d^2 \tag{9.99}$$

However, the PSD of the $S_{\phi,e}$ is formed with PSDs of both compared generators and in some instances with an additional noise. Consequently, the measured output phase noise is given by

$$S_{\phi,e}(f) = \frac{S_{\phi,1}(f) + S_{\phi,2}(f) + S_{v,add}(f)\left(\frac{K_{o,2}}{s}\right)^2}{|1 + G(s)|^2} \tag{9.100}$$

where the PLL open-loop gain $G(s)$ is given by

$$G(s) = \frac{K_d K_{o,2} K_A F(s)}{s} \tag{9.101}$$

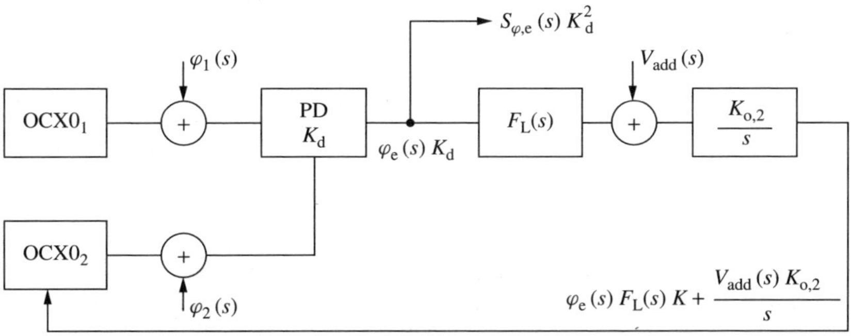

Figure 9.23 The block diagram of the set for the phase-noise measurement of precision oscillators.

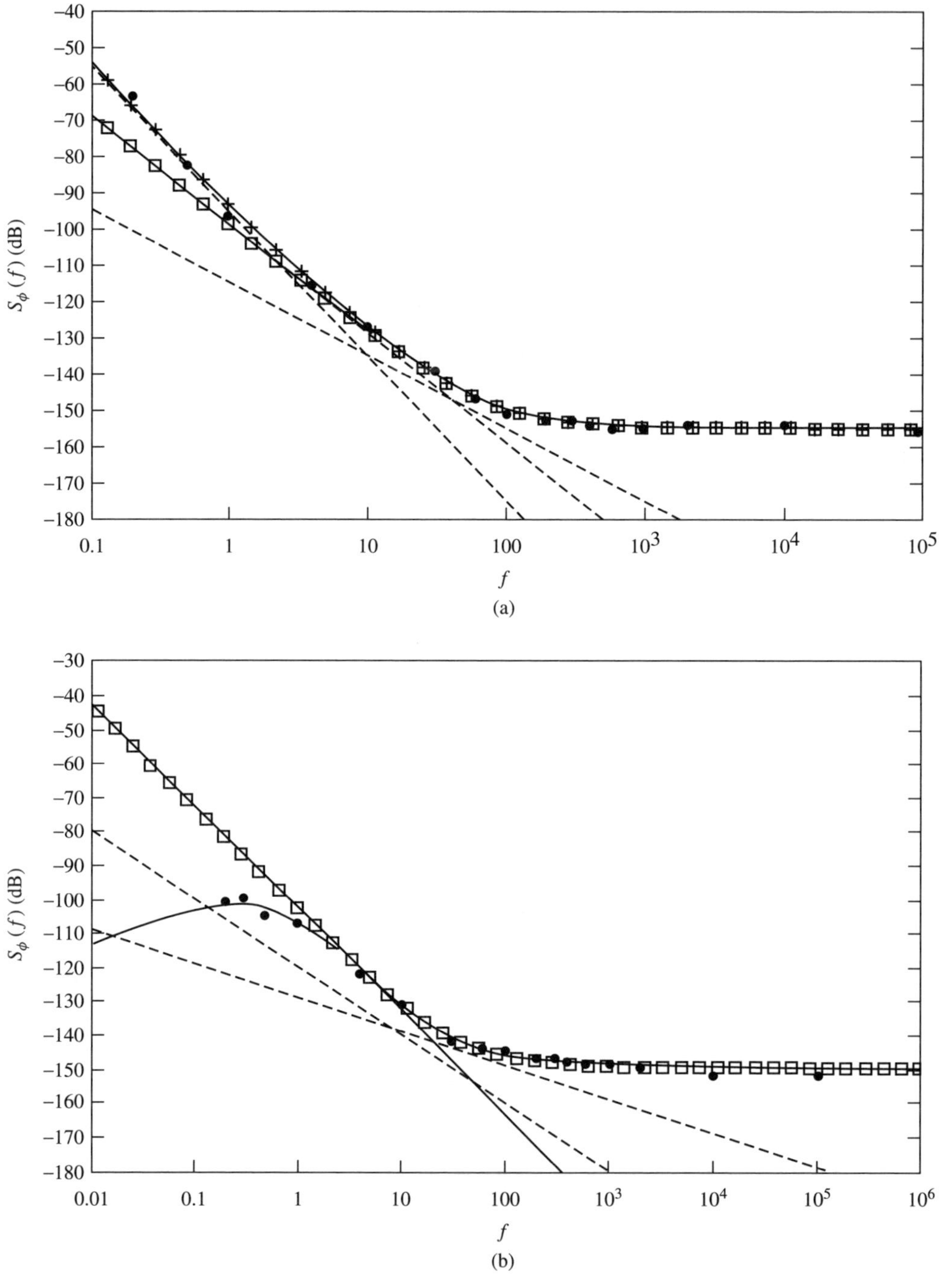

Figure 9.24 Examples of the measured phase-noise PSD of two different combinations of 10-MHz oscillators: (a) H-P 10811 versus H-P 10811 (the effective $\zeta \approx 0.2$) and (b) Hradec Kralove # 63 versus # 138 (the effective $\zeta \approx 1.2$), with asymptotes $1/f^3$, $1/f^2$, $1/f^1$.

Generally, we use a second-order loop of type 2 with the natural frequency in the range of 0.1 to 1 Hz. However, the damping factor changes in accordance with the loop gain K. Table 9.2 reveals that the oscillator gain of the tuned oscillator may change substantially, and with it the damping factor and the shape of the measured PSD in the vicinity of 1-Hz frequency (see Fig. 9.24(a) and (b)).

9.5 PLL NOISE BANDWIDTH

One of the most important properties of PLLs is the loop bandwidth B_L

$$B_L = \int_0^\infty |H(j\omega)|^2 \, df \quad [\text{Hz}] \tag{9.102}$$

Its introduction is in close connection with applications in communications in which the received signal is superimposed with large, generally white noise. In instances in which the noise of the local oscillator is small, the SNR in the receiver is dominated by the spurious input noise with the PSD $S_{\phi,\text{in}}$

$$\frac{1}{\text{SNR}} = \frac{P_n}{P_s} = \langle S_{\text{out}}{}^2(t) \rangle = S_{\phi,\text{in}} B_L \tag{9.103}$$

Noise bandwidth of the first- and second-order loops is summarized in Tab. 9.8. The normalized bandwidth of the second- and third-order loops is compared in Fig. 9.25.

9.6 APPENDIX: PROPERTIES OF RING OSCILLATORS

The trend to combine the whole PLL on one IC chip is met with monolithic VCOs, particularly in the low gigahertz ranges. We encounter LC versions, relaxation circuits, and nowadays ring oscillators [38, 40–42]. Since designers and customers are interested, of all the properties, mostly in the phase noise and time or radian (degrees) jitter, we shall discuss these properties briefly.

The ring oscillator can be considered as a black box as depicted in Fig. 9.26. The oscillating conditions require fulfilment of the condition (4.1), that is,

$$\omega\tau = \pi(2k+1); \quad \tau = n\sum_1^n \tau_i \tag{9.104}$$

Table 9.8 Noise bandwidth of first- and second-order loops

PLL type	PLL order	Loop filter	Noise bandwidth B_L (Hz)
1	First	No filter	K/4
1	Second	RC	K/4
1	Second	RRC; RCC	$(K/4)[1+(KT_2)^2/KT_1]/(1+KT_2)$
2	Second	Integrating	$(K/4)[1+(KT_2)^2/KT_1]/KT_2 = (\omega_n/2)(\zeta + 1/4\zeta)$

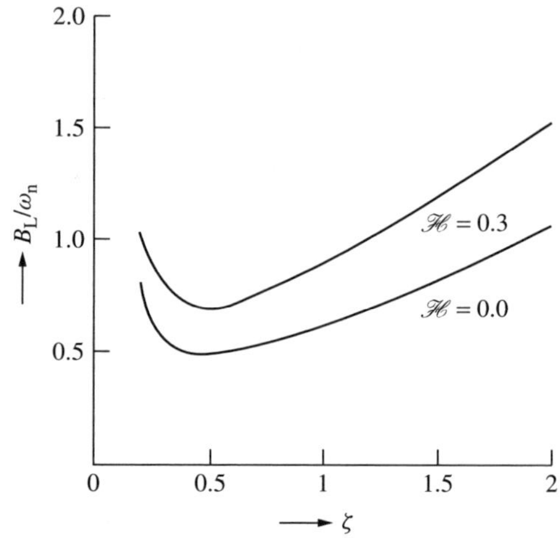

Figure 9.25 PLL noise bandwidth for the high gain of the second- and third-order loops versus the damping factor.

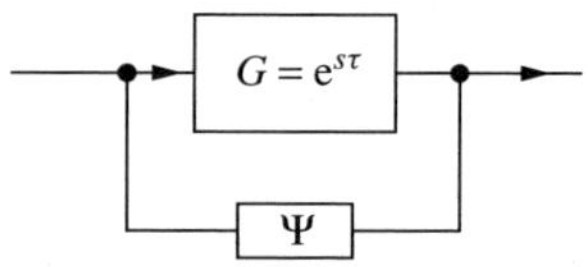

Figure 9.26 The ring oscillator as an oscillating black box.

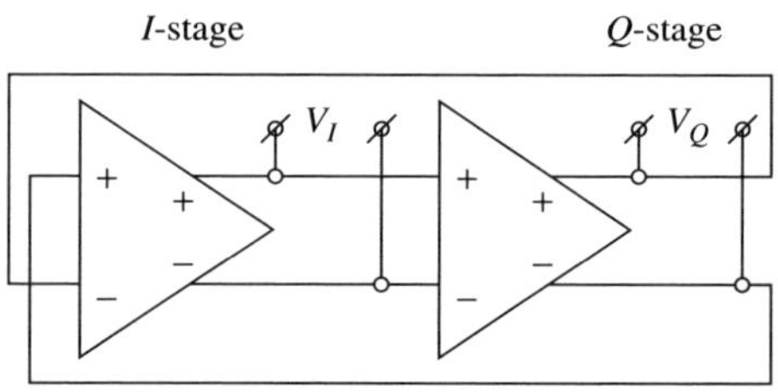

Figure 9.27 The ring oscillator quadrature outputs (© IEEE, 2002).

For the fundamental frequency, we put $k = 0$, and we get for the "resonant" frequency

$$f_o = \frac{\tau}{2} \tag{9.105}$$

Evidently, the ring oscillator will oscillate for the period $2N$ times the stage delay (cf. Fig. 9.27). A short investigation of the noise properties may start with frequency fluctuations

$$(\omega_o + \Delta\omega_n) = \frac{\pi}{\tau} \tag{9.106}$$

from which the phase noise is

$$\Delta\phi = \Delta\omega_n \tau \tag{9.107}$$

Introduction of the above relation into the gain, G, of the oscillating loop leads to

$$G = \mathrm{e}^{-\mathrm{j}\omega\tau} = \mathrm{e}^{-\mathrm{j}(\omega_\mathrm{o}+\Delta\omega)\tau} \tag{9.108}$$

which may be simplified to

$$G \approx \mathrm{e}^{-\mathrm{j}\omega_\mathrm{o}\tau}(1 - \mathrm{j}\Delta\omega\tau) \approx \mathrm{e}^{-\mathrm{j}\omega_\mathrm{o}\tau}\frac{1}{1 + \mathrm{j}\Delta\omega\tau} \tag{9.109}$$

However, the last factor in the above relation resembles the phase fluctuations in an LC resonant circuit (cf. (3.34)). If we put

$$\frac{2\Delta\omega Q}{\omega_\mathrm{o}} = \Delta\omega\tau \tag{9.110}$$

we arrive at the effective quality factor Q

$$Q_\mathrm{eff} = \frac{\omega_\mathrm{o}}{2}\tau = \frac{\pi}{2} \tag{9.111}$$

At this point we can investigate the influence of the additive phase fluctuations due to the electronics, which is a source of additional frequency fluctuations (cf. (9.82)), that is,

$$\phi = \frac{Q\Delta\omega}{\omega_\mathrm{o}} \tag{9.112}$$

After application of (9.83), we get for the PSD

$$S_\phi(f) = \left(\frac{a_\mathrm{e}}{f} + \frac{2kT}{P_r} + a_\mathrm{o}\right)\frac{f_\mathrm{o}^2}{f^2 {Q_\mathrm{ef}}^2} + \frac{a_\mathrm{e}}{f} + a_\mathrm{o} \tag{9.113}$$

Note that we have neglected the frequency fluctuation term of the "bulk circuit" a_r.

Introduction of the conservative values for a_e and a_o

$$a_\mathrm{e} \approx 10^{-11\pm1} \quad \text{and} \quad a_\mathrm{o} \approx 10^{-16\pm1} \tag{9.114}$$

reveals for the h-coefficients of ring oscillators

$$\begin{aligned} h_{-1} &= \frac{10^{-11\pm1}}{(\pi/2)^2}; \quad h_0 = \frac{10^{-16\pm1}}{(\pi/2)^2}; \\ h_1 &= \frac{10^{-11\pm1}}{f_0^2}; \quad h_2 = \frac{10^{-16\pm1}}{f_0^2} \end{aligned} \tag{9.115}$$

Table 9.9 Properties of several ring oscillators

f_o [GHz]	h_{-1}	h_o	h_2	K_o [MHz/V]	References
0.232	$10^{-11.6}$	$10^{-16.6}$	–	–	[40]
1	10^{-12}	10^{-16}	–	280	[41]
0.9	10^{-11}	–	10^{-30}	606	[38]
2	–	10^{-18}	–	–	[42]

Several data computed from the published measurements are summarized in Tab. 9.9. The expected corner frequency between flicker and white frequency noise is

$$f_{cor} \approx \frac{h_{-1}}{h_0} \approx 10^4 \text{ [Hz]} \tag{9.116}$$

Example 9.10

Evaluate the expected h-coefficients of a ring oscillator for the output frequency $f_o = 1\,\text{GHz}$

$$h_{-1} = 10^{-31\pm1}; \quad h_0 = 10^{-15.4\pm1}; \quad h_1 = h_{-1}; \quad h_2 = 10^{-33\pm1} \tag{9.117}$$

In gigahertz frequency range, one often prefers to have information about jitter instead of phase noise (see Sections 9.3.4.1 and 9.3.4.2).

Example 9.11

Let us investigate the jitter generated in a PLL-DDFS system using a ring VCO oscillator with $f_o = 1\,\text{GHz}$ and the natural frequency $f_n = 10\,\text{kHz}$ with the PSD in the pass band $S_{\varphi,L} \approx 10^{-5}$ [cf. 40]

$$\langle\Delta\phi^2\rangle = 10^{-6} * 10^4 + 10^{18} * 10^{-16} * \int_{10^4}^{\infty} \frac{1}{f^2}\,\mathrm{d}f =$$

$$= 10^{-2} + \frac{1}{2} * \frac{100}{10^{-4}} \doteq 10^{-2} \tag{9.118}$$

from which the phase jitter is

$$\langle\Delta\varphi\rangle \doteq 0.1 \text{ rad} \quad \text{or} \doteq 6^\circ$$

and the respective time jitter is

$$\langle\Delta \text{t}\rangle \doteq 100 \text{ [ps]}$$

Note that major contribution or the overall jitter is generated in the pass band of the PLLs.

REFERENCES

[1] R.F. Voss, "1/f (flicker) noise: a brief review", *Annual Frequency Control Symposium*, 1979, Proceedings, pp. 40–46.
[2] J.B. Johnson, "Thermal agitation of electricity in conductors", *Phys. Rev.*, **32**, 97–109, 1928.
[3] H. Nyquist, "Thermal agitation of electric charge in conductors", *Phys. Rev.*, **32**, 110–113, 1928.
[4] S. Lezter and N. Webster, "Noise in amplifiers", *IEEE Spectrum*, **August**, 67–75, 1970.
[5] B.M. Olivier, "Thermal and quantum noise", *Proc. IEEE*, **53**, 436–454, 1965 (Reprinted by Gupta p.129).
[6] J.B. Johnson, "Electronic noise: the first two decades", *IEEE Spectrum*, **8**, 42–46, 1971 (Reprinted by Gupta p.17).
[7] W. Schottky, *Phys. Rev.*, **28**, 74, 1926.
[8] J. Bernamont, *Ann. Phys. [Paris]*, **7**, 71, 1937.
[9] A. Papoulis, *Probability, Random Variables and Stochastic Processes*. New York: McGraw-Hill, 1965.
[10] *Proc IEEE*, Special issue on Frequency Stability, **54**, February, 103–338, 1966.
[11] J. Rurman, "Characterization of phase and frequency instabilities in precision sources: fifteen years of progress", *Proc. IEEE*, **66**, 1048–1075, 1978 (also in 9.11).
[12] V.F. Kroupa, *Frequency Stability: Fundamentals and Measurements*, New York: IEEE Press, 1983.
[13] V.F. Kroupa, *Frequency Synthesis: Theory, Design et Applications*. London: Ch. Griffin, 1973.
[14] V.F. Kroupa, ed. *Direct Digital Frequency Synthesizers*. New York: IEEE Press, 1999.
[15] V.F. Kroupa, J. Pavlovec and L. Sojdr, "Noise in standard frequency sources", *Digest of the Conference on Precision Electromagnetic Measurements*, Braunschweig, 1980, pp. 147–151.
[16] M.M. Driscoll and T.D. Merrell, "Spectral performance of frequency multipliers and dividers", *1992 IEEE Frequency Control Symposium, Proc.*, May 1992, pp. 193–200.
[17] F.L. Walls and C.M. Felton, "Low noise frequency synthesis", *Proc. of the 41st Annual Frequency Control Symposium*, May 1987, pp. 512–518.
[18] M.R. McClure, "Residual phase noise of frequency dividers", *Microwave J.*, **March**, 124–130, 1992.
[19] W.F. Egan, *Frequency Synthesis by Phase Lock*. 2nd ed. New York: John Wiley, 2000.
[20] V.F. Kroupa, Jitter and phase noise in frequency dividers, IEEE Trans. IM, **50**(5), 1241–1243, 2001.
[21] W.A. Edson, "Noise in oscillators", *Proc. IRE*, **August**, 1454–1466, 1960.
[22] K. Kurokawa, "Noise in synchronized oscillators", *IEEE Trans.*, **MTT-16**, 234–240, 1968.
[23] F.L. Walls and A.E. Wainwright, "Measurement of the short-term stability of quartz crystal resonators and the implications for crystal oscillator design and applications", *IEEE Trans.*, **IM-24**, 15–20, 1975.
[24] G. Marianneau, J.J. Gagnepain and J. Uebersfeld, "Bruit de phase et d'amplitude des resonateurs et oscillateurs a quartz", *Proc. Int. Symp. in Telecommun, Lanion*, October 1977, France, pp. 240–244.
[25] V.F. Kroupa, "Flicker frequency noise in BAW and SAW quartz resonators", *IEEE Trans.*, **UFFC-35**, 406–420, 1988.
[26] D.B. Leeson, "A simple model of feedback oscillator noise spectrum", *Proc. IEEE*, **February**, 329–330, 1966.
[27] T.E. Parker, "Precision surface-acustic-wave (SAW) oscillators", *IEEE Trans. UFFC*, **35**(3), 342–364, 1988.
[28] F.L. Walls, "The quest to understand and reduce 1/f noise in amplifiers and BAW quartz oscillators", *Proc. 9th European Frequency and Time Forum*, Besancon, France, 1995, pp. 227–240.

[29] HP 1081D/1081E Crystal oscillators (© 1991 H-P Comp).
[30] E. Cantor and M. Vaish, "Statistical results from high volume production of ultra stable quartz oscillators", *Proc. of EFTF and IEEE Int. Fr. Control Symp.* Besancon, April 1999, pp. 358–365.
[31] C. O'Connor, "Tracking advances in VCO technology", *Microwaves & RF*, **July** 53–65, 2002.
[32] E. Hegazi, H. Sjöland and A. Abidi, "A filtering technique to lower LC oscillator Phase noise", *IEEE J. Solid-State Circuits*, **36**, 1921–1930, 2001.
[33] P. Vancorenland and M.S.J. Steyaert, "A 1.57-GHz fully integrated very low-phase-noise quadrature VCO", *IEEE J. Solid-State Circuits*, **37**, 653–656, 2002.
[34] F. Herzel, M. Pierschel, P. Weger and M. Tiebout, "Phase noise in a differential CMOS voltage-controlled-oscillator for RF applications", *IEEE Trans. Ciruits Syst. II*, **47**(1), 2000, 11–15.
[35] W.M.Y. Wong, P.S. Hui, Z. Chen, K. Shen, J. Lau, P.C.H. Chan and P.-K. Ko, "A wide tuning range gated varactor", *IEEE J. Solid-State Circuits*, **35**, 773–779, 2000.
[36] P. Andreani and H. Sjöland, "Tail current noise suppression in RF CMOS VCOs", *IEEE J. Solid-State Circuits*, **37**, 342–348, 2002.
[37] J. van der Tang, P. van de Ven, D. Kasperkovitz and A. van Roermund, "Analysis and design of an optimally coupled 5-GHz quadrature LC oscillator", *IEEE J. Solid-State Circuits*, **37**, 657–661, 2002.
[38] M. Thamsirianunt and T.A. Kwasniewski, "CMOS VCO's for PLL frequency synthesis in GHz digital mobile radio communications", *IEEE J. Solid-State Circuits*, **32**, 1511–1524, 1997.
[39] V.F. Kroupa, "Low-noise microwave-frequency synthesizers: design principles", *IEEE Proceedings-H*, **130**, 483–488, 1983 (Reprinted in [9.14]).
[40] A. Hajimiri and T.H. Lee, "A general theory of phase noise in electrical oscillators", *IEEE J. Solid-State Circuits*, **33**(2), 179–194, 1998.
[41] S.-J. Lee, B. Kim and K. Lee, "A fully integrated low-noise 1-GHz frequency synthesizer design for mobile communication application", *IEEE J. Solid-State Circuits*, **32**(5), 760–765, 1997.
[42] C. Vaucher and D. Kasperkovitz, "A wide-band tuning system for fully integrated satellite receivers", *IEEE J. Solid-State Circuits*, **33**(7), 987–997, 1998.
[43] J.D. van der Tang, D. Kasperkovitz and A. van Roermund, "A 9.8–11.5 GHz quadrature ring oscillator for optical receviers", *IEEE J. Solid-State Circuits*, **37**(3), 438–442, 2002.

10

Digital PLLs (Sampled Systems)

With the advent of integrated circuits, we face a widespread use of digital techniques replacing analog systems. Their advantages are small size, small power consumption, and low prices on one hand and reliability, dependability, stability, and so on, on the other. The major feature is the digital or quantized transfer of information. The situation is not completely new since some control systems or regulation devices, in the past, had been working on discrete instructions (cf. [1]).

The advantages of the digital operation were soon recognized for applications in phase-locked systems, particularly in connection with frequency synthesis and modern communications in high-megahertz and low-gigahertz ranges. The situation being such, we encounter digital phase-locked loops (DPLLs) replacing the earlier analog system (analog phase-locked loops – APLLs) discussed up to now [2, 3].

Nowadays, DPLLs frequently find applications in synchronous communications as MF demodulators or auxiliary channel carriers, as generators of clock frequencies, often for all the received channels, and others. In the following sections we shall discuss the problem of sampled PLLs as well as all digital PLLs with the inevitable theoretical introduction.

We have already encountered the sampling PD in which the information is supplied to the memory capacitors, and without any rigorous proof we have stated that for information about PLL stability, it was sufficient to replace the sampling PD merely with a time delay. However, this approach is not sufficient for evaluation of DPLL systems that require a much more rigorous approach with the assistance of the z-transform.

10.1 FUNDAMENTALS OF THE z-TRANSFORM

In our discussion of the z-transform, we shall start from the Fourier transform, proceed with the investigation of the sampled systems [4], and in this way avoid the errors and misunderstandings that are often encountered. We shall start with the open-loop time response $g(t)$ subjected to the sampling function $h(t)$ (see Fig. 10.1).

$$g(t) = h(t) \otimes g(t) \tag{10.1}$$

Phase Lock Loops and Frequency Synthesis V.F. Kroupa
 ISBN: 0-470-84866-9

Figure 10.1 Open-loop time response $g(t)$ subjected to switching.

In the frequency domain, the time convolution may be replaced with the multiplication of the respective frequency spectral components

$$G(s) = H_h(s)\hat{G}(s) \tag{10.2}$$

where the sample and hold function for the duration time T is represented in the Laplace transform as

$$H_h(s) = \int_0^T e^{-st}\,dt = \frac{1 - e^{-sT}}{s} \tag{10.3}$$

In the case in which $g(t)$ is a slowly varying function of time, we can approximate its Laplace transform with the expansion

$$\hat{G}(s) = \int_0^\infty g(t)e^{-st}\delta(t - nT)\,dt = \sum_{n=0}^{\infty} G(nT)e^{-snT} \tag{10.4}$$

Now let us revert to Fig. 8.10 and find out the Laplace transform of the output voltage $v_2(t)$

$$V_2(s) = \int_0^\infty v_2(t)e^{-st}\,dt \tag{10.5}$$

In accordance with Fig. 8.11, the output voltage $v_2(t)$ is proportional to $\phi_e(t)$, a slowly varying function, and is formed with narrow rectangles reproduced in Fig. 8.13. Consequently, we can write the above relation as an infinite sum

$$V_2(s) = \int_0^T v_1(0)e^{-st}\,dt + \int_T^{2T} v_1(T)e^{-st}\,dt + \cdots + \int_{nT}^{(n+1)T} v_1(nT)e^{-st}\,dt + \cdots \tag{10.6}$$

which may be simplified into

$$V_2(s) = \sum_{n=0}^{\infty} \int_{nT}^{(n+1)T} v_1(nT)e^{-st}\,dt \tag{10.7}$$

Since all $v_1(nT)$ are nearly constant, we arrive, after integration, at

$$V_2(s) = \sum_{n=0}^{\infty} v_1(nT) \frac{e^{-snT} - e^{-s(n+1)T}}{s} \tag{10.8a}$$

$$= \frac{1 - e^{-st}}{s} \sum_{n=0}^{\infty} v_1(nT) e^{-snT} \tag{10.8b}$$

$$= T e^{-sT/2} \frac{\sin h(sT/2)}{sT/2} \sum_{n=0}^{\infty} v_1(nT) e^{-snT} \tag{10.8c}$$

After replacement of the first factor in the rhs of the relation (10.8b), with the transfer function $H_h(s)$ and the sum with $\hat{V}_1(s)$, we get

$$V_2(s) = H_h(s) \hat{V}_1(s) \tag{10.9}$$

Next we proceed with the introduction of a new variable

$$z = e^{sT} \tag{10.10}$$

Evidently, we can write

$$V_1(z) = \sum_{n=0}^{\infty} v_1(nT) z^{-n} \tag{10.11a}$$

$$= v_1(0) + v_1(T) z^{-1} + v_1(2T) z^{-2} + \ldots \tag{10.11b}$$

After realizing that relation (10.10) makes possible a conformal mapping of the plane s into the plane z, that is, every point in one plane has a corresponding point in the other plane, we arrive at

$$V_1(z) = \hat{V}_1(s) \tag{10.12}$$

However, there is one substantial difference that is best understood by mapping the imaginary axis from the plane s into the plane z; that is, by introducing the equality

$$z = e^{j\omega T} \tag{10.13}$$

we get

$$\begin{aligned} \omega &= 0 \quad z = 1 \\ \omega &= \frac{\pi}{T} \quad z = -1 \\ \omega &= \frac{2\pi}{T} \quad z = 1 \end{aligned} \tag{10.14}$$

and so on.

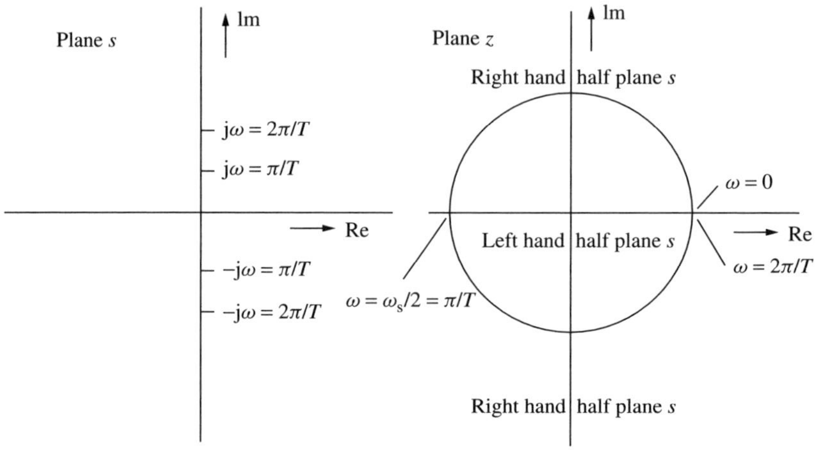

Figure 10.2 Mapping of the plane s in the plane z; the imaginary axis is mapped on the periphery of the unit circle in the plane z in the counterclockwise circulation.

We see that the imaginary axis is mapped on the periphery of the unit circle in the plane z in a counterclockwise circulation of rate $2\pi kT\,(k = 0, 1, \ldots)$ (see Fig. 10.2). From the above discussion and using relation (10.12) and $s = \mathrm{j}\omega$, we have

$$\hat{V}_1(\mathrm{j}\omega) = \sum_{n=0}^{\infty} v_1(nT)\mathrm{e}^{-\mathrm{j}n\omega T} \tag{10.15}$$

We see that $\hat{V}_1(\mathrm{j}\omega)$ is a periodic function with sampling frequency

$$\omega_\mathrm{s} = \frac{2\pi}{T} \tag{10.16}$$

as shown in Fig. 10.3.

Since $\hat{V}_1(s)$ in eq. (10.15) depends on the sample values of $v_1(t)$ it seems reasonable that $\hat{V}_1(s)$ should be related in some way to $V_1(s)$, the Laplace transform of $V_1(t)$. This, in fact, is the case. It can be shown that [4]

$$\hat{V}_1(s) = \frac{1}{T}\sum_{n=-\infty}^{\infty} V_1(s - \mathrm{j}n\omega_\mathrm{s}) \tag{10.17}$$

In the case where spectral components ω_n exceed half of the sampling frequency (cf. Fig. 10.3(c)), the output spectrum of the sampler is distorted. The consequence of this is that the spectra around individual harmonics of the sampling frequency overlap – we face the process of aliasing. After replacing s with $\mathrm{j}\omega$ we have

$$\hat{V}_1(\mathrm{j}\omega) = \frac{1}{T}\sum_{n=-\infty}^{\infty} V_1(\mathrm{j}(\omega - n\omega_\mathrm{s})) \tag{10.18}$$

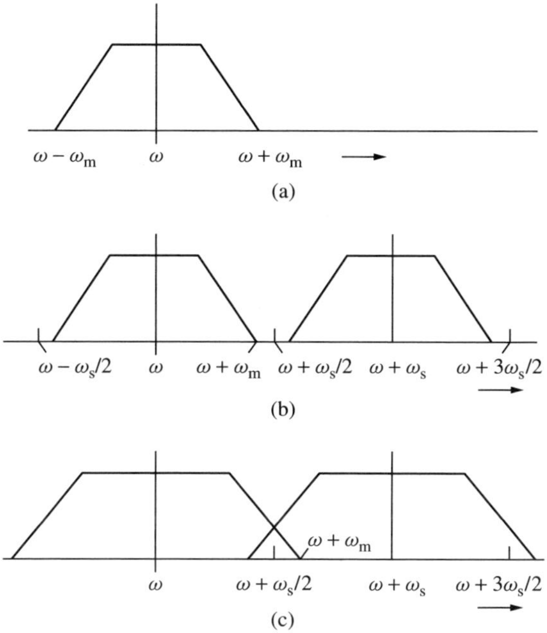

Figure 10.3 (a) The output spectrum around the carrier – nonsampled; (b) the output spectrum after sampling in the case that all modulation frequencies ω_m are smaller than $\omega_s/2$; and (c) the output spectrum after sampling in the case that some modulation frequencies ω_m are larger than $\omega_s/2$.

As long as the frequency of the modulation components in the base band is smaller than $\omega_s/2$ (cf. Fig. 10.3(a)), that is,

$$\omega_m < \omega_s/2 \tag{10.19}$$

then the output spectrum of the sampler in accordance with (10.8) and (10.18) is

$$V_3(j\omega) = \left| \frac{\sin(\pi\omega/\omega_s)}{\pi\omega/\omega_s} \right| \cdot |V_1(j\omega)| \tag{10.20}$$

Note that the above relation and the assumption (10.19) were applied in Chapter 3 in the discussion of PLLs with time delays (eq. (3.42)).

Next, we shall mention some other properties of the z-transform. Let us assume a function $f(t)$

$$f(t) = 0 \quad \text{for} \quad t < 0 \tag{10.21}$$

and introduce a delayed function $f(t - kT)$

$$f(t - kT) = 0 \quad \text{for} \quad t - kT < 0 \tag{10.22}$$

where k is a positive integer. With the assistance of (10.11), we have for the string of samples $f(nT)$

$$F(z) = \sum_{n=0}^{\infty} f(nT)z^{(-n)} \tag{10.23}$$

$$F(z) = f(0) + f(T)z^{-1} + \cdots \tag{10.24}$$

However, reverting to the sampling in the original intervals reveals for the shifted samples

$$\begin{aligned} F_d(z) &= \sum_{n=0}^{\infty} f[(n-k)T]z^{-n} \\ &= f(-kT) + f(1-k)Tz^{-1} + \cdots + f(-T)z^{-(k-1)} \qquad (10.25) \\ &\quad + f(0)z^{-k} + f(T)z^{-(k+1)} + f(2T)z^{-(k+2)} + \cdots \qquad (10.26) \end{aligned}$$

But with the assumption (10.22), we can simplify the above relation as follows:

$$F_d(z) = f(0)z^{-k} + f(T)z^{-(k+1)} + f(2T)z^{-(k+2)} + \cdots \tag{10.27}$$

$$F_d(z) = z^{-k}[f(0) + f(T)z^{-1} + f(2T)z^{-2} + \cdots] \tag{10.28}$$

$$F_d(z) = z^{-k}F(z) \tag{10.29}$$

Example 10.1

Let us find the transfer function of the network in Fig. 10.4 in the z-transform. For the voltage on the capacitor we have the differential equation

$$\frac{v_2 - v_1}{R} + C\frac{dv_2}{dt} = 0 \tag{10.30}$$

which can be rearranged into

$$v_2 + RC\frac{dv_2}{dt} = v_1 \tag{10.31}$$

and after replacing differentiation with differences

$$\frac{dv_2}{dt} \approx \frac{v_2(nT) - v_2[(n-1)T]}{T} \tag{10.32}$$

we finally arrive at

$$v_2(nT) + \frac{RC}{T}v_2(nT) - \frac{RC}{T}v_2[(n-1)T] = v_1(nT) \tag{10.33}$$

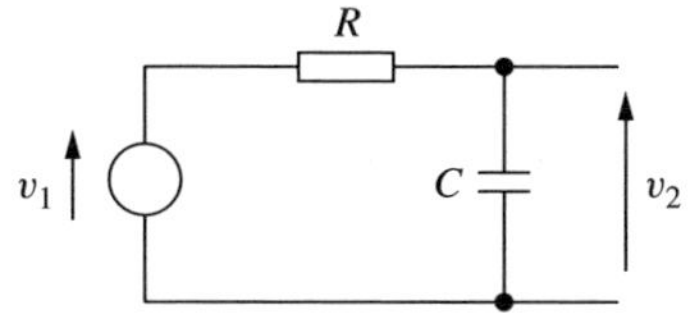

$1 << RC$

$$\frac{V_2(s)}{V_1(s)} \approx \frac{1}{RC}\frac{1}{s}$$

(a)

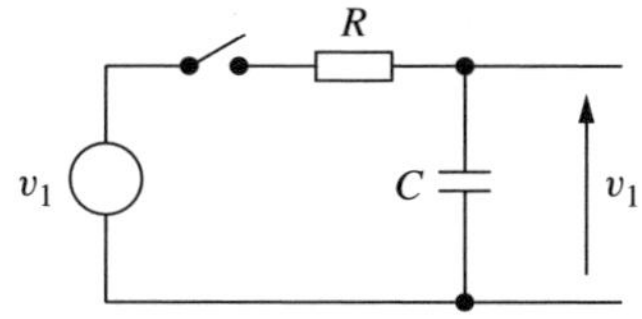

$1 << RC/T$

$$\frac{V_2(z)}{V_1(z)} \approx \frac{T}{RC}\frac{1}{1-z^{-1}}$$

(b)

Figure 10.4 Simple integrating circuit RC: (a) the continuous model and (b) the sampling or switching model.

The summation over all n from zero to infinity and application of the relation (10.23) provides the following z-transform:

$$V_2(z)\left(1+\frac{RC}{T}\right)-\frac{RC}{T}V_2(z)z^{-1}=V_1(z) \tag{10.34}$$

from which the corresponding transfer function is easily computed:

$$H(z)=\frac{V_2(z)}{V_1(z)}=\frac{1}{1+\dfrac{RC}{T}-\dfrac{RC}{T}z^{-1}} \tag{10.35}$$

In instances where $1 \ll RC/T$, the above relation simplifies into

$$H(z)=\frac{T}{RC}\frac{1}{1-z^{-1}} \tag{10.36}$$

where the second factor on the rhs is the z-transform model of the integrator.

Now we shall revert to the conformal mapping of the plane s into plane z. By introduction of s

$$s=s_r+\mathrm{j}s_i=s_r+\mathrm{j}\omega \tag{10.37}$$

we arrive with the assistance of (10.13) at

$$z=\mathrm{e}^{sT}=\mathrm{e}^{s_rT}\cdot\mathrm{e}^{\mathrm{j}\omega T} \tag{10.38}$$

with the absolute value of z

$$|z|=\mathrm{e}^{s_rT} \tag{10.39}$$

However, the period T is always positive. Consequently, for negative s_r we have

$$|z| < 1; (s_r < 0) \tag{10.40}$$

and the whole left-half plane s is mapped inside the unit circle, and for $|z| = 1$, on its periphery (cf. Fig. 10.2). This property is used for appreciating the stability of the systems investigated with the assistance of the z-transform. We can state the following theorem: *A stable system with the transfer function H(z) must have poles inside the unit circle in the plane z.*

Example 10.2

Investigate the behavior of the accumulator *modulo Y*, used in direct digital frequency synthesizers (DDFS) (cf. Chapter 12), in the z-transform representation. The block diagram is shown in Fig. 10.5. At each clock pulse the content of the accumulator is increased by the integer X and reset each time the accumulator exceeds Y; however, the next "repetition" period starts with a remainder in the accumulator but not with the zero state. In accordance with (10.27), we have

$$q(z) = Xz^{-B} + Xz^{-(B-1)} + \cdots + Xz^0 \tag{10.41}$$

where

$$B = \text{integer}\left(\frac{Y}{X}\right) \tag{10.42}$$

Evidently

$$q(z) = X\sum_{n=0}^{B} z^{-n} = X\frac{1 - z^{-B+1}}{1 - z^{-1}} \tag{10.43}$$

which for large B can be simplified to

$$q(z) \approx X\frac{1}{1 - z^{-1}}. \tag{10.44}$$

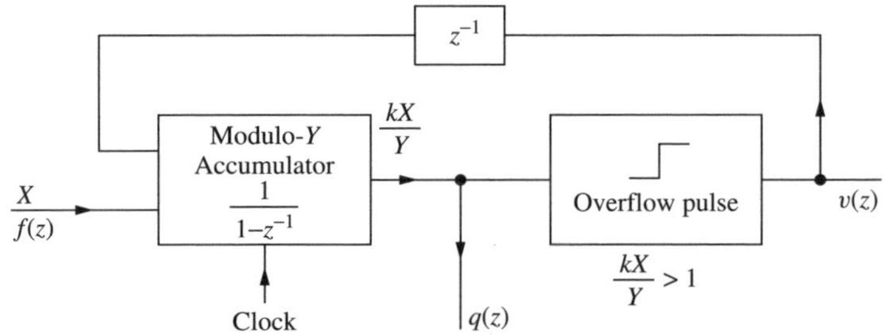

Figure 10.5 An accumulator *modulo Y*, used in DDFS, in z-transform representation.

Since the sum in (10.41) represents integration, we easily conclude that integration in the z-transform is represented, as stated in (10.36), by

$$\frac{1}{1 - z^{-1}} \tag{10.45}$$

Once in the repetition period, a pulse is generated at the output. Symbolically we can write

$$\frac{f(z) - v(z)_{-1}}{1 - z^{-1}} = v(z) - q(z) \tag{10.46}$$

from which the spurious modulation is

$$q(z) = \frac{v(z) - f(z)}{1 - z^{-1}} \tag{10.47}$$

10.2 PRINCIPLES OF DIGITAL FEEDBACK SYSTEMS

The principal block diagram of digital feedback system is shown in Fig. 10.6. To reduce distortion of the demodulated messages, we place at the input a band-pass filter with the pass band approximately

$$B_w \approx \omega_s = \frac{2\pi}{T_s} \tag{10.48}$$

As with the analog system, the open-loop gain can be expressed as

$$G(z) = K_d(z) \cdot D(z) \cdot K_o(z) \tag{10.49}$$

and the corresponding transfer function is

$$H(z) = \frac{G(z)}{1 + G(z)} = \frac{K_d(z)K_o(z)D(z)}{1 + K_d(z)K_o(z)D(z)} \tag{10.50}$$

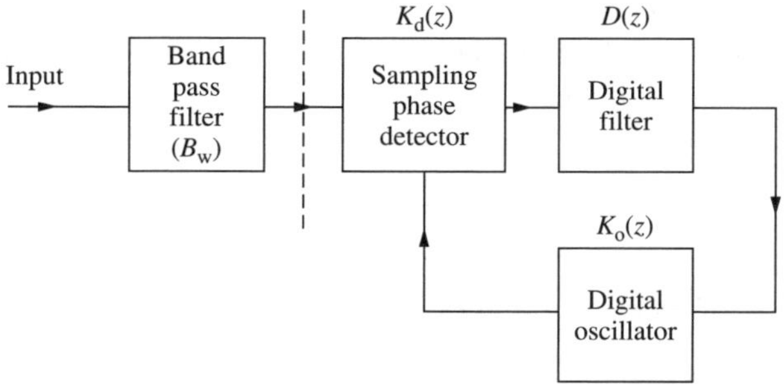

Figure 10.6 Principle block diagram of digital feedback system (DPLL).

In instances in which the demodulated information is taken at the output of the digital PD, the effective transfer function is

$$1 - H(z) = \frac{1}{1 + K_d(z)K_o(z)D(z)} \tag{10.51}$$

As in the case of analog systems, we introduce the first-order loop when $D(z) = 1$, and higher-order loops in instances where $D(z)$ is a function of z – most often we encounter second-order loops. In the following sections, we shall discuss, first, properties of the individual digital blocks.

10.3 MAJOR BUILDING BLOCKS OF DPLLs

We shall now examine the basic parts of digital PLLs: the phase detectors, the digital filters, and the digital oscillators, illustrated in Fig. 10.6.

10.3.1 Digital Phase Detectors

There are several types of digital phase detectors, of which only a few will be discussed in the following sections.

10.3.1.1 Sampling phase detectors

This simple flip-flop circuit provides two functions: *set* and *clear*. The principle block diagram is shown in Fig. 10.7(a). The pulse generators supply output signals on the zero transitions from, for example, the negative to the positive polarity to the gating flip-flop. A pulse at its set input S opens the counter and the other at its clear input C closes it. The counted pulses are supplied by the clock generator, the frequency of which is generally much higher than the input frequency f_i, that is,

$$f_c = 2^M f_i \tag{10.52}$$

Consequently, the frequency difference between the pulse rates supplied by the input or reference generator and the number controlled oscillator (NCO) is proportional to the number N of counted pulses

$$N = \frac{|T_i - T_o|}{T_c} = f_c \left| \frac{1}{f_i} - \frac{1}{f_o} \right| \tag{10.53}$$

With the assistance of (10.52), we get

$$N = 2^n f_o \frac{f_o - f_i}{f_i f_o} \approx 2^n T_i |f_o - f_i| \tag{10.54}$$

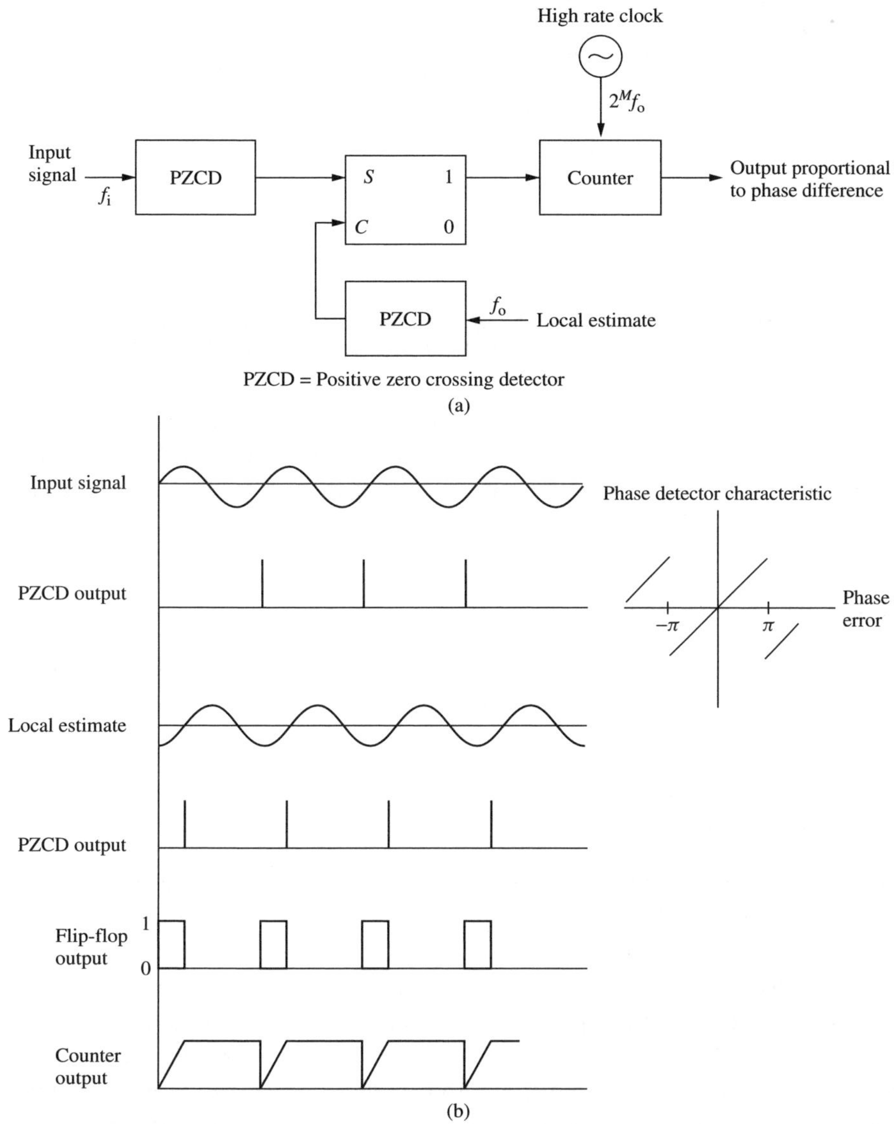

Figure 10.7 The simple flip-flop circuit (a) as a phase detector (b) (Reproduced from W.C. Lindsey and C.M. Chie, *Phase-Locked Loops*, IEEE Press, 1985 by permission of IEEE, 2002 [3]).

from which the phase difference is

$$\Delta\phi = 2\pi T_i \Delta f = 2\pi \frac{N}{2^n} \tag{10.55}$$

10.3.1.2 The "Nyquist rate" PD

The block diagram is shown in Fig. 10.8. It is based on the multiplication of the input signals to the phase detector (cf. eq. (1.4)).

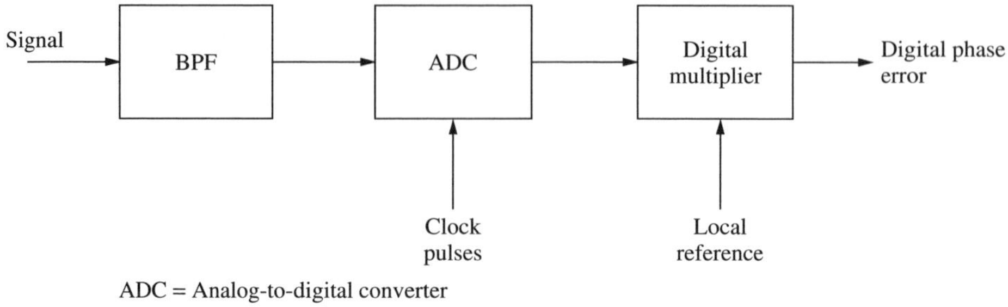

Figure 10.8 "Nyquist rate" sampling phase detector (Reproduced from W.C. Lindsey and C.M. Chie, *Phase-Locked Loops*, IEEE Press, 1985 by permission of IEEE, 2002 [3]).

A band-pass filter is placed at the input of the DPLL to prevent aliasing; thereafter, the signal is fed to an analog-to-digital convertor (ADC) and then digitally multiplied with the *Local Reference* (VCO) output. The required phase difference is obtained in this way.

10.3.1.3 Zero crossing PD

The positive zero crossing detector is one of the simplest phase detectors (PDs) encountered in DPLLs. The principle is illustrated in Fig. 10.9. Another type of PD registers every zero crossing.

10.3.1.4 The lag-lead PD

This PD is characterized by a simple binary operation. The output indicates whether the local reference leads or lags the input signal (see the rhs in Fig. 10.10). We can use the popular phase-frequency detector, often designated as a phase comparator, as a binary PD. However, the rectangular output wave must be used for the generation of pulses. Since the quantization of the output signal is rough, a smoothing or sequential filter is introduced. It is made up of three counters and two OR gates. The designation "sequential" includes the fact that the output is not a linear function of a number of the input pulses. It works as a weighting circuit with a changing time difference between individual input pulses.

Pulses from both leading and lagging outputs are stored in counters N (till the number N) and similarly in the counter M where

$$N < M < 2N \tag{10.56}$$

The operation may result in two situations:

1. After simultaneous resetting of all three counters, one of the N counters overflows and supplies a pulse used for correction of the NCO and provides resetting of all three counters.

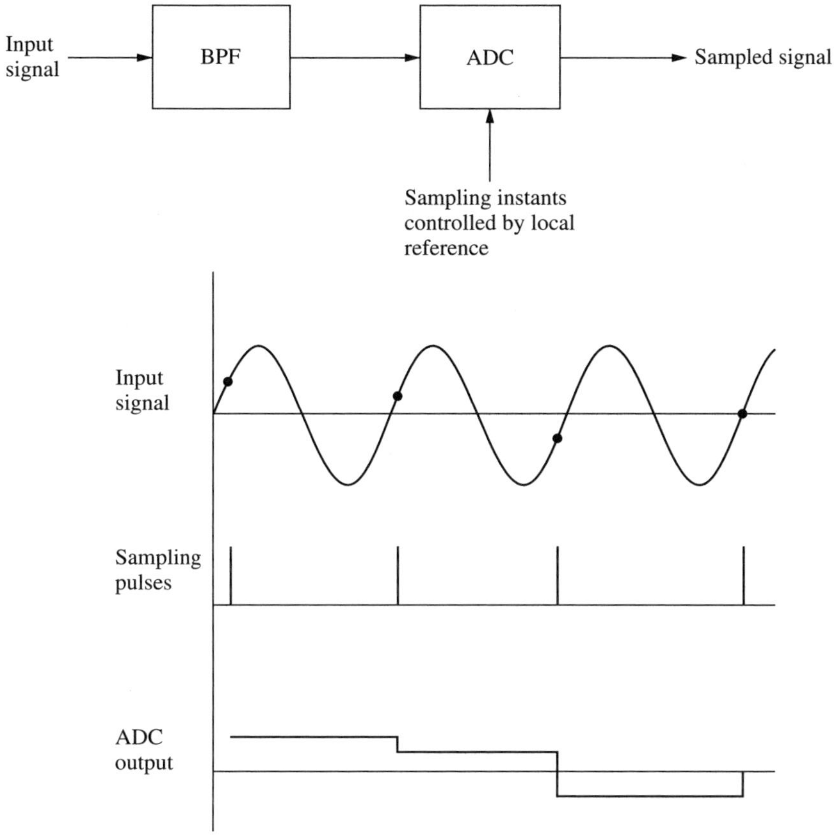

Figure 10.9 Positive zero crossing phase detector (Reproduced from W.C. Lindsey and C.M. Chie, *Phase-Locked Loops*, IEEE Press, 1985 by permission of IEEE, 2002 [3]).

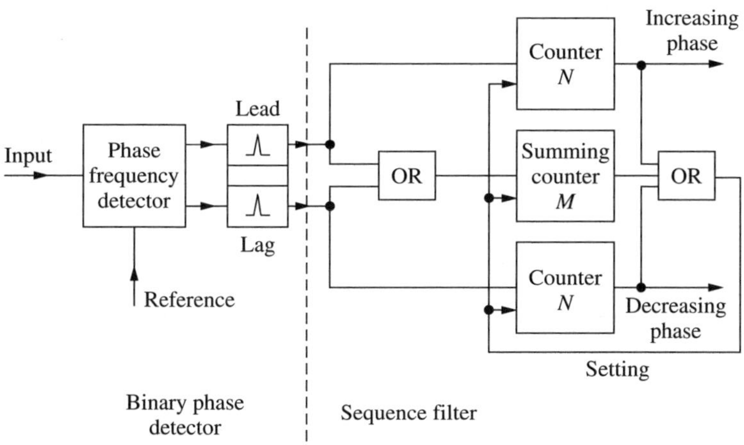

Figure 10.10 Binary phase detector with a sequential filter.

2. The summing counter overflows sooner than any of the N counters. At this situation it commands, via the OR gate, resetting of all three counters, and no output signal is generated; this is the case when no phase correction is needed.

10.3.2 Digital Filters

As in analog PLLs, in DPLLs also we introduce a filter for improving their function (cf. Fig. 10.6). At the same time, the loop order is increased. Our investigation will start with the second-order type 2. In this case the transfer function of the loop filter in the Laplace transform is easily changed into z-transform by a simple rearrangement

$$F(s) = \frac{1 + sT_2}{sT_1} = \frac{T_2}{T_1} + \frac{1}{sT_1} \tag{10.57}$$

We see that the filter is formed by the sum of two terms, and in accordance with the block algebra (Fig. 1.5b), it can be realized as a branched system. Finally, with the assistance of Tab. 10.1 we change (10.57) into the z-transform.

$$F(z) = K_1 + \frac{K_2}{1 - z^{-1}} \tag{10.58}$$

where $K_1 = T_2/T_1$ and $K_2 = 1/T_1$. The block diagram is shown in Fig. 10.11. Addition of the integration branch reveals the DPLL of the third order (see Fig. 10.12).

10.3.3 Digital Oscillators

Digitally controlled oscillators (DCOs) are based on programmable frequency dividers. One arrangement is presented in the block diagram in Fig. 10.13. Its input is supplied with the clocking pulses that are stored in the counter memory (Q) and compared with a predetermined value W set in the comparing circuit. In the instant when $Q = W$,

Table 10.1 Table of Laplace and z-transforms

$f(t)$	$F(s)$	$F(z)$	$f(nT)$
–	1	z^{-0}	$\delta(nT)$
–	e^{-kTs}	–	$\delta[(n-k)T]$
1	$\frac{1}{s}$	$\frac{1}{1-z^{-1}}$	1
t	–	$\frac{Tz^{-1}}{(1-z^{-1})^2}$	nT
$\frac{t^2}{2!}$	$\frac{1}{s^3}$	$\frac{T^2z^{-1}(1+z^{-1})}{2(1-z^{-1})^3}$	$\frac{(nT)^2}{2!}$
e^{-at}	$\frac{1}{s+a}$	$\frac{1}{1-e^{-aT}z^{-1}}$	e^{-anT}
$e^{-at} - e^{-bt}$	$\frac{b-a}{(s+a)(s+b)}$	$\frac{e^{-aT} - e^{-bT}}{(1-e^{-aT}z^{-1})(1-e^{-bT}z^{-1})}$	$e^{-anT} - e^{-bnT}$

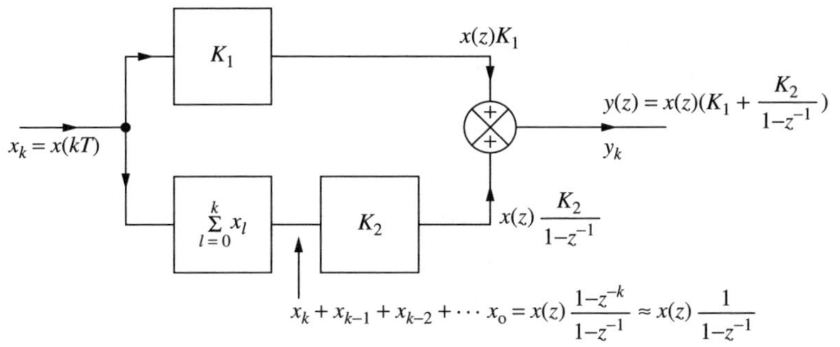

Figure 10.11 Lag-lead filter used in type 2 PLLs in the z-transform notation.

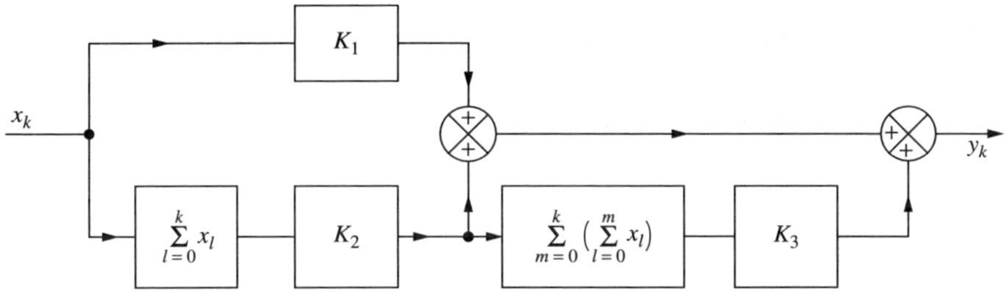

Figure 10.12 Lag-lead filter used in type 2 PLLs with an integrating RC section in the z-transform notation.

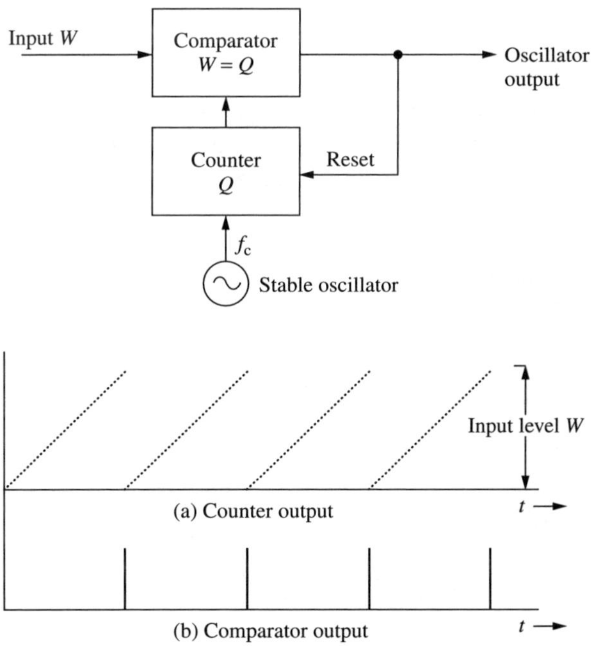

Figure 10.13 Block diagram of a typical digitally controlled oscillator (Reproduced from W.C. Lindsey and C.M. Chie, *Phase-Locked Loops*, IEEE Press, 1985 by permission of IEEE, 2002 [3]).

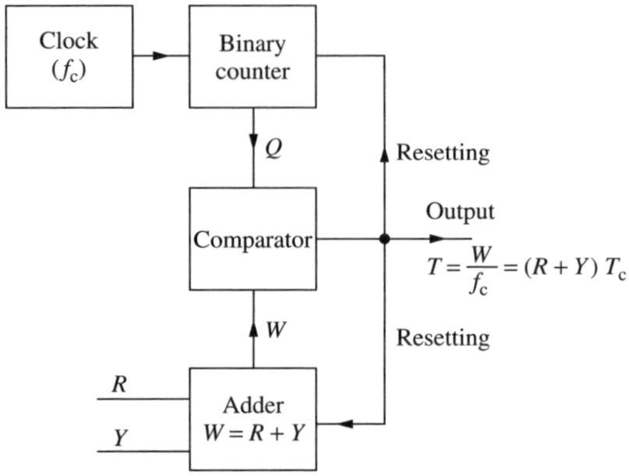

Figure 10.14 A more detailed arrangement of the digitally controlled oscillator.

the summation is stopped and the counter reset to zero. Evidently, the output period T_o is equal to

$$T_o = W \cdot T_{\text{clock}} \tag{10.59}$$

and is easily changed by resetting the number W.

A more detailed arrangement is shown in Fig. 10.14. The clocking frequency is approximately R times higher than the desired output frequency $f_o = 1/T_o$. The content of the binary counter is always compared with the preset value in the adding circuit

$$W = R + Y \tag{10.60}$$

where the number Y is supplied by the loop filter of the DPLL. As soon as Q is equal to W, the comparator is activated, and its output pulse resets the binary counter as well as the adder; however, the controlling adder is set back to R and to the corrected value Y. The output period is

$$T = W/f_c = (R + Y)T_c = T_o + YT_o/R \tag{10.61}$$

with sensitivity

$$\frac{\Delta T_o}{T_o} = \frac{Y}{R} \tag{10.62}$$

Example 10.3

Evaluate the respective phase sensitivity for $R = 512$, $Y = 1$

$$\Delta\phi = \Delta T_o f_o 2\pi = \frac{2\pi}{512} \approx 0.01\,\text{rad} \tag{10.63}$$

or

$$\Delta\phi = 0.7^\circ \tag{10.64}$$

Another arrangement of digital oscillators is illustrated in Fig. 10.15 where, in accordance with the outputs of the sequential filter, the clock pulses are either added to or removed from the counter set to a fixed number R.

What remains is the derivation of the digital oscillator transfer function, knowledge of which is needed for evaluation of properties of DPLLs. To this end we rearrange the block diagram in Fig. 10.6 into the basic DPLL system depicted in Fig. 10.16. Note that both phase detector gain, $g(.)$, generally nonlinear, and the equivalent noise samples, η_k, are functions of the actual design parameters of the loop. By considering the phase difference to be small, as in analog counterparts, the application of the Taylor expansion on the slowly varying phase detector output gives

$$g(\Phi_k) \doteq g'(0)\Phi_k \tag{10.65}$$

For simplification of the whole process we can also neglect, without any loss of generality, the noise components η_k and consider the transfer function of the digital filter, $D(z)$, to be constant. By assuming, further, that the input phase deviation

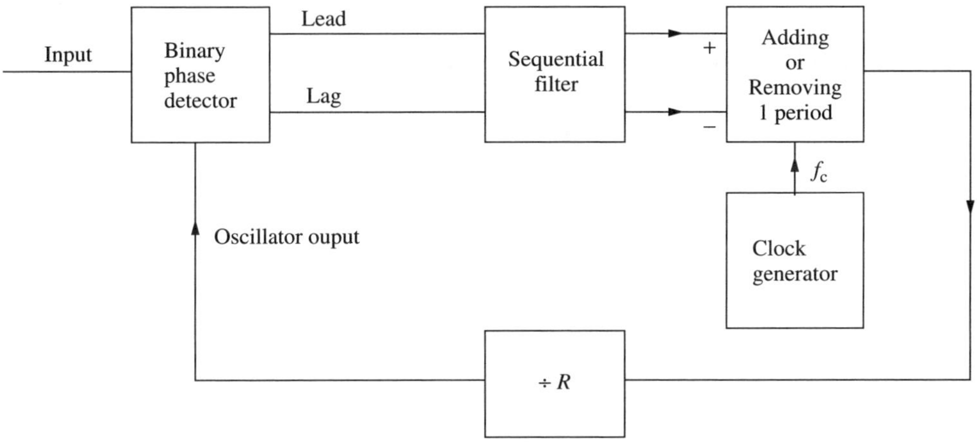

Figure 10.15 Digital oscillator in DPLL with a sequential filter.

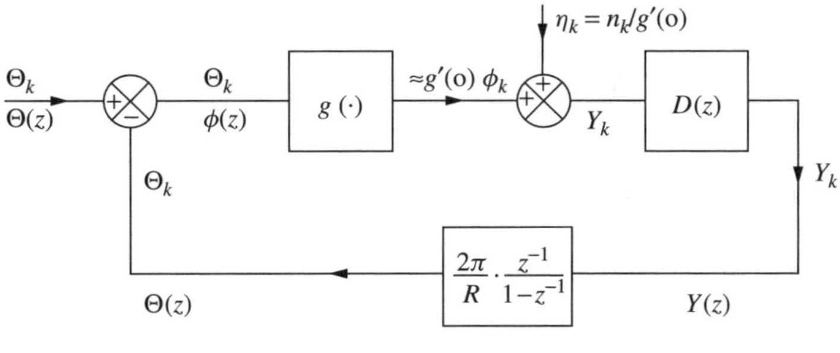

Figure 10.16 The basic arrangement of a digital phase lock loop.

$\Theta(kT) = \Theta_k$ is compensated by the deviation $\hat{\Theta}_k$, we have

$$\hat{\Theta}_0 = 0 \tag{10.66}$$

$$\Phi_o = \Theta_0 \tag{10.67}$$

$$y_0 \neq 0 \tag{10.68}$$

However, a positive nonzero y_0 changes the division factor R into W_0 (see Fig. 10.14)

$$W_0 = R - Y_0 \tag{10.69}$$

and consequently, in accordance with (10.61), changes the output frequency ω_0 into

$$\omega_{01} \doteq \omega_0(1 + Y_0/R) \tag{10.70}$$

During one period, the input phase $\hat{\Theta}_0 = 0$ changes into

$$\hat{\Theta}_1 = (\omega_{01} - \omega_0)T_0 \doteq 2\pi Y_0/R \tag{10.71}$$

After comparing the input phase Θ_1 we find the phase Φ_1

$$\Phi_1 = \Theta_1 - \hat{\Theta}_1 \tag{10.72}$$

and new y_1 with the corresponding Y_1 corrects the output frequency into

$$\omega_{02} \doteq \omega_0(1 + Y_1/R) \tag{10.73}$$

and phase difference into

$$\hat{\Theta}_2 = \frac{2\pi}{R}(Y_0 + Y_1) \tag{10.74}$$

Generalization reveals

$$\hat{\Theta}_k = \frac{2\pi}{R}\sum_{r=0}^{k-1} Y_r \tag{10.75}$$

The above relation can easily be rewritten as

$$\hat{\Theta}_k = \hat{\Theta}_{k-1} + \frac{2\pi}{R}Y_{k-1} \tag{10.76}$$

or in the z-transform symbolic form

$$\hat{\Theta}(z) = z^{-1}\hat{\Theta}(z) + z^{-1}\frac{2\pi}{R}Y(z) \tag{10.77}$$

and from this we get the transfer function

$$\frac{\hat{\Theta}(z)}{Y(z)} = \frac{2\pi}{R} \cdot \frac{z^{-1}}{1 - z^{-1}} \tag{10.78}$$

where the second factor on the rhs is generally designated as the oscillator transfer function.

10.4 DIGITAL PHASE-LOCKED LOOPS (DPLLs)

Solution of DPLLs in the closed form, in contradiction to analog systems, is only possible after application of some significant simplifications. The major obstacle is the nonlinearity of sampling PDs and digital convertors using a rather small number of bits. Consequently, the following discussion may serve for demonstration and not for a rigorous application.

10.4.1 Digital Phase-locked Loops of the First Order

After reverting to eq. (1.10) and replacing the derivation with differences, we have

$$\frac{\Phi_{k+1} - \Phi_k}{T} = K \sin \Phi_k + \Delta\omega \tag{10.79}$$

where the loop constant K is a factor of the gain of the phase detector gain K_d and the oscillator gain derived in the previous sections ((10.65) and (10.73))

$$K_\mathrm{d} \approx g'(0) \tag{10.80}$$

$$K_\mathrm{o} = \frac{\omega}{R} \tag{10.81}$$

Evidently, relation (10.79) can be written as

$$\Phi_{K+1} - \Phi_k = -K_1 \sin \Phi_k + \Delta\omega T \tag{10.82}$$

where we have introduced

$$K_1 = TK_\mathrm{d}K_\mathrm{o} = Tg'(0)\frac{\omega}{R} = \omega G_1 \tag{10.83}$$

and combined all the DC constants into G_1.

Evidently, the arrangement in Fig. 10.16 can be simplified to the block diagram depicted in Fig. 10.17. Introduction of the DPLL transfer function $H(z)$, in analogy with (1.29), reveals

$$H(z) = \frac{K_1\dfrac{z^{-1}}{1 - z^{-1}}}{1 + K_1\dfrac{z^{-1}}{1 - z^{-1}}} = \frac{K_1}{z - 1 + K_1} \tag{10.84}$$

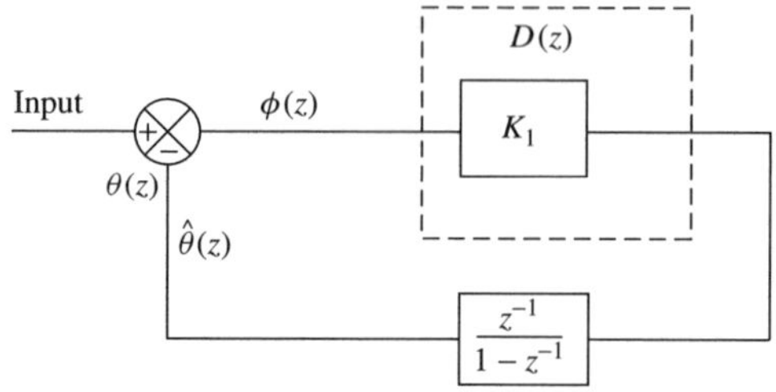

Figure 10.17 A generalized block diagram of the digital phase lock loop.

and

$$1 - H(z) = \frac{z - 1}{z - 1 + K_1} \tag{10.85}$$

Example 10.4

Solution of the first-order digital PLL (DPLL) will start with applications of eqs. (10.8), (10.10), and (10.17). For the latter, we shall limit the sum to $n = 0$ only. In this case we have

$$H(s) = H_h(s)\hat{H}(s) = \mathrm{e}^{-sT/2}\,\mathrm{Sinh}(sT/2)\frac{K_1}{\mathrm{e}^{sT} - 1 + K_1} \tag{10.86}$$

The first approximation will reveal

$$\begin{aligned} H(s) &\approx \mathrm{e}^{-sT/2}\frac{K_\mathrm{d}K_\mathrm{o}T}{1 + sT - 1 + K_\mathrm{d}K_\mathrm{o}T} = \mathrm{e}^{-sT/2}\frac{K}{s + K} \\ &\approx \frac{1}{1 + sT/2} \cdot \frac{K}{s + K} \end{aligned} \tag{10.87}$$

Effectively, the first-order loop was changed to the second-order PLL. However, when the sampling frequency is rather low, the approximation reveals a second-order loop with a time delay

$$H(s) \approx \mathrm{e}^{-sT/2}\,\mathrm{Sinh}(sT/2) \cdot \frac{K}{s^2T/2 + s + K} \tag{10.88}$$

10.4.2 Digital Phase-locked Loops of the Second Order

Replacement of the loop filter $D(z)$ in Fig. 10.17 with that in Fig. 10.11 requires introduction of the relations (10.58) and (10.83), that is,

$$D(z) = K_1 + \frac{K_2 z}{z - 1} \tag{10.89}$$

The resulting transfer function $H(z)$ is

$$H(z) = \frac{D(z)}{z - 1 + D(z)} \tag{10.90}$$

and with the assistance of (10.89) we arrive at the final form

$$H(z) = \frac{(K_1 + K_2)z - K_1}{(z-1)^2 + (K_1 + K_2)z - K_1} \tag{10.91}$$

Example 10.5

The solution of the second-order DPLL was given by Gardner in [5]. First, we change the denominator of (10.91) as shown:

$$(z-1)^2 + (K_1 + K_2)z - K_1 = (z-\alpha)^2 - \beta^2 \tag{10.92}$$

where we have introduced

$$-2\alpha = -2 + K_1 + K_2 \tag{10.93}$$

$$\alpha^2 + \beta^2 = 1 - K_1 \tag{10.94}$$

Next, we compute the roots of the rhs of (10.92):

$$z_{1,2} = \alpha \pm \mathrm{j}\beta \tag{10.95}$$

With the assistance of eq. (10.10), we provide the reverse transformation in the Laplace transform

$$\alpha + \mathrm{j}\beta = \mathrm{e}^{(s_r + \mathrm{j}s_i)T} = \mathrm{e}^{s_r T}[\cos(s_i T) + \mathrm{j}\sin(s_i T)] \tag{10.96}$$

By using Fig. 4.2 we have

$$s_r = -\omega_n \zeta \quad \text{and} \quad s_i = \omega_n \sqrt{1-\zeta^2} \tag{10.97}$$

and for values α and β the following constants:

$$\alpha = \mathrm{e}^{-\omega_n \zeta T} \cos \omega_n T \sqrt{1-\zeta^2} \tag{10.98}$$

$$\beta = \mathrm{e}^{-\omega_n \zeta T} \sin \omega_n T \sqrt{1-\zeta^2} \tag{10.99}$$

After introduction of the results just found into (10.93) and (10.94), we compute the constants K_1 and K_2

$$K_1 = 1 - \alpha^2 - \beta^2 = 1 - \mathrm{e}^{-2\omega_n \zeta T} \tag{10.100}$$

$$K_2 = 1 + \alpha^2 + \beta^2 - 2\alpha = (1-\alpha)^2 + \beta^2 \tag{10.101}$$

For stable systems, the roots z_1 and z_2 must be inside the unit circle in the plane z (cf. Fig. 10.2), which requires the validity

$$\alpha^2 + \beta^2 < 1 \tag{10.102}$$

But with the assistance of eqs. (10.99) and (10.100), we find out that the condition is fulfilled in the same way, since

$$\alpha^2 + \beta^2 = \mathrm{e}^{-2\omega_n \zeta T} \tag{10.103}$$

In this context, we point out root conditions for stability, which are sufficient and necessary:

$$-1 < z_1 z_2 < 1 \tag{10.104}$$

$$-(1 + z_1 z_2) < -(z_1 + z_2) < (1 + z_1 z_2) \tag{10.105}$$

Example 10.6

Let us investigate the sampled second-order loop (10.88) in accordance with the procedure used in Example 10.4:

$$H(s) = H_h \hat{H}(s) \approx e^{-sT/2} \frac{(K_1 + K_2)(1 + sT) - K_1}{(sT)^2 + (1 + sT)(K_1 + K_2) - K_1}$$

$$\approx \frac{1}{1 + sT/2} \cdot \frac{sT(K_1 + K_2) + K_2}{(sT)^2 + sT(K_1 + K_2) + K_2} \tag{10.106}$$

Comparison with the normalized second-order loop (2.24)

$$H(\sigma) = \frac{1 + \sigma 2\zeta}{\sigma^2 + 2\sigma\zeta + 1} \tag{10.107}$$

reveals for the natural frequency ω_n and the damping factor ζ

$$\omega_n = \frac{\sqrt{K_2}}{T} \qquad \zeta = \frac{1}{2} \frac{K_1 + K_2}{\sqrt{K_2}} = \frac{\omega_n}{2} T_{2,\mathrm{APLL}} \tag{10.108}$$

$$T_{2,\mathrm{APLL}} = \frac{2\zeta}{\omega_n} = \frac{K_1 + K_2}{\sqrt{K_2}} \cdot \frac{T}{\sqrt{K_2}} = \frac{K_1 + K_2}{K_2} T \tag{10.109}$$

and the factor κ is

$$\kappa = \frac{T}{2} \cdot \frac{K_2}{(K_1 + K_2)T} = \frac{1}{2} \frac{K_2}{K_1 + K_2} < 0.5 \tag{10.110}$$

Introduction of κ into (3.48) reveals that the loop is unconditionally stable, as already proved in Example 10.5.

10.5 TRANSIENT RESPONSE EVALUATION FOR STEADY AND PERIODIC CHANGES OF INPUT PHASE AND FREQUENCY

As long as we are justified in neglecting the samples of noise $\{\eta_k\}$, we get for the phase the changes at the *PD* output due to the input differences $\{\Theta_k\}$ (cf. relation (1.30))

$$\Phi(z) = [1 - H(z)]\Theta(z) \tag{10.111}$$

Table 10.2 Laplace transforms and z-transforms of various uniformly sampled phase inputs

Input phase	$f(t)$	$F(s)$	$F(z)$	$f(nT)$
Phase step	$\Delta\phi$	$\Delta\phi\dfrac{1}{s}$	$\Delta\phi\dfrac{z}{z-1}$	$\Delta\phi$
Frequency step	$\phi = \Delta\omega t$	$\Delta\omega\dfrac{1}{s^2}$	$\Delta\omega T\dfrac{z}{(z-1)^2}$	$\Delta\omega nT$
Frequency ramp	$\Delta\dot{\omega}\dfrac{t^2}{2}$	$\Delta\dot{\omega}\dfrac{1}{s^3}$	$\Delta\dot{\omega}\dfrac{T^2}{2}\dfrac{z(z+1)}{(z-1)^3}$	$\Delta\dot{\omega}\dfrac{(nT)^2}{2}$
Phase modulation	$\Delta\phi\sin(\omega_m t)$	$\dfrac{\Delta\phi\omega_m}{s^2+{\omega_m}^2}$	$\dfrac{\Delta\phi z\sin\omega_m T}{z^2-2z\cos\omega_m T+1}$	$\Delta\phi\sin(n\omega_m T)$
Frequency modulation	$\dfrac{\Delta\omega}{\omega_m}\cos(\omega_m t)$	$\dfrac{(\Delta\omega/\omega_m)s}{s^2+{\omega_m}^2}$	$\dfrac{(\Delta\omega/\omega_m)z(z-\cos\omega_m T)}{z^2-2z\cos\omega_m T+1}$	$\dfrac{\Delta\omega}{\omega_m}\cos(n\omega_m T)$

Table 10.3 Limiting time functions in accordance with the Laplace and z-transforms

$\lim f(t)$	$\lim\,[sF(s)]$	$\lim\,[F(z)]$
$t \to 0$	$s \to \infty$	$z \to \infty$
$t \to \infty$	$s \to 0$	$\lim\limits_{z\to 1}[(1-z^{-1})F(z)]$

which can be rearranged for second-order loops with the assistance of (10.91) and (10.92) into

$$\Phi(z) = \frac{(z-1)^2}{(z-\alpha)^2+\beta^2}\Theta(z) \tag{10.112}$$

After introduction of the step or periodic change relations from Tab. 10.2, we can find the respective differential equations and compute the appropriate transient and steady states – the latter with the assistance of Tab. 10.3.

Lindsey and Chie [3] pointed out that digital and analog results did not practically differ in the case where coefficients K_1 and K_2 met conditions (10.101) and (10.102) as long as $\omega_n T \leq 0.1$.

10.6 LOOP NOISE BANDWIDTH OF DIGITAL PLLs

As with analog PLLs, we shall also introduce the noise bandwidth B_L in DPLLs. In the case where noise samples η_k (cf. Fig. 10.16) are stationary [6] with the zero mean value, we can write for the variance $\{\phi_k\}$ the following relation:

$$\sigma_\Phi^2 = \frac{1}{2\pi j}\int\limits_{|z|=1} H(z)H(z^{-1})z^{-1}\frac{\hat{R}_n(z)}{2P}\,dz \tag{10.113}$$

where $\hat{R}_n(z)$ is the autocorrelation of the input noise in the z-transform and P is the input power. By assuming the white noise has a constant power spectral density (PSD)

$$S_n(f) = N_0 \tag{10.114}$$

we get, in cases where the pass band of the input filter is $2B_w$,

$$\frac{\hat{R}_n(z)}{2P} = {\sigma_n}^2 = 2N_0 B_w \tag{10.115}$$

The situation being such, we find for the one-sided noise bandwidth B_L of the DPLL

$$\frac{B_L}{B_w} = \frac{1}{2\pi j} \int_{|z|=1} H(z)H(z^{-1})z^{-1}\,dz \tag{10.116}$$

In this case, the variance σ_Φ^2 will be

$$\sigma_\Phi^2 = \frac{N_0 B_L}{P} = \frac{P_N}{P} \tag{10.117}$$

Note the agreement with the analog PLL (9.113).

The integral in (10.116) is evaluated in accordance with the residues theorem ([7], Section 7.7.2), and for DPLLs of the first order we have

$$\frac{B_L}{B_w} = \frac{K_1}{2 - K_1} \tag{10.118}$$

and for second-order DPLLs we have

$$\frac{B_L}{B_w} = \frac{2}{[4/(r+1)] - K_1} - 1 \tag{10.119}$$

where we have introduced

$$r = 1 + K_2/K_1 \tag{10.120}$$

REFERENCES

[1] J.R. Ragazzini and G.F. Franklin, *Sampled-Data Control Systems*. New York: McGraw-Hill, 1958.

[2] G.S. Gill and C. Gupta, "First order discrete phase-locked loop with applications to demodulations of angle-modulated carrier", *IEEE Trans. Commun.*, **COM-25**, 454–462, 1972.

[3] W.C. Lindsey and C.M. Chie, *Phase-Locked Loops*. New York: IEEE Press, 1985.

[4] E.J. Angelo Jr., "Tutorial introduction to digital filtering", *Bell Syst. Tech. J.*, **60**(7), 1499–1546, 1981.

[5] W.C. Lindsey and C.M. Chie, "A survey of digital phase-locked loop", *Proc. IEEE*, **69**, 410–431, 1981 (Reprinted in [3]).

[6] A.V. Oppenheim and R.W. Schafer, *Digital Signal Processing*. Englewood Cliffs, N.J.: Prentice-Hall, 1975.

[7] G.A. Korn and T.M. Korn, *Mathematical Handbook*. New York: McGraw-Hill, 1958.

11

PLLs in Frequency Synthesis[1]

As a result of the steadily increasing congestion in communication bands of the electromagnetic spectrum, efforts are being made to utilize new ranges, particularly microwave and optical frequencies, to better exploit existing frequency allocations. The latter tasks have led to improvements in communication techniques [spurious sideband (SSB) modulation, multiplexing, digital systems, spread spectrum application, cellular global mobile services (GMS), etc.]. In this connection, perfection of frequency generators in transmitter exciters and local oscillators in receivers has been attempted, in most cases with the assistance of frequency synthesizers, that is, devices changing a given standard frequency, f_s, into another – the desired – output frequency, f_x.

This approach is preferred to single-frequency oscillators (Xtals) because of larger tuning ranges, better short- and long-term stabilities (lower output noise), microprocessor-aided tuning or programming, and many other advantages.

The present situation of frequency synthesis techniques in the whole range of electromagnetic waves, from audio and RF to the visible frequencies, is shown in Fig. 11.1. The lesson one learns from this figure is that there exist different approaches in both hardware and software solutions to particular frequency synthesizers. Nevertheless, one basic principle is common to all the systems.

As already mentioned, the task of frequency synthesis is to generate an arbitrary frequency, f_x, from a given reference or standard frequency, f_s, that is, to solve equation

$$f_x = \xi_x f_s \tag{11.1}$$

where the factor ξ_x is the so-called normalized frequency.

The solution of eq. (11.1) is always based on algorithms for step-by-step approximations of real numbers – as will be explained in the following sections of this chapter. In instances where the factor ξ_x is a rational number, that is, a ratio of two

[1] An adapted and updated version of pp. 1–26 of Kroupa: *Direct Digital Frequency Synthesizers*, IEEE Press, 1999 by permission of IEEE, 2002.

Phase Lock Loops and Frequency Synthesis V.F. Kroupa
 ISBN: 0-470-84866-9

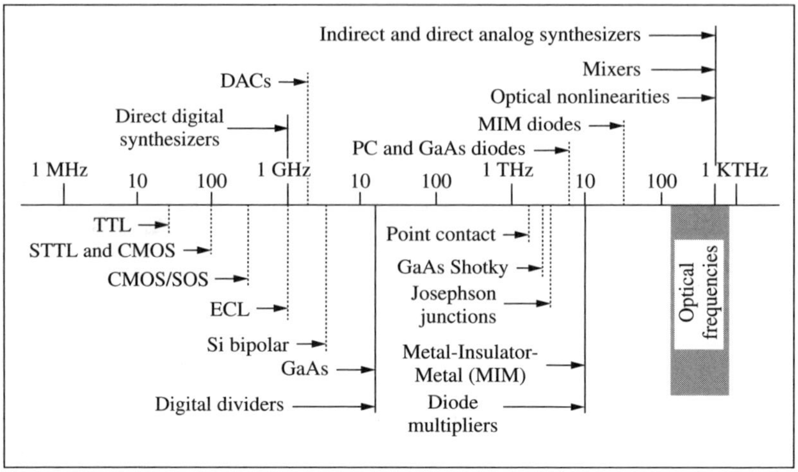

Figure 11.1 The state of the art of frequency synthesis from the lowest frequency ranges to the optical frequencies (Reproduced from V.F. Kroupa, ed. *Direct Digital Frequency Synthesizers*. New York: IEEE Press, 1999 by permission of IEEE, 2002).

mutually prime integers

$$\xi_x = \frac{X}{Y} \tag{11.2}$$

the output frequency and the reference frequency will be in harmonic relation. However, in a more general case, ξ_x can belong to a much broader set of real numbers, and in this case we generally stop the approximation process as soon as the remainder falls below a prescribed or acceptable value.

Practical realizations of different types of frequency synthesizers are discussed in several books [1–6]. However, Direct Digital Frequency Synthesizers (DDFS), which are of most interest in this chapter, were mentioned by Gorski-Popiel [2] nearly 30 years ago, and recently by Crawford, Golberg, and Kroupa [7–9]. Otherwise, there are a lot of papers and literature about new developments in the field of IC designs of DDFS – Application Notes and Technical Data.

11.1 HISTORICAL INTRODUCTION

Defining particular events in the never-ceasing flow of time is as old as mankind itself. Daily alternations of light and dark, seasonal weather variations, and so on were among the first periodic occurrences experienced by men. However, the introduction of the term "frequency," that is, the inverse value of the duration of one period or cycle

$$f = \frac{1}{T_c} \tag{11.3}$$

is rather recent.

On the contrary, the term "angular velocity," connected with the rotation of the Earth, wheels, and so on, is much older. Nevertheless, the physics of both phenomena are the same since the parametric equations of the uniform circular motion (cf. Fig. 11.2) are given by

$$x = r \times \cos(\omega t) \quad \text{and} \quad y = r \times \sin(\omega t) \tag{11.4}$$

where r is the radius and the angular velocity ω is connected with the period of motion

$$\omega = \frac{2\pi}{T_c} \tag{11.5}$$

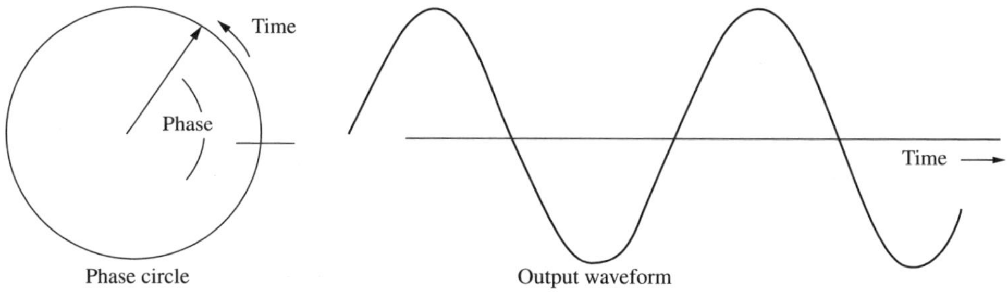

Figure 11.2 Uniform circular motion and its projection on the horizontal time axis.

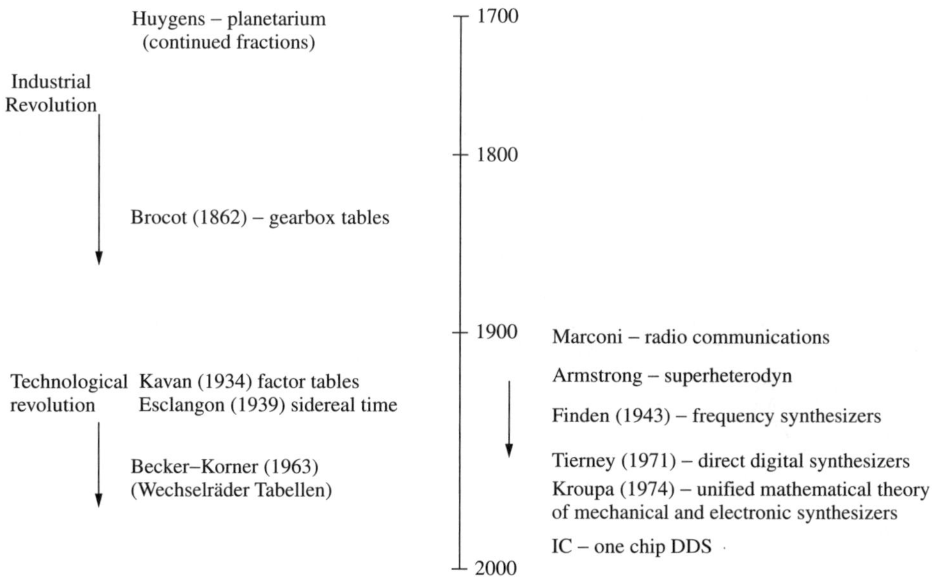

Figure 11.3 The past and present situation of mechanical and electronic synthesizers (Reproduced from V.F. Kroupa, ed. *Direct Digital Frequency Synthesizers*. New York: IEEE Press, 1999 by permission of IEEE, 2002).

Evidently, devices for changing the angular velocity of the driving engine into another angular velocity of working wheels can be considered as single-frequency mechanical synthesizers resting on the same mathematical design principles. The past and present situation of the mechanical and electronic synthesizers is shown schematically in Fig. 11.3.

11.2 GEARBOXES

Long before the Industrial Revolution started, mechanical engineers had to solve the problem of changing the input angular velocity ω_s (i.e., some given *input revolutions/second, minute, etc.*) into a desired output angular velocity ω_x, or in other words, had to solve the following basic equation

$$\omega_x = \xi_x \times \omega_s \tag{11.6}$$

by taking into account several restrictions that will be discussed later – an example would be the gear ratio of the compound train shown in Fig. 11.4.

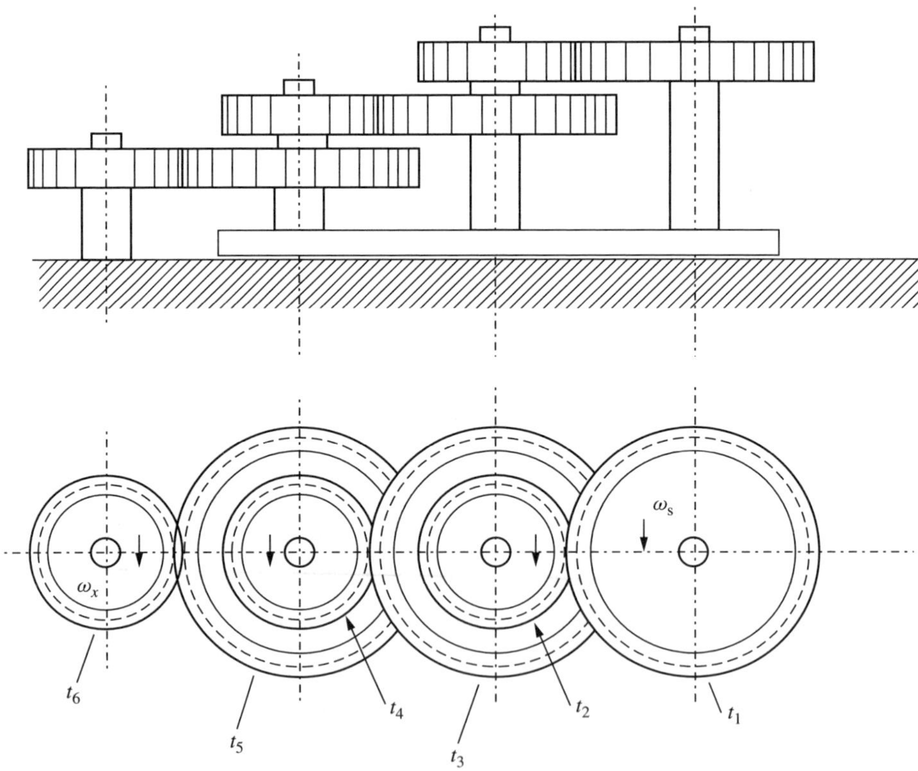

Figure 11.4 Compound gear train (Reproduced from V.F. Kroupa, ed. *Direct Digital Frequency Synthesizers*. New York: IEEE Press, 1999 by permission of IEEE, 2002).

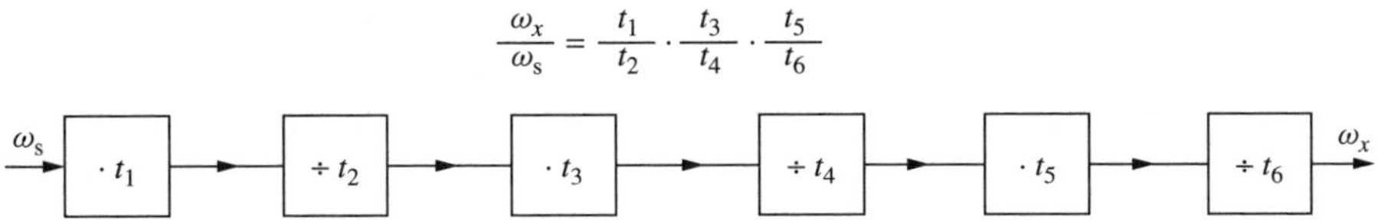

Figure 11.5 The block diagram circuit of the compound gear train (Reproduced from V.F. Kroupa, ed. *Direct Digital Frequency Synthesizers*. New York: IEEE Press, 1999 by permission of IEEE, 2002).

The normalized output angular velocity, in accordance with eq. (11.6), is

$$\xi_x = \frac{\omega_x}{\omega_s} = \frac{t_1}{t_2} \times \frac{t_3}{t_4} \times \frac{t_5}{t_6} \tag{11.7}$$

where t_1 to t_6 are the number of teeth on individual wheels. Factors in the numerator represent multiplication and those in the denominator represent dividers (cf. the block diagram in Fig. 11.5).

The difficulty with the solution of the gearbox equation is that only wheels with a limited number of teeth are practicable.

Further, experience has taught mechanical engineers that a mere expansion of the ratio ξ_x into a continued fraction (first used by Huygens for the design of his planetarium [10]) generally does not work: either the approximation error or the number of teeth was too large.

Because of the lack of an appropriate theory, mechanical engineers compiled tables (e.g., [11] or [12]) that made it possible to solve nearly all practical cases with sufficient accuracy.

11.3 FREQUENCY SYNTHESIS

The aim of frequency synthesis is to generate an arbitrary frequency, f_x, from a given standard frequency, f_s, that is, to solve equation

$$f_x = \xi_x \times f_s \tag{11.8}$$

in such a way that a practical device could be manufactured accordingly (note the similarity with eq. (11.6)).

The term "frequency synthesis" was coined in the early 1940s when the first measuring generators were devised, the variable output frequency of which was always harmonically related to a submultiple of the driving frequency f_s [13, 14].

11.3.1 Single-frequency Synthesizers

In the simplest case, ξ_x is a fraction formed by small relatively prime integers, that is,

$$\xi_x = X_1/Y_1 \tag{11.9}$$

In this case the synthesizer is reduced merely to a chain of one frequency divider and one multiplier. If X_1 and Y_1 in eq. (11.9) are products of small prime numbers, the synthesizer may be realized by a chain of frequency multipliers and dividers as in Fig. 11.5. Factorization of integers X_1 and Y_1 is performed either with the assistance of tables [15] or by using a simple algorithm not recalled here.

11.3.2 Multiple Output Frequency Synthesizers

"Tunable" frequency synthesizers with many output frequencies in a desired frequency range were required in the standard time and frequency laboratories for measurement purposes. But soon uses in communications followed, where applications as transmitter exciters and as local oscillators in receivers were appreciated. For simplicity of operation in these first devices, the variable output frequency was set decimally. In some instances, the setting between the smallest frequency steps was provided with an incoherent interpolating oscillator. These first direct frequency synthesizers, manufactured in the 1960s and the early 1970s (e.g., Hewlett-Packard type 5100) were bulky, with a rather good spectral purity and phase stability (e.g., Fluke type 645).

11.4 MATHEMATICAL THEORY OF FREQUENCY SYNTHESIS

In the early years, the trial and error method was the workhorse for designers of frequency synthesizers. However, soon the author of this book [1] showed that the actual frequency synthesis process was nothing but a step-by-step approximation of real numbers. Only the approximation process, in the physical world of frequency synthesis, is subjected to limitations caused by the required frequency range, the hardware used, the desired spectral purity, the switching speed, the acceptable cost, and so on.

11.4.1 Approximation of Real Numbers

The situation being such, we shall repeat here, in short, all important theorems for approximation of real numbers [16].

1. Cantor series families
2. Lueroth series families
3. Cantor product expansions
4. Continued fraction approximations
5. Modulo-N approximations.

In the following sections, we shall verify the advantages and disadvantages of their application in frequency synthesis.

11.4.1.1 Cantor series approximations

Before the widespread use of digital circuits, Cantor series expansions provided the general mathematical model for single as well as for multiple output frequency synthesizers. Their principle is as follows.

For arbitrarily chosen integers $y_1, y_2, \ldots (\geq 2)$, we can approximate any real number ξ_o with the assistance of the algorithm

$$\begin{aligned} x_i &= \text{integer}\,(\xi_o) \quad (i = 0, 1, 2, \ldots) \\ \xi_{i+1} &= (\xi_i - x_i) \times y_{i+1} \end{aligned} \tag{11.10}$$

into the so-called *Cantor series*

$$\xi_o = x_o + \sum_{i=1}^{n} \frac{x_i}{y_1 y_2 \ldots y_i} \tag{11.11}$$

with the approximation error

$$R_n \approx \frac{1}{y_1 y_2 \ldots y_n} \tag{11.12}$$

In instances where all divisors y_i are the same, that is,

$$y_i = g \tag{11.13}$$

the Cantor series is changed into the so-called *g-dic* or *systematic fraction*

$$\xi_o = \sum_{i=0} \frac{x_i}{g^i} \tag{11.14}$$

Earlier "tunable" frequency synthesizers were based mainly on decimal fractions, that is, on systematic fractions with the base, $g = 10$. However, effective suppression of spurious signals generated by the mixing process required some modifications (discussed in detail in [1]).

On the other hand, DDFSs are based on systematic fractions with the base, $g = 2$. Since the useful normalized frequency is always $\xi_x < 0.5$, because of the Nyquist phenomenon, we have in this case the following mathematical model

$$\xi_x = \frac{a_1}{2} + \frac{a_2}{2^2} + \frac{a_3}{2^3} + \cdots \tag{11.15}$$

where a_1 is equal to 0 and the other a_i equal either to 1 or 0. The approximation error cannot exceed

$$R_n \leq \frac{1}{2^{n+1}} \tag{11.16}$$

We shall see later that unlike other approximation theorems, the convergence of the above series is rather slow.

11.4.1.2 Modified Engel series approximations

From all the Lueroth series providing a number of possibilities for approximation of real numbers [16], only the Engel series, in the particular form modified by the author, was used for purposes of frequency synthesis [17].

By assuming that the normalized output frequency $\xi_x = \gamma_o$ is smaller than one, the expansion of the modified Engel series is given by

$$\gamma_o = \sum_{i=1}^{n} \frac{(-1)^{i-1}}{p_1 p_2 \dots p_i} \tag{11.17}$$

where

$$\begin{aligned} \gamma_1 &= \frac{1}{1-\gamma_o} \qquad p_1 = \text{integer } (\gamma_1) \\ \gamma_2 &= \frac{\gamma_1}{\gamma_1 - p_1} \qquad p_2 = \text{integer } (\gamma_2) \end{aligned} \tag{11.18}$$

and so on till $p_n \dots$

Note that the partial divisors, p_i, are steadily increasing, which is the condition for convergence. At the same time the remainder is very small, that is,

$$R_n < \frac{1}{p_n^2} \tag{11.19}$$

Practical frequency synthesizers, using this mathematical model, have been built (see again [17]).

11.4.1.3 Continued fraction approximations

As already mentioned, this was the first theorem used in "frequency synthesis." The advantage of this theorem is that it provides the best approximations of real numbers, that is, with fractions formed by the smallest integers in the numerators and denominators for a given approximation error [18, 19]. The difficulty is the uniqueness of the *continued fractions*, which limits applications. Expansion of any real number ξ_0 is found by application of *Euclid's theorem*

$$\xi_0 = b_0 + \frac{1}{\xi_1} \quad \xi_1 \geq 1; \quad \xi_1 = b_1 + \frac{1}{\xi_2} \quad \xi_2 \geq 1 \tag{11.20}$$

By combining the above results, we get

$$\xi_0 = b_0 + \cfrac{a_1}{b_1 + \cfrac{a_2}{b_2 + \cfrac{a_3}{\vdots}}} \tag{11.21}$$

The nth-order approximation is often written as [16, 18]

$$\frac{A_n}{B_n} = [b_o, b_1, \ldots, b_n] \tag{11.22}$$

where A_n and B_n are relatively prime and are found by the recurrence formulae

$$A_{-2} = 0, \quad A_{-1} = 1, \quad A_n = A_{n-1}b_n + A_{n-2} \tag{11.23}$$

$$B_{-2} = 1, \quad B_{-1} = 0, \quad B_n = B_{n-1}b_n + B_{n-2} \tag{11.24}$$

and the remainder is

$$|R_n| = \left| \frac{(-1)^n}{B_n(B_n\xi_{n+1} + B_{n-1})} \right| \le \frac{1}{B_n B_{n+1}} < \frac{1}{B_n^2} \tag{11.25}$$

In some instances where one or more partial quotients are equal to 1, we can apply modifications for speeding up the expansion

$$\xi_0 = b_0 \pm \frac{1}{\xi_1} \quad \xi_1 \ge 2; \quad \xi_1 = b_1 \pm \frac{1}{\xi_2} \quad \xi_2 \ge 2 \tag{11.26}$$

and so on, all $b_0, b_1, \ldots$, being integers equal to or larger than 2. In general, continued fraction can be expressed as

$$\xi_0 = b_0 + \cfrac{a_1}{b_1 + \cfrac{a_2}{b_2 + \cfrac{a_3}{\vdots}}} \tag{11.27}$$

where all a_k are equal to 1 in the first case or some are equal to -1 in the *shortened continued fraction expansion*, which is important for appreciation of spurious signals in some types of DDFS.

Finally, the *series expansions* theorem, based on the shortened continued fractions in expansions [19, 20], provides a tool for appreciation of spurious modulation in DDFS.

$$\begin{aligned}\xi_0 &= A_0 + \frac{1}{B_1} - \frac{a_1a_2}{B_1B_2} + \cdots + (-1)^{n-1}\frac{a_1a_2\ldots a_n}{B_n - 1B_n} \\ &= A_0 + \sum_{r=1}^{n}(-1)^{r-1}\left(\frac{a_1a_2\ldots a_r}{B_{r-1}B_r}\right)\end{aligned} \tag{11.28}$$

Example 11.1

Design a generator of the sidereal time from the mean solar time. Their relation is given approximately by

$$\xi_x \approx \frac{366.24219430}{365.24219430} \tag{11.29}$$

Table 11.1 Continued fraction expansion of the normalized frequency of ξ_x in relation (11.29)

n	b_n	A_n	B_n	R_n
0	0	0	1	–
1	9	1	9	-10^{-3}
2	1	1	10	2.7×10^{-4}
3	35	36	359	4.8×10^{-6}
4	1	37	369	2.8×10^{-6}
5	1	73	728	-1.0×10^{-6}
6	1	110	1,097	3.2×10^{-7}
7	1	183	1,825	-1.8×10^{-7}
8	1	293	2,922	5.8×10^{-9}
9	19	5,750	57,343	-1.1×10^{-10}
10	2	11,793	117,608	3.6×10^{-11}

Source: © IEEE, 1999. Reproduced from V.F. Kroupa, ed. *Direct Digital Frequency Synthesizers*. New York: IEEE Press, 1999 by permission of IEEE, 2002.

Table 11.2 Shortened continued fraction expansion of the normalized frequency of ξ_x in relation (11.29)

n	a_n	b_n	A_n	B_n	R_n
0	1	0	0	1	–
1	−1	10	1	10	2.38×10^{-4}
2	−1	37	37	369	2.79×10^{-6}
3	−1	3	110	1,097	3.17×10^{-7}
4	−1	3	293	2,922	5.85×10^{-9}
5	1	20	5,750	57,343	-1.17×10^{-10}
6	−1	3	174,951	174,951	-1.73×10^{-13}

Let the input frequency be 100 kHz in the UTC (Coordinated Universal Time) timescale and the output frequency be in the 10-kHz range of the sidereal timescale. In this case, the normalized frequency is

$$\xi_x = 10.02737909350795/100$$

The expansion into a continued fraction is reprinted in Tab. 11.1. Note that its slow convergence is caused by the consecutive b_ns being equal to 1. The expansion until $1/\xi_4$ is

$$\xi_o = \cfrac{1}{9 + \cfrac{1}{1 + \cfrac{1}{35 + \cfrac{1}{\xi_4}}}} \tag{11.30}$$

In Tab. 11.2 the shortened expansion is recalled.

11.4.1.4 Modulo-N approximations

The principle of modulo-N approximations is depicted in Fig. 11.6. At each cycle or pulse of the clock, or reference frequency, a number, k, which represents a phase

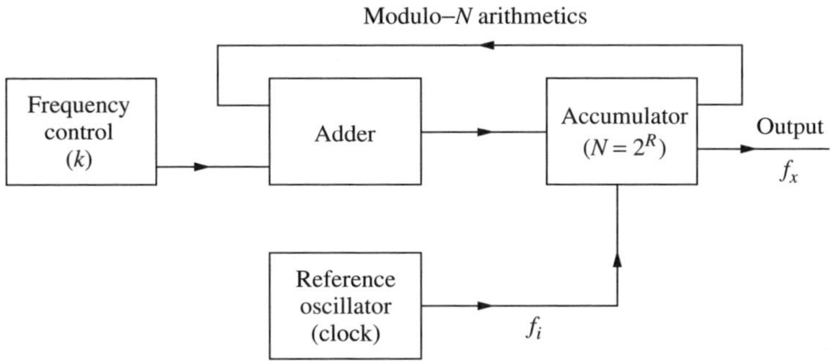

Figure 11.6 Principle block diagram of overflowing modulo-N device (Reproduced from V.F. Kroupa, ed. *Direct Digital Frequency Synthesizers*. New York: IEEE Press, 1999 by permission of IEEE, 2002).

increment, is added to that already stored in the accumulator. Every time the accumulator overflows, an output pulse is generated and a new period of accumulation is started, however, with a *modulo-N remainder* (smaller than k) in the accumulator memory. Finally, after N clock periods the remainder stored in the accumulator is zero; one fundamental period of the system ends and a new one starts.

Example 11.2

Let us synthesize a normalized frequency

$$\xi_x = 3/16$$

In this example, we use a 4-bit accumulator ($R = 4$) and put $k = 3$ into the frequency control circuit (see Fig. 11.6). The stored numbers in the accumulator will be successively

0,3,6,9,12,15,"18"	with the remainder 2 after the first overflow
2,5,8,11,14,"17"	with the remainder 1 after the second overflow
1,4,7,10,13,"16"	with the remainder 0 after the third overflow
0,3,6, and so on.	

Figure 11.7(a) illustrates the accumulation process in a graphic form and Fig. 11.7(b) the corresponding output pulse train. We easily deduce that generally

$$NT_i = \sum_{r=1}^{k} T_{xr} = kT_{xo} \tag{11.31}$$

where T_{xr} is the duration of actual output periods in Fig. 11.7(b) and T_{xo} is the idealized equidistant separation of the output pulses – see Fig. 11.7(c). As a consequence, the output pulse rate has the mean frequency

$$f_{xo} = \frac{k}{N} f_i \tag{11.32}$$

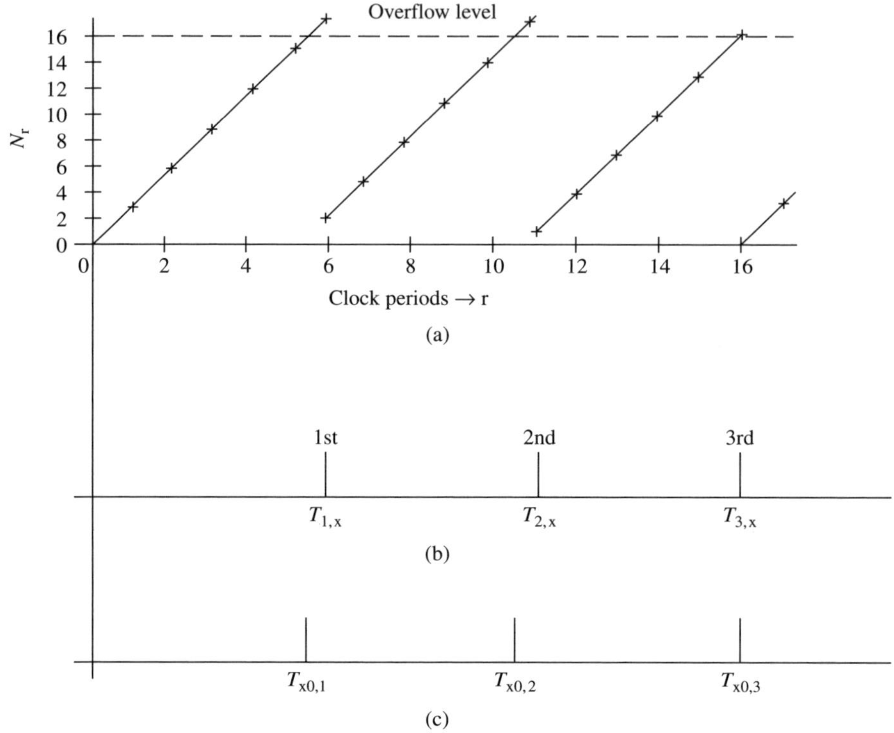

Figure 11.7 Generation of a pulse train with a normalized frequency $\xi_x = 3/16$ in a modulo-N system: (a) accumulator contents N_r; (b) modulated; and (c) idealized output pulse train (Reproduced from V.F. Kroupa, ed. *Direct Digital Frequency Synthesizers*. New York: IEEE Press, 1999 by permission of IEEE, 2002).

which can easily be changed by reprogramming the "word" k in the *frequency control* (steering memory) block (see Fig. 11.6). Further, in practical devices, N is generally equal to a large power of 2. In this way, we can arrive at very small frequency steps and at a simple hardware for the accumulators since the smallest frequency step is for $\Delta k = 1$

$$\Delta f_{xo} = \frac{1}{N} f_i \quad \text{where} \quad N = 2^R (R = 24, 32, 48, etc.) \tag{11.33}$$

Examination of the output pulse train in Fig. 11.7(b) reveals that individual periods T_{xr} are not equal. In the following text, we shall investigate this deficiency. By starting with zero in the accumulator, we arrive after m_1 clock pulses at its first overflow. As a consequence, we have the remainder

$$m_1 k - N < k \tag{11.34a}$$

After the second overflow, we get

$$m_2 k - 2N < k \tag{11.34b}$$

and so on (cf. Fig. 11.7(a) and Example 11.2) until

$$m_n k - nN = 0 \tag{11.34c}$$

Generalization reveals

$$mk - rN < k \quad (r < k) \tag{11.35}$$

from which

$$m - r\frac{N}{k} = \varepsilon \tag{11.36}$$

where

$$1 > \varepsilon \geq 0 \tag{11.37}$$

Since m is an integer, the condition $\varepsilon = 0$ is met when

$$\text{integer}\left(m - r\frac{N}{k}\right) = 0 \tag{11.38}$$

or

$$m = \text{integer}\left(r\frac{N}{k}\right) \equiv N \tag{11.39}$$

Now, after reverting to eq. (11.36) and multiplying it by the clock period T_i, we get with the assistance of (11.31) and (11.37) the timing error

$$mT_i - rT_{xo} \leq T_i \tag{11.40}$$

and finally the modulation function [9]

$$s(t_r) = T_i\left[r\frac{N}{k} - \text{integer}\left(r\frac{N}{k}\right)\right] \tag{11.41}$$

Evidently, the spurious phase time modulation does not exceed one clock period T_i. As a consequence, modulo-N arithmetic also provides a mathematical model for DDFS meeting the requirement for the *quasiperiodic* omission of pulses in the steering pulse train. Only the output pulses are on the places of the "swallowed" pulses [17, 21].

11.5 DIRECT DIGITAL FREQUENCY SYNTHESIZERS

DDFSs with a single output frequency (based on the continued fraction expansions or the modified Engel series) are required only in rare instances [e.g., 17, 20, 21]. However, devices with variable output frequencies are encountered more often and DDFSs, based on the IC technology, simplify the hardware considerably and at the same time provide very small tuning steps. These frequency synthesizers are based on

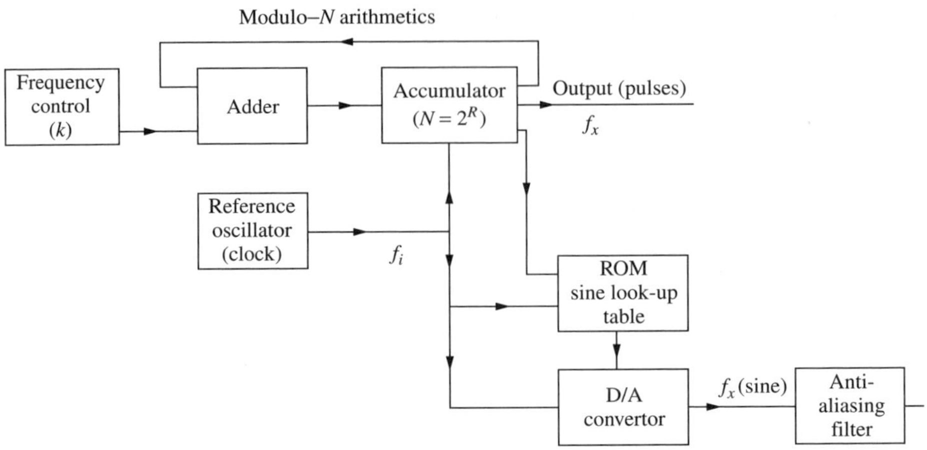

Figure 11.8 Block diagram of the Modulo-N DDFS (Reproduced from V.F. Kroupa, ed. *Direct Digital Frequency Synthesizers*. New York: IEEE Press, 1999 by permission of IEEE, 2002).

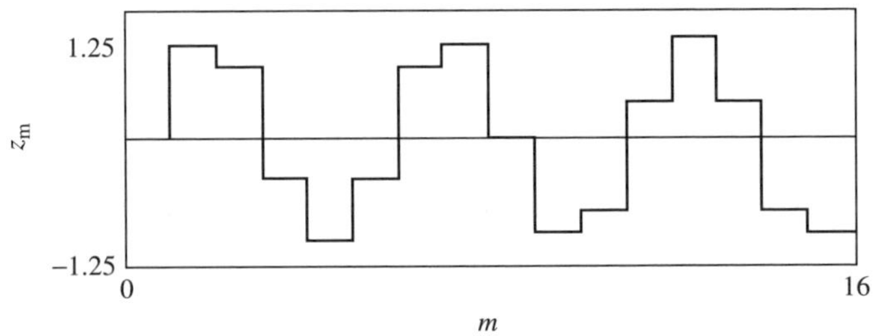

Figure 11.9 The output sine wave for the normalized frequency $\xi_x = 3/16$ (Reproduced from V.F. Kroupa, ed. *Direct Digital Frequency Synthesizers*. New York: IEEE Press, 1999 by permission of IEEE, 2002).

modulo-N arithmetics. At present they cover a range from "DC" to some 500 MHz; however, recently clock rates beyond 1 or 2 GHz have been reported.

The principle can be easily deduced from the basic block diagram in Fig. 11.8. Note that it differs from Fig. 11.6 only by means of the generation of the output sine wave, that is, by ROM sine lookup table, digital to analog converter, and antialiasing filter.

In instances where the normalized frequency ξ_x is a ratio of small integers, the output sine wave is a staircase wave (see Fig. 11.9). In instances where the denominator in (11.2) is a very large number, generally a large power of 2, the output frequency steps are very small since the accumulator capacity is very large (cf. (11.33)).

11.5.1 Spurious Signals in DDFSs

The major difficulty with all DDFSs are spurious signals accompanying the desired carrier. In some instances these spurious signals are quite large. Therefore, knowledge

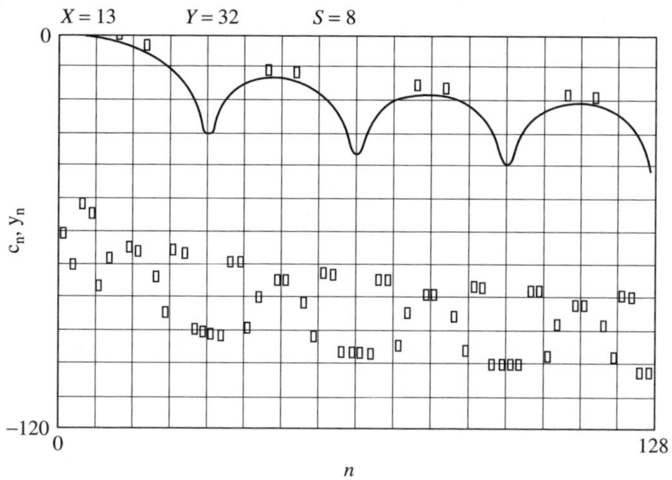

Figure 11.10 Example of the output spectrum of the *DDFS* without antialiasing filter (Reproduced from V.F. Kroupa, ed. *Direct Digital Frequency Synthesizers*. New York: IEEE Press, 1999 by permission of IEEE, 2002).

of their sources and expected amplitudes and frequencies may be useful both for the designers as well as for the users.

Since all DDFSs are based on the sampling process, the output spectrum extends to multiples of the sampling frequency (cf. relation (10.18) and Fig. 11.10). This problem is solved by introduction of the antialiasing filters with the cutoff frequency smaller than the Nyquist frequency, that is, smaller than one-half of the sampling frequency.

However, in some instances the application of a pass-band filter makes it possible to put the useful range to higher output frequencies.

Note that beyond the inherent undesired signals in DDFSs (to be computed in the closed form), the IC circuits are plagued with leaking signals, spurious mixer products, and other inevitable nonidealities.

11.5.1.1 Spurious signals in DDFSs with rectangular output waves

A simple frequency synthesizer based on Fig. 11.6, which is supplemented with some sort of Digital to Analog Convertor (DAC), has a rectangular output with the time modulation

$$\begin{aligned} s(t_r) &= T_c\left[r\frac{Y}{2X} - \text{integer}\left(r\frac{Y}{2X}\right)\right] \\ &= T_c\left[r\left(A_o + \frac{a_1}{B_1} - \frac{a_1 a_2}{B_1 B_2} + \cdots\right) - \text{integer}\left(r\frac{Y}{2X}\right)\right] \end{aligned} \tag{11.42}$$

where the series expansion of $Y/2X$ is based on the shortened continued fraction theorem. Investigation of the above series reveals superpositions of sawtooth waves with amplitudes, T_c, T_c/B_1, T_c/B_2, ..., T_c/B_{m-1} and periods A_1T_c, A_2T_c, ..., $A_nT_c = YT_c$. The largest and second largest spurious levels in decibel measure are

$$a_{1,m_1} \approx 20 \log \left(\frac{X}{m_1 Y} \right); \quad (m_1 = 1, 2, \ldots) \tag{11.43}$$

and

$$a_{2,m_2} \approx 20 \log \left(\frac{X}{m_2 B_1 Y} \right); \quad (m_2 = 1, 2, \ldots) \tag{11.44}$$

11.5.1.2 Spurious signals in DDFSs with sine wave lookup tables

The effective way for the reduction of spurious signals in "tunable" DDFSs is by changing the phase stored in the accumulator into a sine wave. The situation is depicted in Fig. 11.8. Nevertheless, even here we encounter several sources of spurious signals that are illustrated schematically in Fig. 11.11.

11.5.1.2.1 Truncation of sine values in sine lookup tables Generally, the sine values are only rational numbers with infinite decimal or binary places.

$$\sin x = \pm \left(\frac{a_1}{2} + \frac{a_2}{2^2} + \cdots + \frac{a_{s-1}}{2^{s-1}} + \frac{a_s}{2^s} + \frac{a_{s+1}}{2^{s+1}} + \cdots \in \infty. \right) \tag{11.45}$$

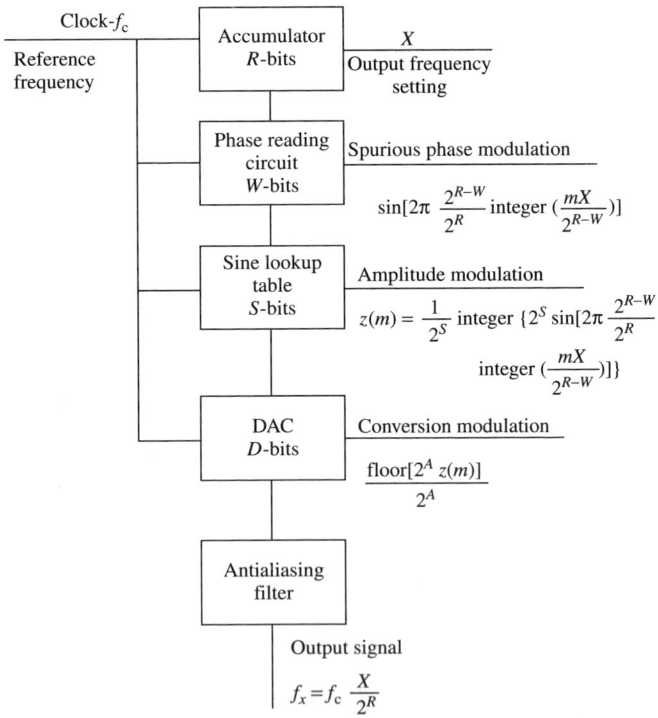

Figure 11.11 Sources of spurious signals in DDFSs (Reproduced from V.F. Kroupa, ed. *Direct Digital Frequency Synthesizers*. New York: IEEE Press, 1999 by permission of IEEE, 2002).

However, to keep the memory of sine lookup tables in reasonable dimensions, we must cut the stored words to, say, S bits only. The consequence is a spurious amplitude modulation of the output signal.

$$\varepsilon(m) = \sin(2\pi m\zeta_x) - 2^{-s} \text{ integer } [2^s \sin(2\pi m\zeta_x)] \tag{11.46}$$

By assuming that the error signal $\varepsilon(m)$ has a uniform distribution in the interval from $-1/2^S$ to $+1/2^S$, we get for its variance

$$\sigma_e^2 \approx \tfrac{1}{3}2^{-2s} \tag{11.47}$$

However, computation of the respective mean "Power Spectral Density (PSD)" (spurious levels) performed in [9] revealed

$$S(n) \approx \frac{1}{3Y}2^{-2s} \tag{11.48}$$

With the assistance of the computer simulations, we get an even lower bit value, that is,

$$S(n) \approx \frac{1}{6Y}2^{-2s} \tag{11.49}$$

or in decibel measure

$$S(n) = -7.8 - 6 \times S - 10\log(Y) \pm 10 \quad [\text{dB}] \tag{11.50}$$

See the spectrogram in Fig. 11.12. Note that when the frequency step is smaller than 1 Hz (cf. relation (11.33)), we face a real power spectral density

$$S(f) = -7.8 - 6 \times S - 10\log(f_{\text{clock}}) \pm 10 \quad [\text{dB}] \tag{11.51}$$

In some instances we encounter odd harmonics with amplitudes

$$a_h = 2 \times \frac{2^{-s}}{\pi} \times \frac{1}{h} \qquad (h = 3, 5, \ldots) \tag{11.52}$$

from which the level of the third harmonic of the output frequency in decibel measure becomes [9]

$$20 \times \log\left(\frac{a_3}{a_1}\right) \approx -6 \times S - 13.5 \quad [\text{dB}] \tag{11.53}$$

11.5.1.2.2 Truncation of the accumulator bits The high resolution of DDFSs requires a large modulo-N accumulator capacity (e.g., 2^R where $R = 24, 32, 48, \ldots$ bits, cf. eq. (11.33)). As a consequence, it is impossible to use all the stored phase information for generation of the output sine wave. Generally, we refer only to some W most

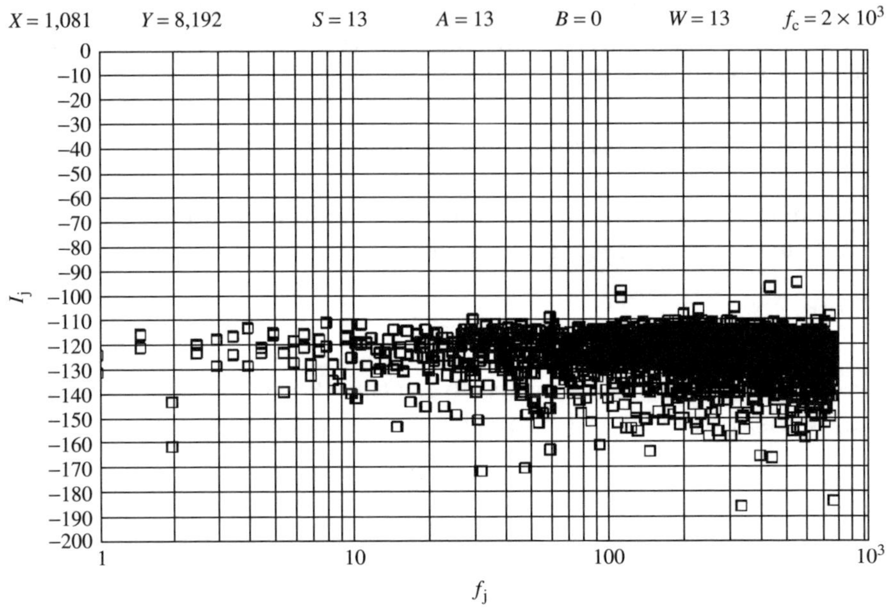

Figure 11.12 Spurious signals for $\xi_x = 1081/2^{13}$, for $S = A = 13$ with no accumulator bits neglected.

significant bits (MSB) from the R bit large accumulator, and neglect all the remaining B bits

$$W = R - B \tag{11.54}$$

The consequence is a phase modulation of the output sine wave. A good estimation of the level of the largest spurious signals due to the truncation of the accumulator bits provides the following relation in decibel measure [9]

$$20\log\left(\frac{a_{sp}}{a_{car}}\right) \approx -6W - 20\log(r); \quad (r = 1, 2, \ldots) \tag{11.55}$$

where the largest r_{max} cannot exceed 2^{B-1} (see the spectral diagram in Fig. 11.13).

In a properly designed DDFS, the level of the expected largest spurious signals generated by the truncation of the accumulator and sine lookup tables should be approximately the same. Having this condition in mind, we get after comparison of the relations (11.53) and (11.55)

$$-6W \approx -6S - 13.5 \tag{11.56}$$

from which

$$W \geq S + 2 \tag{11.57}$$

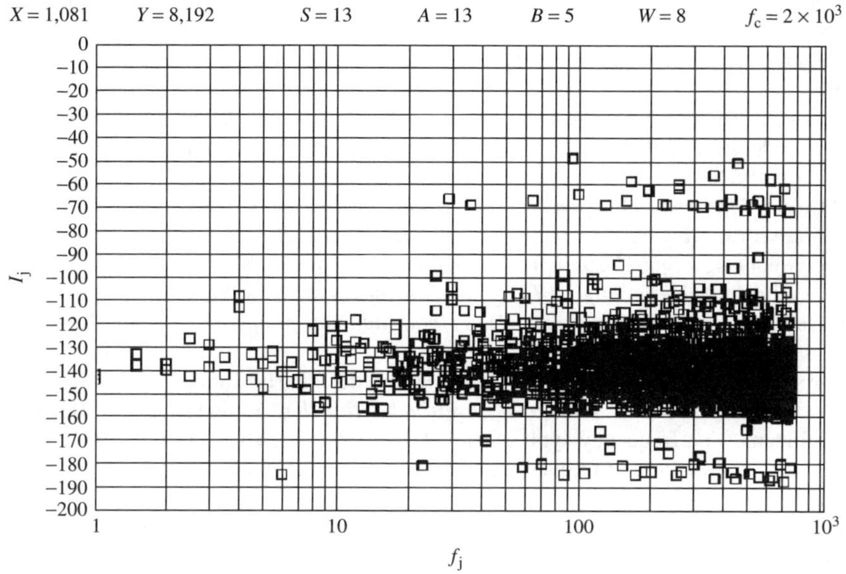

Figure 11.13 Spurious signals for $\xi_x = 1081/2^{13}$, for $A = S = 13$ with 5 least significant bits (LSB) in the accumulator neglected.

11.5.2 Phase and Background Noise in DDFSs

In all frequency generators, we encounter both amplitude and phase or frequency variations that cause the so-called frequency instability defined generally by the respective PSD, as discussed in Chapter 9. The problem is not fully understood in such complicated devices as DDFSs. We often encounter the statement that their output phase noise is that of the reference or clock oscillator reduced by the square of the normalized frequency. However, this may not be true since nearly all building blocks are noise generators (cf. Fig. 11.8). The difficulty about estimation of the output DDFS noise is its dependency on the normalized frequency set, on the number of accumulator, phase, sine, and DAC bits used. Therefore, very few reliable measurements have been published. The situation being such, estimation for the output DDFS output noise may be carried out with the assistance of (9.59), (11.50), or (11.51), and (11.55)

$$S_{\varphi,\mathrm{DDFS}}(f) \approx S_{\varphi,\mathrm{c}}(f) * \zeta_x^2 + S_{\varphi,\mathrm{div}} + S(n)_{\mathrm{DAC}} + S(n)_{\mathrm{W}} \tag{11.58}$$

In addition, a lot of spurious mixer products are often present.

REFERENCES

[1] V.F. Kroupa, *Frequency Synthesis: Theory, Design et Applications*. London: Ch. Griffin; New York: John Wiley, 1973.

[2] J. Gorski-Popiel, *Frequency Synthesis: Techniques and Applications*. New York: IEEE Press, 1975.
[3] V. Manassewitsch, *Frequency Synthesizers, Theory and Design*. New York: Wiley, 1976 and 1980.
[4] W.F. Egan, *Frequency Synthesis by Phase Lock*. 2nd ed. New York: Wiley 1981 and 2000.
[5] U.L. Rohde, *Digital PLL Frequency Synthesizers, Theory and Design*. Englewood Cliffs: Prentice Hall, 1983.
[6] R. Stirling, *Microwave Frequency Synthesizers*. Englewood Cliffs: Prentice Hall, 1987.
[7] J.A. Crawford, *Frequency Synthesizer Design Handbook*. Boston/London: Artech House, 1994.
[8] B.-G. Goldberg, *Digital Techniques in Frequency Synthesis*. New York: MacGraw-Hill, 1996.
[9] V.F. Kroupa, ed. *Direct Digital Frequency Synthesizers*. New York: IEEE Press, 1999.
[10] Ch. Huygens, *Opuscula posthuma. Descriptio automati planetarii*. Leiden: Lugduni Batavorum, 1703.
[11] A. Brocot, *Le Calcul des Rouages par Approximation*. Paris, 1862.
[12] F. Becher and A. Korner, *Pfauter Wechselraeder Tabellen*. Ludwigsburg: F. Becher, 1963.
[13] H.J. Finden, "The frequency synthesizer", *J. IEE*, Part III, **90**, 165–180, 1943.
[14] M. Boella, "Generatore di frequenze campione per misure di alta precisione", *Alta Frequenza*, **14**, 183–194, 1945.
[15] J. Kavan, Tabula omnibus a 2 usque ad 256000 numeris integris omnes divisores primos preabens. Stara Dala, 1934, Czechoslovakia. Reprinted during the World War II by MacMillan as: Factor Tables (Prime Factors of all numbers up to 256000).
[16] O. Perron, *Irrationalzahlen*. 3rd ed. Berlin: de Gruyter, 1947.
[17] V.F. Kroupa, "Approximating frequency synthesizers", *IEEE Trans. Inst. Meas.*, **IM-23**, 521–524, 1974 (Reprinted in [9] pp 23–26).
[18] O. Perron, *Die Lehre von den Kettenbruechen*, Teubner, Leibzig 1912, Stuttgart, 1954.
[19] H.S. Wall, *Analytic Theory of Continued Fractions*, D. Van Nostrand Company, Inc. New Jersey, 1948.
[20] V.F. Kroupa, Spectra of pulse rate frequency synthesizers, *Proc. IEEE*, **December**, 1680–1682, 1979.
[21] G. Becker, Quasiperiodic frequency synthesis, *Proceedings of the 26th Annual Frequency Control Symposium* **June** 279–291, 1972.

12
PLLs and Digital Frequency Synthesizers

Phase locked loops (PLLs) make it possible to generate accurate frequencies in the high megahertz and low gigahertz ranges (cf. Fig. 11.1). In principle, they are based on PLLs with a divider in the feedback path discussed earlier (Fig. 2.12). In practice, we encounter integer-N, fractional-N, and higher order fractional-N (Σ-Δ modulation divider systems) setups.

Many of these PLLs direct digital frequency synthesizers (DDFS) systems find use in modern communications such as cellular Global System for Mobile Communications (GSM), in data transfer systems, and many other applications. Generally, good spectral purity is desired and in many instances fast switching from one channel to another is required. Some of the critical parameters are leakage of the reference signal and its harmonics, the duration of the switching time from one channel to the other, and the PLL background noise.

12.1 INTEGER-N PLL DIGITAL FREQUENCY SYNTHESIZERS

These frequency synthesizers are based on the divide-by-N PLL configuration discussed in Chapter 2. In connection with Fig. 2.12, the divider-N reduces the VCO frequency f_o, to be equal to the reference frequency f_r, evidently the division factor N is

$$N = \frac{f_o}{f_r} \tag{12.1}$$

Since the smallest step is

$$\Delta N = 1 \tag{12.2}$$

the channel spacing is just equal to the reference frequency f_r. However, there are some other difficulties: small natural frequency ω_n due to the generally large division

Phase Lock Loops and Frequency Synthesis V.F. Kroupa
© 2003 John Wiley & Sons, Ltd ISBN: 0-470-84866-9

factors N (see relation (2.49)) and consequently, long switching times on one hand and large phase noise (mainly of the PD) on the other.

Digital dividers make possible construction of the PLL – programmable frequency synthesizer in the IC form. However, to reduce power consumption, they are combined with the fast fixed prescalers followed by a dual modulus arrangement. A typical arrangement is shown in Fig. 12.1.

There are services in which longer switching times may be tolerated [1]. In such a case, only the spectral properties are important and deserve investigation.

12.1.1 Spurious Signals in Integer-N PLL Digital Frequency Synthesizers

In integer-N PLL-DDFSs, a major source of spurious signals is the PD with a leaking reference frequency and its harmonics. Another problem has been discussed in connection with the properties of phase frequency detectors (PFD), in Section 8.4.4.3, namely, the problem of the dead zone. If not dealt with, the up–down switching around the zero phase starts to be random and source of the undesired noise [2]. One suggested remedy was a shunting resistor $R_{s,leak}$ (cf. Fig. 2.11). The result is that the charge on the loop capacitor C, is not constant. The capacitor is discharged practically during the whole period with a very small current. However, its lost charge

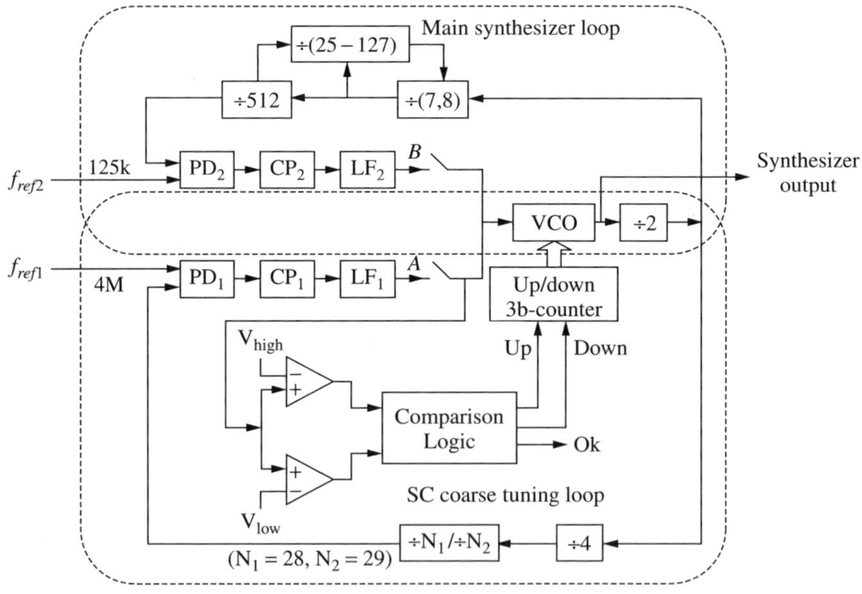

Figure 12.1 Integer-N programmable digital frequency synthesizer (Digital-PLL) in the IC form with the fast fixed prescaler followed by a dual modulus arrangement. (Reproduced from T.-H. Linn and J.T. Kaiser, "A900-MHz 2.5 mA CMOS Frequency synthesizer with an automatic SC Tuning Loop," *IEEE J. Solid-State Circuits*, **36**, March 2001, pp. 424–430, 2001 by permission IEEE [1].)

is compensated in a very short time τ, by the switching current I_P. The situation is described by the following relation for the change of the charge on the capacitor C:

$$\Delta q = I_P \tau = C E_C (1 - e^{-T_r/R_s C}) \tag{12.3}$$

where E_C is the mean value of the voltage on the capacitor. An unimportant simplification reveals change of the time-dependent voltage $\Delta v_2(t)$ with the period T_R

$$\Delta v_2(t) = \frac{\Delta q}{C} = E_C \frac{t}{R_s C} \tag{12.4}$$

Evidently we face a sawtooth wave during the reference period T_r. Its expansion into individual harmonics reveals

$$\Delta v_2(t) = E_C \frac{T_r}{R_s C} \frac{1}{\pi} \left[\frac{\pi}{2} + \Sigma \frac{1}{n} \sin(n\omega_r t) \right] \tag{12.5}$$

Since

$$\Delta \omega(t) = \Delta v_2(t) K_o \tag{12.6}$$

we get after integration of (12.5) the phase noise components

$$\varphi(t) = E_C \frac{T_r}{R_s C} \sum_1 \frac{K_o}{\pi n^2 \omega_r} \cos(n\omega_r t) \tag{12.7}$$

The VCO output signal is phase modulated. However, since the phase modulation is small, we can simplify the modulated output signal

$$V_{out} \sin[\omega_o t - \varphi(t)] = V_{out}[\sin \omega_o t - \varphi(t) \cos \omega_o t] \tag{12.8}$$

with phase noise amplitudes (cf. (12.7))

$$\varphi_n = \frac{2 E_C K_o}{R_s C} \frac{1}{(n\omega_r)^2} \tag{12.9}$$

and powers

$$S_\varphi \left(f_n = \frac{n}{T_r} \right) = -20 \log \varphi_n \tag{12.10}$$

The DC sawtooth wave component is compensated by the current pulse recharging the capacitor C with the switching time τ:

$$\tau = \frac{E_C}{R I_P} T_r \tag{12.11}$$

Evidently we face a short current pulse at the PD output with amplitude I_P and a compensating sawtooth wave with the amplitude

$$I_R = 2I_P \frac{\tau}{T_R} = 2\frac{E_C}{R} \tag{12.12}$$

The corresponding change of the VCO frequency is

$$\Delta\omega_o = \Delta v_2 K_o = E_C K_o \frac{T_r}{RC} \tag{12.13}$$

Example 12.1

Let us evaluate the reference spurious signals in the following PLL arrangement

Output frequency	$f_o = 3.6\,[\text{GHz}]$
Input frequency	$f_i = 26\,[\text{MHz}]$
Reference frequency	$f_r = 0.4\,[\text{MHz}]$
Divider	$N = 9.10^3$
Loop natural frequency	$f_n = 30\,[\text{kHz}]$
Damping factor (2nd order loop)	$\zeta = 0.7$
PD current	$I_P = 8.10^{-4}\,[\text{A}]$
Main capacitor	$C = 1.5.10^{-9}\,[\text{F}]$
Main resistor	$R_2 = 15\,[\text{k}\Omega]$
Leaking resistor	$R_s = 10\,[\text{M}\Omega]$
Oscillator gain	$K_o = 200\,[\text{MHz/V}]$

From (12.10) we get for the first spurious component (f_{spurious}) due to the reference frequency,

$$S_\varphi(0.4\ \text{MHz}) = \left(\frac{2\cdot 1\cdot 2\cdot 10^8}{10^7\cdot 10^{-9}\cdot(2\pi\cdot 10^5)^2}\right)^2 = 1.6\cdot 10^{-5} \tag{12.14}$$

Note, that a compensating signal of the same amplitude but of the opposite sign is present, which enlarges the above value by 6 dB. Consequently in dB measure we have

$$S_\varphi(fr_1) = 10\log(1.6\cdot 10^{-5}) + 6 \approx -40\ [\text{dB}] \tag{12.15}$$

and for higher harmonics of the leaking reference signal

$$S_\varphi(fr_n) \approx -40 - 6n\ (n \geq 1)\ [\text{dB}] \tag{12.16}$$

By assuming a third-order PLL, we can investigate the properties of the leaking spurious signal in the plot illustrated in Fig. 12.2. Owing to the PLL transfer function $H(jx)$ the first reference spurious signal can be reduced to the −80 dB/Hz level (see Fig. 12.2(a)). Note the simulation of the phase noise measurement with a spectral analyzer that can even conceal the spurious signal as demonstrated in Fig. 12.2(b) (cf. Fig. 12 in [3])

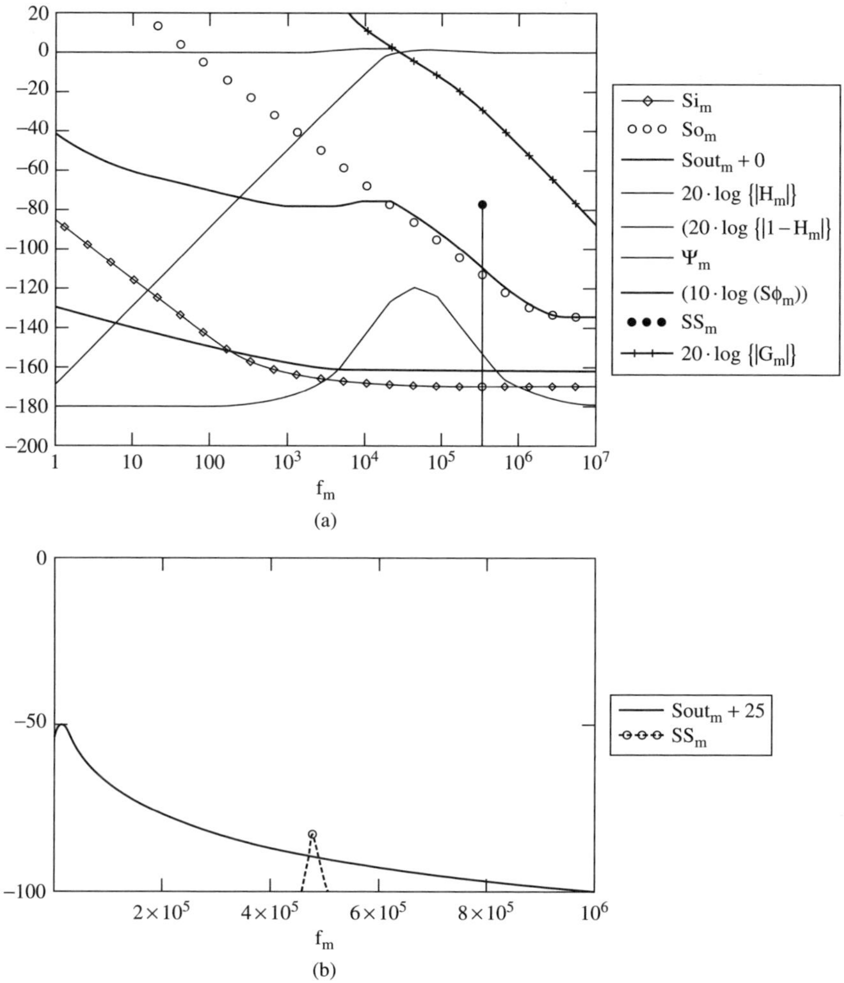

Figure 12.2 (a) Investigation of the leaking spurious signal in a third-order PLL: the characteristics from the top indicate: PLL transfer loop characteristics $|H_{\text{eff}}(f)|^2$, $|1 - H_{\text{eff}}(f)|^2$ and $|G_{\text{eff}}(f)|^2$, phase noise power spectral density (PSD) of the open loop VCO (○), PSD of the output signal (–), PSD of the reference signal (Si_m) (♦), PSD of the loop noise (–), the phase margin (Ψ_m), and finally the level of the spurious signal (SS_m) (●); and (b) simulation of the output phase noise measurement of the same PLL with the assistance of an analyzer with the RBW 300 Hz.

12.1.2 Background Noise in Integer-N PLL Digital Frequency Synthesizers

We shall now evaluate the thermal noise voltages generated in the filter resistors R and $R_{\text{S,leak}}$ (Fig. 12.3)

$$e_{\text{n},2}^2 \approx 4kTR_2 \quad \text{and} \quad e_{\text{n},R_\text{s}}^2 \approx 4kTR_\text{s} \tag{12.17}$$

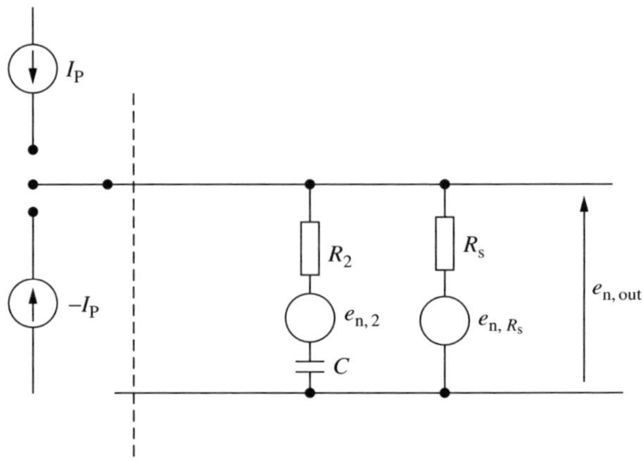

Figure 12.3 Noise generators in the loop of an integer-N PLL digital frequency synthesizer.

The output noise voltage is

$$\begin{aligned} e_{n,out} &\approx \frac{e_{n,2} + e_{n,R_s}}{R_2 + R_s + \dfrac{1}{sC}} \cdot \left(R_2 + \frac{1}{sC}\right) \\ &\approx (e_{n,2} + e_{n,R_s}) \cdot \frac{1 + sT_2}{1 + sT_s} \end{aligned} \tag{12.18}$$

For very low frequencies, the output noise voltage is

$$e_{n,out} \approx (e_{n,2} + e_{n,R_s}) \approx \sqrt{4kT(R_2 + R_s)} \tag{12.19}$$

whereas for large frequencies it is

$$e_{n,out} \approx (e_{n,2} + e_{n,R_s})\frac{T_2}{T_s} \tag{12.20}$$

and the PSD of the output phase noise is approximated

$$S_\varphi(f) \approx 4kT(R_2 + R_s)\left(\left|\frac{1 + sT_2}{1 + sT_s}\right|\right)^2 \tag{12.21}$$

12.2 FRACTIONAL-N PLL DIGITAL FREQUENCY SYNTHESIZERS

The fractional-N frequency synthesizer is similar to the divide-by-N PLL. However, with the assistance of an auxiliary divider, the output frequency of the VCO is not restricted to integral multiples of the reference signal only but can be also locked to the fractional multiples resulting in a substantial reduction of the frequency in the useful bandwidth of the PLL without reduction of its pass band (i.e. without reduction of the loop natural frequency f_n). The consequence is that these frequency synthesizers have both high frequency resolution and shorter settling time, two essential requirements for modern applications, for example, in mobile radio sets and so on. The major difficulties are spurious signals generated by the *fractional-N divider* systems.

These PLL systems began to be popular in connection with the HP 3325 Synthesizer/Function generator [4–6].

The idea and operation of the PLL fractional-N frequency synthesizer will be explained with the assistance of Fig. 12.4. The principle is in *quasiperiodic* removing (swallowing) of pulses from the VCO pulse train. In practice, pulse suppression is provided by *two-mode* dividers, the operation of which has been discussed in Section 8.5.2.3 (cf. Fig. 12.1).

For a better understanding of the problem, we assume that the auxiliary DDFS is a simple divide-by-F, which divides the reference frequency f_r. Considering the PLL without pulse swallowing, the period T_{rN}, at the divider-N output is equal to the N period, T_o, of the VCO frequency

$$T_{rN} = T_o N \tag{12.22}$$

However, after each overflowing of the divider-F one pulse is missing in the VCO train T_o and the corresponding period of T_{rN} is prolonged by one T_o. Since the periods of the signals applied to the inputs of PD, T_r and T_{rN}, are very close to each other, we can write for the repetition period due to the auxiliary branch

$$T_{rN} F \approx T_r F \approx T_o NF + T_o \tag{12.23}$$

Finally, in the locked state of the respective PLL, both periods T_r and T_{rN} are equal. Consequently, the fundamental repetition period of the signal supplied to the feedback

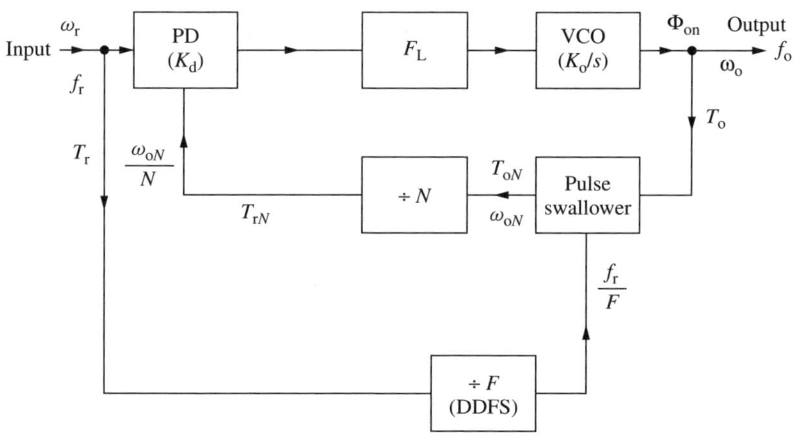

Figure 12.4 Block diagram of the fractional-N frequency synthesizer. (Reproduced from V.F. Kroupa, *Direct Digital Frequency Synthesizers*, by permission IEEE Press, 1999, John Wiley & Sons, Inc., 2002 [7]).

port of the phase detector is exactly

$$T_o NF + T_o = T_r F \tag{12.24}$$

and the effective division ratio is easily computed as

$$N_{\text{eff}} = \frac{T_r}{T_o} = N + \frac{1}{F} \tag{12.25}$$

For a later estimation of the spurious signals, introduced by the fractional-N behavior, we shall investigate a solution provided from the frequency domain point of view [7, 8].

12.2.1 Spurious Signals in Fractional-N PLL Digital Frequency Synthesizers

The above discussed fractional-N division introduces spurious frequency and consequently also a phase modulation. Even in instances of the quasiperiodic omission of pulses in a pulse rate [9] we encounter large spurious signals.

> ***Example 12.2***
>
> Let us consider a simple case with the periodic suppression of a single pulse, that is, with the quasiperiodic omission of each eighth pulse ($q_1 = 8$) as illustrated in Fig. 12.5. One easily arrives at the conclusion that the process generates a sawtooth wave with a maximum amplitude of nearly 2π.

12.2.1.1 Quasiperiodic omission of pulses

The pulse subtractor in Fig. 12.6. quasiperiodically removes pulses in the T_o train at each overflowing of the divider-F as shown in Fig. 12.5. The time-dependent frequency can be expressed with the assistance of δ-functions as

$$\omega_{oN}(t) = \omega_o - 2\pi\delta(t - FT_r) \tag{12.26}$$

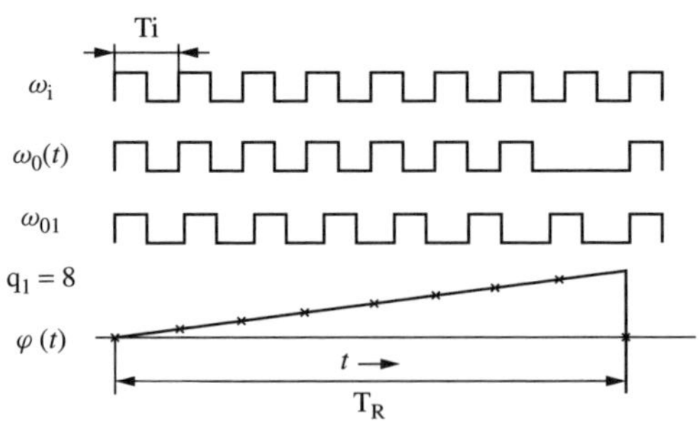

Figure 12.5 Spurious phase modulation generated by quasiperiodic omission of each eighth pulse ($q_1 = 8$). (Reproduced from V.F. Kroupa, *Direct Digital Frequency Synthesizers*, by permission IEEE Press, 1999, John Wiley & Sons, Inc., 2002 [7]).

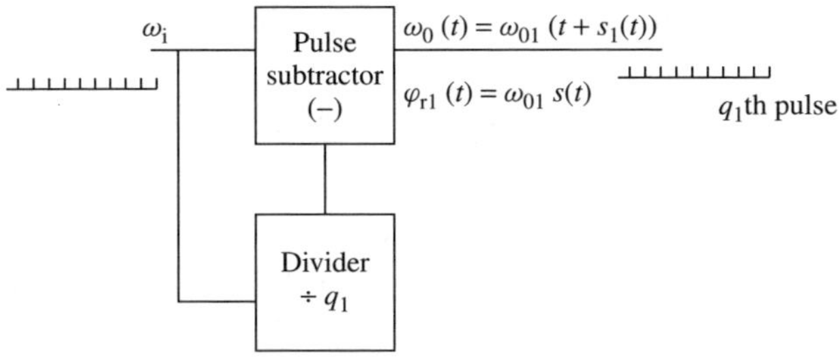

Figure 12.6 Idealized arrangement for "swallowing" of each q_1th pulse from a periodic pulse train. (Reproduced from V.F. Kroupa, *Direct Digital Frequency Synthesizers*, by permission IEEE Press, 1999, John Wiley & Sons, Inc., 2002 [7]).

Now we can compute individual spurious components due to the second term on the rhs of the above relation. On the timescale of the T_o we may consider the s_k burst as δ-functions placed at kT_r distances:

$$s(t) = \sum_{k=1}^{Y} s_k \delta(t - kT_r) \tag{12.27}$$

For the evaluation of complex amplitudes of the individual spurious components, we apply the Fourier series expansion

$$\Delta\hat{\omega}_n \approx \frac{2\pi}{YT_r} \int_0^{YT_r} s_k \delta(t - kT_r)\, e^{-j\frac{2\pi}{YT_r}nt}\, dt \tag{12.28}$$

After integrating the above relation we get

$$\Delta\omega_n = |\Delta\hat{\omega}_n| = \frac{2\pi}{YT_r} \left| \sum_{k=1}^{Y} s_k\, e^{-j2\pi\frac{nk}{Y}} \right| \tag{12.29}$$

The frequency $\omega_{oN}(t)$ at the output of the pulse subtractor is given by

$$\omega_{oN}(t) = \omega_{oN,\text{mean}} + \sum_{n=1}^{\infty} 2\Delta\omega_n \cos\left(2\pi \frac{n}{YT_r} t + \Psi_n\right) \tag{12.30}$$

with the fundamental modulation frequency

$$\omega_{\text{mod}} = \frac{2\pi}{YT_r} \tag{12.31}$$

and the steady state component given by the zero term ($n = 0$) in (12.28), that is,

$$\Delta\omega_o = \frac{2\pi}{YT_r} \sum_{k=1}^{k=Y} s_k = 2\pi \frac{X}{YT_r} \tag{12.32}$$

With this result and relation (12.25) we can compute the effective division factor:

$$N_{\text{eff}} = \frac{T_{\text{r}}}{T_{\text{o}}} = N + \frac{X}{Y} \tag{12.33}$$

Further, after integrating the time-dependent term in (12.30) we get for the amplitudes of the spurious phase modulation components (of the sinusoidal waves)

$$\phi_n \approx \Delta\omega_n \frac{YT_{\text{r}}}{2\pi n} = \frac{1}{n} \sum_{k=1}^{Y} s2_k \, \mathrm{e}^{-\mathrm{j}\pi \frac{nk}{Y}} \tag{12.34}$$

and their respective powers

$$S(n) = \tfrac{1}{2} |\phi_n|^2 \tag{12.35}$$

Note that the spurious phase exhibits discrete components with spurious frequencies

$$f_{\text{spurious}} = n \frac{f_{\text{r}}}{Y} \tag{12.36}$$

12.2.1.2 Phase modulation function in fractional-N PLL digital frequency synthesizers

Another approach for evaluation of the modulation function generated by the process of quasiperiodic omission of pulses can be explained with the assistance of the theory of spurious signals generated in Direct Digital Frequency Synthesizers (DDFS) (cf. [7, 8]).

By reverting to Fig. 12.4, we may consider the *pulse swallower* as a mixer supplied by the output frequency f_{o}, on one hand and by the unfiltered output frequency of the DDFS which is subjected to the time (or phase modulation). The time modulation function in the fractional-N frequency synthesizer is

$$s_{\text{DDFS}}(t_k) = T_{\text{r}} \left[k\frac{X}{Y} - \text{integer}\left(k\frac{X}{Y} \right) \right] \tag{12.37}$$

with the fundamental period YT_{r}.

The shape and periods of the spurious signals were investigated briefly in Chapter 11. For an in-depth analysis, see [7]. Here, we highlight the salient features. With the assistance of the modified continued fraction expansion (relations (11.26) to (11.28) and relations (11.41) and (11.42)) we find out that $s_{\text{DDFS}}(t_k)$ is actually a superposition of sawtooth waves (cf. Fig. 12.7).

By performing the continued fractional expansion of the effective divisor N_{eff} (12.33) we get for the first-order approximation

$$N_{\text{eff}} \approx N + \frac{1}{B_1} \tag{12.38}$$

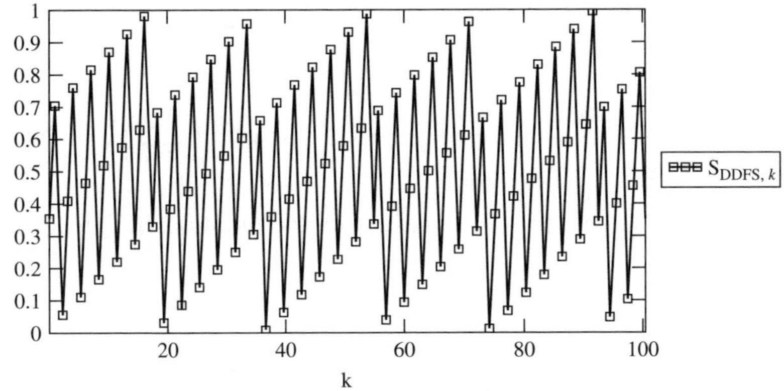

Figure 12.7 Plot of the spurious modulation in fractional-N PLL with $N = 32$ and DDFS set to $X/Y = 45/128$. (Reproduced from V.F. Kroupa, *Direct Digital Frequency Synthesizers*, by permission IEEE Press, 1999, John Wiley & Sons, Inc., 2002 [7]).

Note that the above relation is exactly relation (12.25), only F is replaced by B_1. Evidently we can express the period of the spurious modulation component with the largest amplitude approximately as follows:

$$T_{sp,1} \approx T_o B_1 N \approx B_1 T_r > T_r \tag{12.39}$$

Since B_1 can be anywhere in the range

$$2 = B_1 \le Y \tag{12.40}$$

we find out that $T_{sp,1}$ may exceed the reference period T_r, in some cases even substantially.

However, in accordance with the rule of thumb, the natural loop frequency f_n of the PLL is generally in the range

$$0.01 \times f_r < f_n < 0.1 \times f_r \tag{12.41}$$

and we can conclude that the largest spurious modulation components, introduced by the fractional-N division, may be well in the PLL pass band and may appear, multiplied by N_{eff}, at the VCO output.

The respective spurious phase modulation, with the repetition period T_r at the PD input is

$$\begin{aligned} \varphi_r(t_k) &= \omega_{ro} s_r(t_k) \\ &\approx \frac{2\pi}{N_{eff}} \left[k\frac{X}{Y} - \text{integer}\left(k\frac{X}{Y} \right) \right] \end{aligned} \tag{12.42}$$

Owing to the multiplication properties of PLLs with a divider in the feedback path the peak of the spurious phase in the VCO output might reach nearly 2π.

We shall proceed with a more exact investigation. Spurious phase at the feedback input of the PD is

$$\varphi_{\mathrm{r}} \approx \frac{2\pi}{N}\left[k\left(\frac{1}{B_1} \pm \frac{1}{B_1 B_2} \pm\right) - \text{integer}\left(\frac{X}{Y}\right)\right] \tag{12.43}$$

The amplitude of the largest sawtooth wave with the duration $B_1 T_{\mathrm{r}}$ is approximately

$$\varphi_{N,\mathrm{r}} \approx \frac{2\pi}{N} \tag{12.44}$$

The phase error of the output of the PD is

$$\varphi_{\mathrm{e}} \approx \frac{2\pi}{N} \times K_{\mathrm{di}} \tag{12.45}$$

and the corresponding VCO phase disturbance will be

$$\varphi_{\mathrm{VCO}}(s) \approx \varphi_{\mathrm{e}} Z(s) \frac{K_{\mathrm{o}}}{2\pi / B_1 T_{\mathrm{r}}} \tag{12.46a}$$

$$= \frac{2\pi}{N} K_{\mathrm{di}} K_{\mathrm{o}} \frac{B_1 T_{\mathrm{r}}}{2\pi} Z(s) \tag{12.46b}$$

Now we shall introduce the loop parameters ω_n and ζ (cf. (2.39) to (2.43)):

$$Z(s) = \frac{1 + sRC}{sC} = \frac{1 + sT_2}{sC} \tag{12.47}$$

$$K_{\mathrm{di}} K_{\mathrm{o}} = \omega_n^2 C N \quad \zeta = \omega_n \frac{T_2}{2} \tag{12.48}$$

Consequently,

$$\varphi_{\mathrm{VCO}}\left(\frac{\omega_{\mathrm{r}}}{B_1}\right) \approx 2\pi \omega_n^2 \left(\frac{B_1}{\omega_{\mathrm{r}}}\right) CN \frac{1 + 2\zeta\left(\dfrac{\omega_{\mathrm{r}}}{B_1 \omega_n}\right)}{\dfrac{\omega_{\mathrm{r}}}{B_1} C} \tag{12.49a}$$

$$\approx 2\pi \left(\frac{\omega_n}{\omega_{\mathrm{r}}} B_1\right)^2 \zeta \frac{\omega_{\mathrm{r}}}{B_1 \omega_n}$$

$$\approx 2\pi\zeta \left(\frac{B_1 \omega_n}{\omega_{\mathrm{r}}}\right) \tag{12.49b}$$

We shall illustrate the problem with the following example.

Example 12.3

For reasonable loop values such as:

$$\zeta \approx 1.5$$
$$B_1 \geq 2$$
$$\omega_n/\omega_r \approx 0.1$$

we have for the power of the respective output spurious signals of the largest saw-tooth wave

$$\varphi\left(\frac{\omega_{\mathrm{r}}}{B_1}n\right) \approx \left(2\zeta\frac{B_1\omega_n}{n\omega_{\mathrm{r}}}\right)^2 \leq \left(\frac{0.6}{n}\right)^2 \tag{12.50}$$

which is an impractical, large value.

The above example reveals that Fractional-N PLL-DDFS frequency synthesizers would be spoiled by unwanted spurious signals.

12.2.2 Reduction of Spurious Signals in Fractional-N PLL-DDFSs

Investigation of the information stored in the accumulator of the DDFS, owing to its modulo-Y operation, reveals [7]:

$$s_{\mathrm{DDFS}}(t_k) = T_{\mathrm{r}}\left[k\frac{X}{Y} - \text{integer}\left(k\frac{X}{Y}\right)\right] \tag{12.51}$$

which has the same form as the phase time modulation fed to the feedback input of the PD (cf. relation (12.37)). Consequently, the information stored in the DDFS accumulator can be used with the assistance of a digital to analog convertor (DAC) for compensation of the spurious modulation introduced by the fractional-N operation (see Fig. 12.8).

Investigation of the output phase noise with the assistance of Fig. 12.8 and simplified eq. (9.93) in Chapter 9 leads to

$$\Phi_{\mathrm{o,n}} = \left[\Phi_{\mathrm{r}} - \frac{\Phi_{\mathrm{o}N}}{N} + \frac{V_{\mathrm{PD}}}{K_{\mathrm{d}}} + \Phi_{\mathrm{L}}\right] N_{\mathrm{eff}} \frac{F_{\mathrm{L}}(s)K/(N_{\mathrm{eff}}s)}{1 + F_{\mathrm{L}}(s)K/(N_{\mathrm{eff}}s)} \tag{12.52}$$

where Φ_{L} is the noise generated in the PLL filter. For the spurious signals in the pass band, the above relation simplifies to

$$\Phi_{\mathrm{o,n}} \approx \left[\Phi_{\mathrm{r}} - \frac{\Phi_{\mathrm{o},N}}{N} + \frac{V_{\mathrm{PD}}}{K_{\mathrm{d}}} + \Phi_{\mathrm{L}}\right] N_{\mathrm{eff}} \tag{12.53}$$

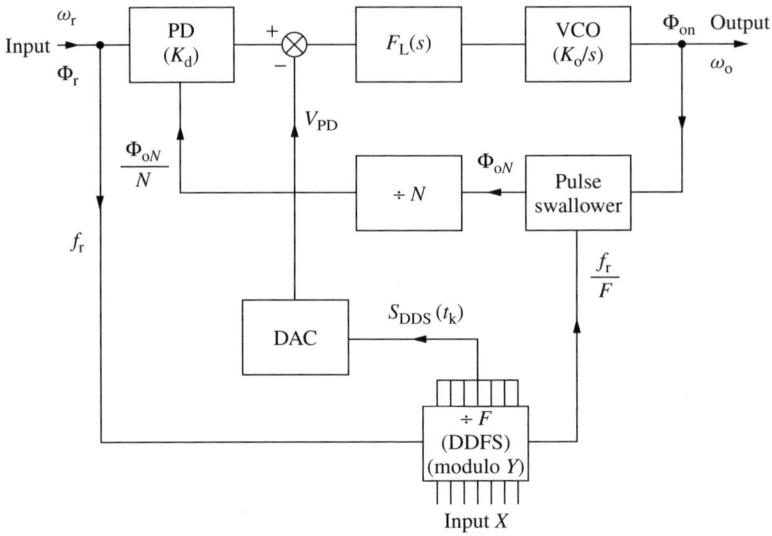

Figure 12.8 Means for compensation of the spurious phase noise in fractional-N PLLs. (Reproduced from V.F. Kroupa, *Direct Digital Frequency Synthesizers*, by permission IEEE Press, 1999, John Wiley & Sons, Inc., 2002 [7]).

Clearly, by putting the output voltage of the DDFS-DAC equal to

$$V_{\mathrm{DAC}} = V_{\mathrm{PD}} = K_{\mathrm{d}} \frac{\Phi_{\mathrm{o}N}}{N} \tag{12.54}$$

we can compensate for the spurious phase modulation, with the assistance of a D-bit DAC, in the loop to the level [7, p. 251].

$$N/S \approx -1.76 - 6.02D\,[\mathrm{dB}] \tag{12.55}$$

Example 12.4

By assuming the natural loop frequency $f_{\mathrm{n}} \geq 10\,\mathrm{kHz}$ (which is a reasonable assumption in fractional-N PLL frequency synthesizers) we may expect the background PSD to be approximately (cf. Fig. 9.18)

$$S_{\phi\mathrm{L}} \approx 10^{-14}$$

Even after assuming a perfect compensation, assisted, for example, with a 16-bit DAC, we cannot achieve such a low noise level since DACs are not perfect devices (nonlinearities, glitches, etc.).

Consequently, this type of compensation will reduce the corresponding spurious level by some 30 to 40 dB only. Therefore, Hewlett-Packard introduced a specially designed analog phase interpolating circuit (API) instead of a DAC [6, p. 125] by using the top five decimal digits from the DDFS for switching compensation currents into the loop. All spurious signals were reduced below the level of −90 dB in the output signal, actually below the loop noise level.

12.3 SIGMA-DELTA FRACTIONAL-N PLL DIGITAL FREQUENCY SYNTHESIZERS

Another effort for reduction of spurious signals in fractional-N frequency synthesizers is based on the application of properties of the sigma-delta quantizers (SDQ) shown in Fig. 12.9. The origin of the arrangement may be found in the Marconi European patent [10].

Theory of the noise or more exactly of the spurious signals in the analog to digital quantizer (ADQ) was discussed by Y. Matsuya [11] and that of the Σ-Δ quantizer chain (SDQ) was published by Miller and Conley [12] and later on by Riley *et al.* [13].

For the spurious "noise," authors of [12] found the following relation indicating the phase noise PSD $\mathscr{L}(f)$ in Σ-Δ fractional-N frequency synthesizers

$$\begin{aligned}\mathscr{L}(f) &\approx \frac{(2\pi)^2}{12 f_{\mathrm{ref}}}[2\sin(\pi f/f_{\mathrm{ref}})]^{2(m-1)}\\ &\approx \frac{(2\pi)^2}{12 f_{\mathrm{ref}}}[2\pi f/f_{\mathrm{ref}}]^{2(m-1)}\ [\mathrm{rad}^2/\mathrm{Hz}]\end{aligned} \tag{12.56}$$

where f_{ref} is equal to input reference frequency f_{r} and m is the number of SDQ stages in the Σ-Δ chain (cf. Fig. 12.10).

However, there are difficulties with the application of this relation:

1. Some authors are tempted to generalize relation (12.48) even for simple systems where $m = 1$ with the expected result that $\mathscr{L}(f) = \text{const.}$, which is not the case as proved in the preceding section and in Fig. 12.11.
2. The value f_{ref} in the denominator might be misleading on one hand when evaluating the spurious level in PLL systems and raises questions about the physical

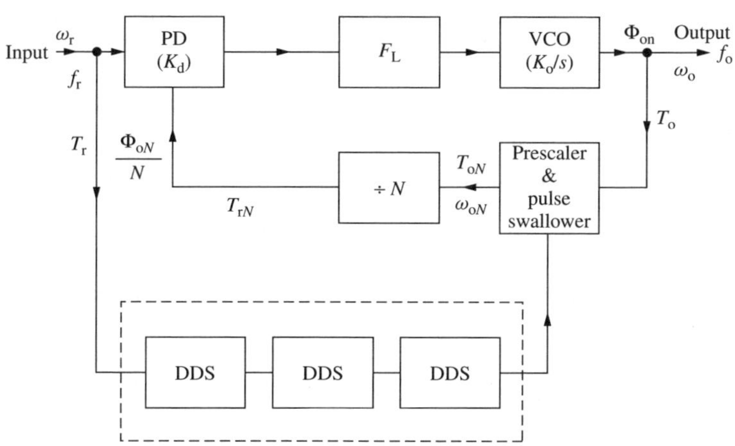

Figure 12.9 Block diagram of the fractional-N frequency synthesizer with the Σ-Δ arrangement. (Reproduced from V.F. Kroupa, *Direct Digital Frequency Synthesizers*, by permission IEEE Press, 1999, John Wiley & Sons, Inc., 2002 [7]).

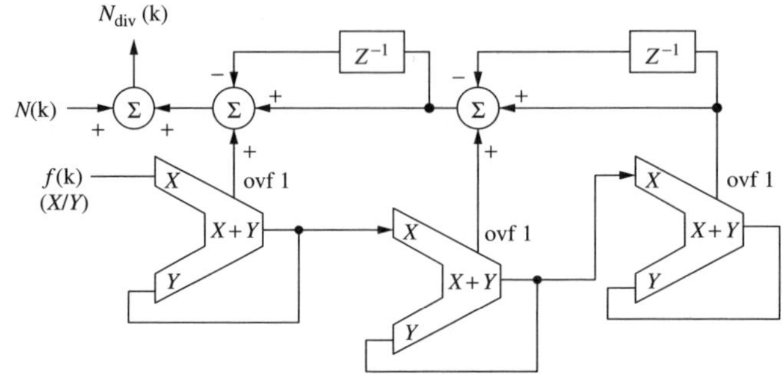

Figure 12.10 Σ-Δ chain investigated by Miller and Conley. (Reproduced from B. Miller and R.J. Conley, "A multiple modulator fractional divider", *IEEE Trans.*, **IM-40**, 578–583, 1991, by permission IEEE [12]).

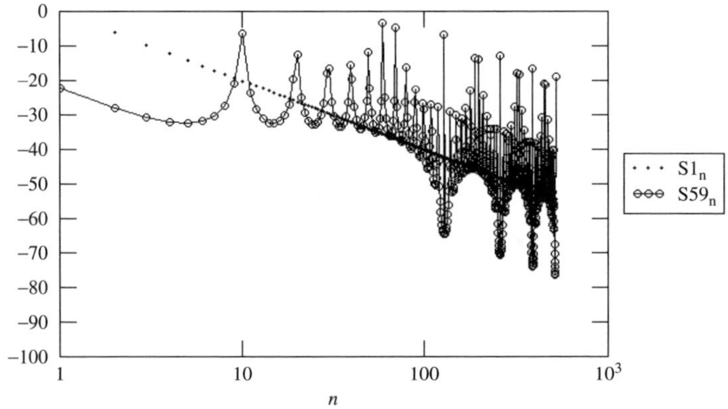

Figure 12.11 Spurious signals S_n due to the simple fractional-N arrangement (without compensation) with the DDFS tuned to X/Y = 1/128(dots) and X/Y = 59/128 (circles) (Reproduced from V.F. Kroupa, J. Štursa, V. Čížek and H. Švandová, "Direct digital frequency synthesizers with the – arrangement in PLL systems", *2001 IEEE International Frequency Control Symposium and PDA Exhibition*. Proceedings, pp. 799–805 [14]).

correctness, since $\mathscr{L}(f)$ in (12.56) is only an envelope of the spurious signals without dimensions.

3. The spurious content is a function of the whole PLL arrangement and not of the f_{ref} only.

12.3.1 Operation of the Σ-Δ Modulator in the *z*-Transform Notation

We shall suppose identical operating word length $R_i (R_i = R)$ of all accumulators and their size $Y_i = Y$, where

$$Y_i = 2^R \tag{12.57}$$

We shall consider a Σ-Δ modulator of the third order ($m = 3$). Its block diagram is illustrated in Figs. 12.10 and 12.12.

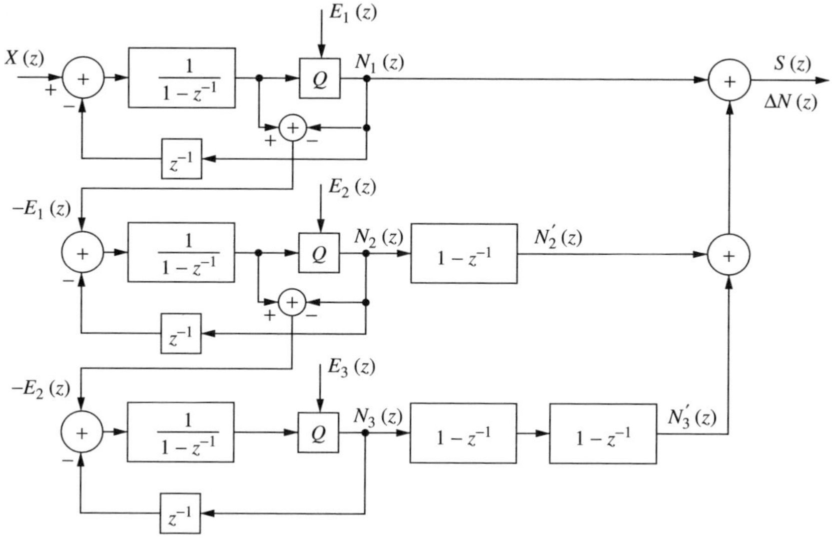

Figure 12.12 Σ-Δ modulator in the z-transform notation. (Reproduced from V.F. Kroupa, J. Štursa, V. Čížek and H. Švandová, "Direct digital frequency synthesizers with the – arrangement in PLL systems", *2001 IEEE International Frequency Control Symposium and PDA Exhibition.* Proceedings, pp. 799–805 [14]).

We shall explain the operation of this modulator in more detail with the model in Fig. 12.12, where we use the z-transform of signals and transfer functions of the involved blocks. Each accumulator produces two output signals. The first one (E_i, $E_i < 1$), is given by the digital accumulation (integration) of the constant input signal x. The "overflow" signal N_i is the second signal and it acquires the value $N_i = 1$ in the case that the accumulator overflows and the value $N_i = 0$ in the other instances. The delay block z^{-1} in the feedback path models the addition modulo 1 of the accumulator (cf. Fig. 10.5). This procedure is the source of overflowing. We can imagine N_i as the output signal of the 1-bit quantizer represented by the block Q of unity gain and by quantization error E_i. The signals $-E_1$ and $-E_2$ are input signals of the subsequent accumulators.

The overflow signal of the first accumulator can be expressed as

$$N_1(z) = (X(z) - N_1(z) \times z^{-1})\frac{1}{1 - z^{-1}} - E_1(z) \tag{12.58}$$

and after a simple modification as

$$N_1(z) = X(z) - E_1(z) \times (1 - z^{-1}) \tag{12.59}$$

The other overflow signals have the analogous form:

$$N_2(z) = -E_1(z) + E_2(z) \times (1 - z^{-1}) \tag{12.60}$$

$$N_3(z) = -E_2(z) + E_3(z) \times (1 - z^{-1}) \tag{12.61}$$

In addition, the overflow signals $N_i(z)$ are subjected to differentiation, that is, are multiplied in the shaping networks (cf. Fig. 12.12) and the results are summed. The contribution of the first accumulator is eliminated by adding the contribution of the second accumulator, and so on. The resulting sum represents the z-transform $S(z)$ of the steering signal s_n

$$S(z) = X(z) + (1 - z^{-1})^3 \times E_3(z) \tag{12.62}$$

The variance of 1-bit quantization $E_{3,n}$, corresponds to

$$\text{var}\ (E_{3,n}) = \frac{1}{12} \tag{12.63}$$

But the ideal mean value of one spectral line is equal to var $\langle (E_{3,n}) \rangle$ divided by $4Y$

$$\langle (E_{3,n}) \rangle = \frac{1}{3 \times Y} \tag{12.64}$$

12.3.2 Solution with the Assistance of the Fourier Series

We shall begin with the arrangement investigated above. With the assistance of Figs. 12.9 and 12.10, we may conclude that the Σ-Δ chain is locked to the reference frequency f_r and generates an output rate of pulse-bursts s_k, which are periodic in rY, where r is an integer ($r = 2$ for a three-stage SDQ; see Fig. 12.13). Unlike the simple fractional-N system, each output burst removes from the T_o pulse-rate s_k pulses (we shall see later that in some instances s_k might be negative, a positive bias removes this difficulty (see Tab. 12.1)) and we can write an equation similar to (12.25), that is,

$$T_{\text{o}N,k} \approx \frac{N + s_k}{N} T_\text{o} \tag{12.65}$$

The process will be explained with the assistance of Fig. 12.9 and Tab. 12.1 where the main feedback path divider is set to $N = 16$ and all three SDQ accumulators to the "normalized frequency" $\xi_x = X/Y = 1/8$. Consequently, each clock signal suppresses s_k VCO pulses (see Fig. 12.11 and Tab. 12.1). Extension of the plot would reveal for the duration of the longest repetition period T_L

$$\begin{aligned} T_\text{L} &= 2YT_{rN,\text{mean}} = 16T_{rN,\text{mean}} \\ &= 16 \times NT_{\text{o}N} = 256T_\text{o} + T_\text{o} \sum_{k=1}^{2Y} s_k \end{aligned} \tag{12.66}$$

Table 12.1 Biased outputs $s2_k$ from the Σ-Δ DDFS system for $X = 1$ and $Y = 2^3 = 8$

$s_k + 3$	3 3 4 2 4 3 2 5 1 5 2 4 2 5 1 4	3 3 4 2 4 3 2 5 1 5 2 ....
k	1 2 3 4 5 6 7 8 9 10 11 12 13 14 15 16	17 18 19 20 21 22 23 24 25 26 27 .....

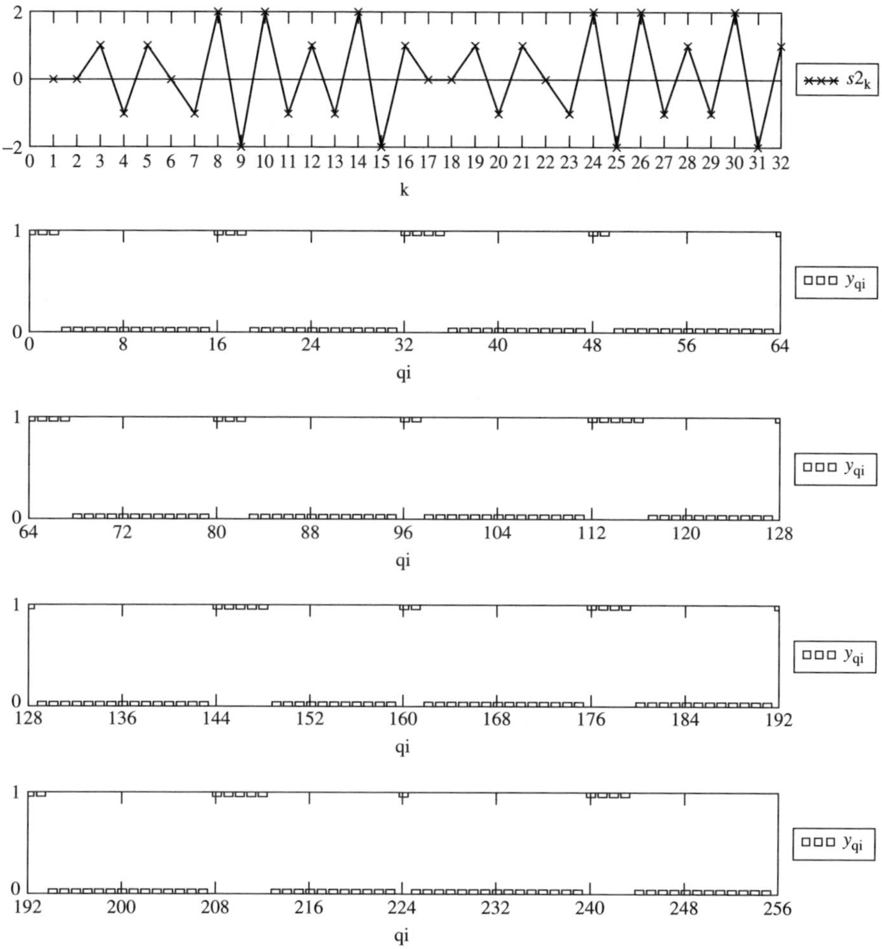

Figure 12.13 Output pulse rate of the fractional-N frequency synthesizer with a set of three DDFSs; with the main divider set to $N = 16$ and all DDFSs set to $X = 1$ and $Y = 2^3 = 8$. (Reproduced from V.F. Kroupa, J. Štursa, V. Čížek and H. Švandová, "Direct digital frequency synthesizers with the – arrangement in PLL systems", *2001 IEEE International Frequency Control Symposium and PDA Exhibition*. Proceedings, pp. 799–805 [14]).

Note that the repetition period in Tab. 12.1 is $2Y$. Another verification of the repetition provides autocorrelation in Fig. 12.14. For other Σ-Δ fractional-N orders see Tab. 12.2. Evidently

$$2YT_r = T_o N 2Y + T_o \sum_{k=1}^{2Y} s_k \tag{12.67}$$

from which we get for the effective frequency modulation in the locked state ($T_{rN,\text{mean}} = T_r$)

$$\frac{1}{T_o} = \frac{N}{T_r} + \frac{1}{2YT_r} \sum_{k=1}^{2Y} s_k \tag{12.68}$$

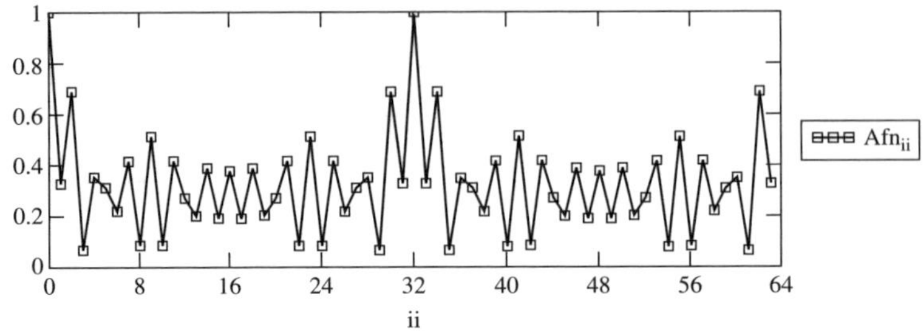

Figure 12.14 Verification of the repetition period of the Σ-Δ PLL discussed in Fig. 12.13 ($N = 16$ and all DDFSs set to $X = 1$ and $Y = 2^3 = 8$) with the assistance of autocorrelation.

Table 12.2 The range Δ_{S_k} was adjusted by considering all possible extreme values of overflow signals $p_{i,n}$. Its knowledge is important for biasing the swallowing process, since only a positive number of pulses can be removed

Order m	Period T_r	Range Δ_{S_k}	Number of levels
1	$1Y$	$\langle 0, 1\rangle$	2
2	$2Y$	$\langle -1, 2\rangle$	4
3	$3Y$	$\langle -3, 4\rangle$	8
4	$4Y$	$\langle -7, 8\rangle$	16

Now, we can compute individual spurious components due to the second term on the rhs. On the timescale of T_o, we may consider the s_k burst as δ-functions placed in the kT_r distances:

$$s(t) = \sum_{k=1}^{2Y} s_k \delta(t - kT_r) \tag{12.69}$$

For the evaluation of the complex amplitudes of individual spurious components, we apply the Fourier series expansion

$$\Delta\hat{\omega}_n \approx \frac{2\pi}{2YT_r} \int_0^{2YT_r} s_k \delta(t - kT_r)\, e^{-j\frac{2\pi}{2YT_r}nt}\, dt \tag{12.70}$$

After integration of the above relation we get

$$\Delta\omega_n = |\Delta\hat{\omega}_n| = \frac{\pi}{YT_r} \left| \sum_{k=1}^{2Y} s_k\, e^{-j\pi\frac{nk}{Y}} \right| \tag{12.71}$$

The frequency $\omega_{oN}(t)$ at the input of the divider is

$$\omega_{oN}(t) = \omega_{oN,\text{mean}} + \sum_{n=1}^{\infty} 2\Delta\omega_n \cos\left(2\pi \frac{n}{2YT_r} t + \Psi_n\right) \tag{12.72}$$

with the fundamental modulation frequency

$$\omega_{\mathrm{mod}} = \frac{2\pi}{2YT_{\mathrm{r}}} \tag{12.73}$$

and the steady state component given by the zero term ($n = 0$) in (12.70), that is,

$$\Delta\omega_{\mathrm{o}} = \frac{2\pi}{2YT_{\mathrm{r}}} \sum_{k=1}^{k=2Y} s_k = 2\pi \frac{X}{YT_{\mathrm{r}}} \tag{12.74}$$

With this result and relation (12.32), we can compute the effective division factor:

$$N_{\mathrm{eff}} = \frac{T_{\mathrm{r}}}{T_{\mathrm{o}}} = N + \frac{X}{Y} \tag{12.75}$$

which is the same as for the first-order fractional-N frequency synthesizer (cf. relation (12.33)). Further, after integrating the time-dependent term in (12.72) we get for the amplitudes of the spurious phase modulation components (of the sinusoidal waves)

$$\phi_n \approx \Delta\omega_n \frac{2YT_{\mathrm{r}}}{2\pi n} = \frac{1}{n} \sum_{k=1}^{2Y} s2_k \, \mathrm{e}^{-\mathrm{j}\pi \frac{nk}{Y}} \tag{12.76}$$

and their respective powers

$$S(n) = \tfrac{1}{2}|\phi_n|^2 \tag{12.77}$$

Note that the spurious phase exhibits discrete components with spurious frequencies

$$f_{\mathrm{spurious}} = n\frac{f_{\mathrm{r}}}{2Y} \tag{12.78}$$

The most advantageous property, from the point of view of the spectral behavior, of these three-stage SDQ fractional-N frequency synthesizers is that the low spurious frequencies have the lowest power, $S(n)$, as is illustrated in Fig. 12.15 and the peak power of $S(n)$ is approximately $n = Y$.

Since in the relations (12.72) and (12.77) we do not encounter the numerator X we witness another important feature, that is, the amplitudes of the spurious phase modulation components $S(n)$ are practically independent of the numerator X in the fractional term of the effective division factor N_{eff} in (12.75). This is illustrated in Fig. 12.16 for three different inputs X, and the same Y in all three accumulators of the SDQ (see Fig. 12.9).

12.3.3 Noise and Spurious Signals in PLL Frequency Synthesizers with Sigma-Delta Modulators (SDQ)

The task of PLL-SDQ systems is to provide fractionality of division factors in the arrangements as simply as possible on one hand and reduce spurious signals, particularly, close to the carrier.

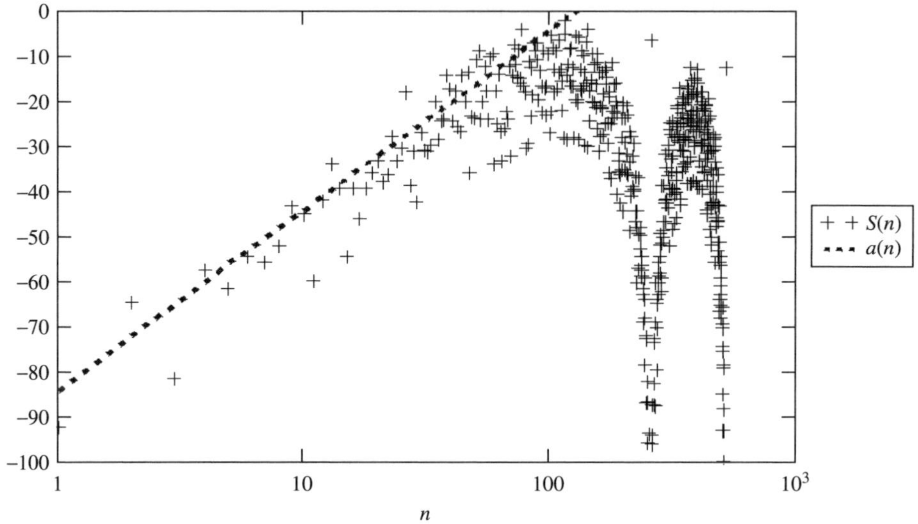

Figure 12.15 Spurious phase modulation signals $S(n)$ due to the three-stage Σ-Δ modulator tuned to $X = 59$ and $Y = 128$; a_n indicates the slope 40 dB/decade. (Reproduced from V.F. Kroupa, J. Štursa, V. Čížek and H. Švandová, "Direct digital frequency synthesizers with the – arrangement in PLL systems", *2001 IEEE International Frequency Control Symposium and PDA Exhibition.* Proceedings, pp. 799–805 [14]).

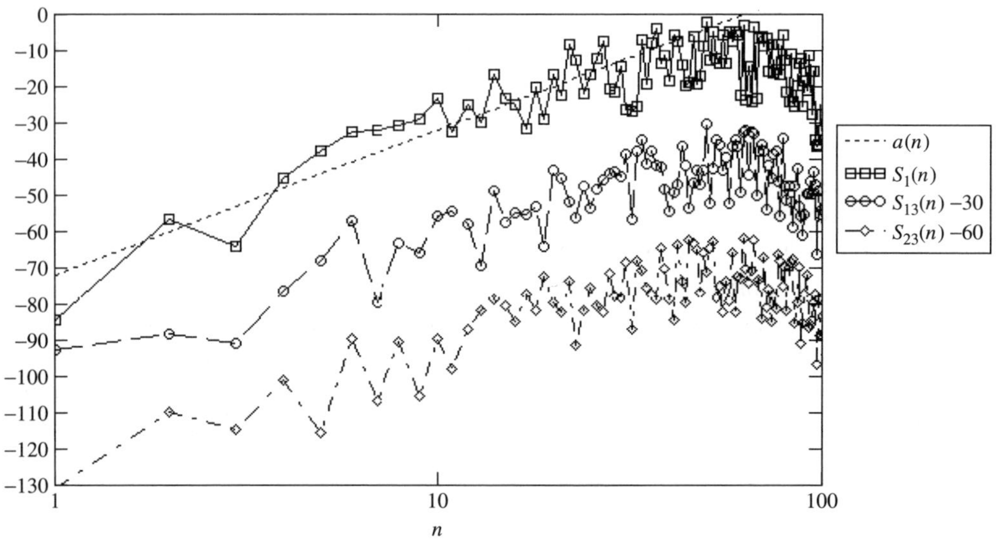

Figure 12.16 Spurious phase modulation signals $S_x(n)$ due to the three-stage Σ-Δ modulator tuned to $X = 1$ (□), $X = 13$ (○) shifted by −30 dB, and $X = 23$ (◇) shifted by −60 dB with $Y = 128$; a_n indicates the slope 40 dB/decade. (Reproduced from V.F. Kroupa, J. Štursa, V. Čížek and H. Švandová, "Direct digital frequency synthesizers with the – arrangement in PLL systems", *2001 IEEE International Frequency Control Symposium and PDA Exhibition.* Proceedings, pp. 799–805 [14]).

12.3.3.1 Random noise and spurious signals

To prevent interference between individual communications channels, the omnipresent noise and all unwanted spurious signals close to the carriers should be as low as possible. In fractional-N frequency synthesizers, we meet both random and discrete "noise" contributions. The question is which of them is more troublesome. By considering a narrow Fourier frequency band Δf, the corresponding spectral content $P(f)$ due to the PSD of the background noise $S_\phi(f)$, and of discrete components $S(n)$, is

$$P(f) = \int_{f_F}^{f_F+\Delta F_F} [S_\phi(f) + S(n)\delta(f_n)]\,\mathrm{d}f \tag{12.79}$$

$$\approx S_\phi(f)\Delta F_f + S(n)$$

from which the respective PSD in the 1 Hz pass band (for one spectral component) is

$$S(f) \approx S_\phi(f_n) + \frac{S(n)}{\Delta F} \tag{12.80}$$

In instances in which $S_\phi(f_n) \ll S(n)$ (which is often the case with the PLL frequency synthesizers), the discrete spurious signals are decisive for the "spectral purity" of the system and the phase noise PSD in (12.80) is actually the envelope of $S(n)$, that is, of the spurious signals.

12.3.3.2 Envelope approximation of the phase spurious signals

From the above discussion and Figs. 12.15 and 12.16, we conclude that in high-frequency PLL synthesizers, with the three-stage Σ-Δ modulator in the steering path, spurious signals exhibit two distinct sets of power: for low n, the power increases with a slope of approximately 40 dB/decade, but for $n > Y$ the power of the largest ones decreases with a slope of −20 dB/decade.

12.3.3.3 Straight line approximation

Using the straight line approximation (see Fig. 12.15)

$$a_n = 40\log\left(\frac{n}{Y}\right); \quad \text{and} \quad b_n = -20\log\left(\frac{2n}{Y}\right) \tag{12.81}$$

we get the following relation for the power envelope W_n of spurious signals:

$$W_n = a_n \quad (\text{for } n < Y); \quad \text{otherwise } W_n = b_n \tag{12.82}$$

See the characteristics $S(n)$ and $W(n)$ in Fig. 12.17. The advantage is that the approximating envelope is very simple, consisting of two straight lines only.

However, investigation of maxima by simulations, cf. (12.76) and (12.77), reveals that they decrease with an increasing number of accumulator bits R in proportion to the accumulator capacity, $Y = 2^R$ [14].

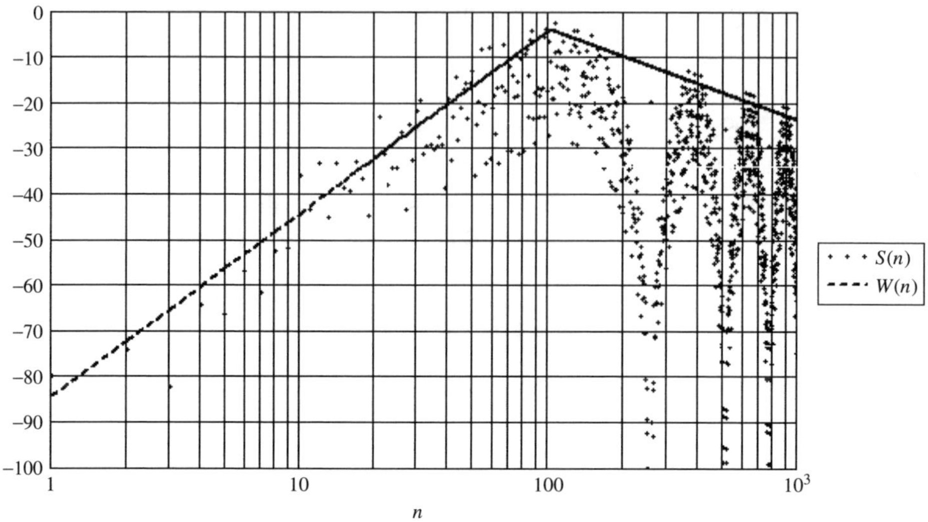

Figure 12.17 Straight line approximation of the spurious signals generated by a Σ-Δ chain ($X = 13$, $Y = 128$). (Reproduced from V.F. Kroupa, J. Štursa, V. Čížek and H. Švandová, "Direct digital frequency synthesizers with the – arrangement in PLL systems", *2001 IEEE International Frequency Control Symposium and PDA Exhibition*. Proceedings, pp. 799–805 [14]).

The situation being such, the behavior of $L(n)$ for small n is also investigated. The result is again inverse proportionality with respect to Y and we get a better straight line approximation:

$$W_n = 40 \log \left(2\frac{n}{Y}\right) + 10 \log \left(4\frac{\pi^2}{Y}\right) \tag{12.83}$$

12.3.3.4 Sine wave approximation

Computer simulations, as illustrated in Figs. 12.15 through 12.17, reveal similarity with relation (12.56). However, we have to introduce important corrections discussed in the Appendix to this chapter:

$$L(n) = 10 \log \left(\frac{(2\pi)^2}{4Y} \left[2 \sin \left(\pi \frac{f_n}{f_r} \right) \right]^4 \left[\sin \left(\frac{\pi f_n / f_r}{\pi f_n / f_r} \right) \right]^2 \right) \tag{12.84}$$

One example is shown in Fig. 12.18.

12.3.4 Spurious Signals in Practical PLL Systems with Sigma-Delta Modulators

For simplicity, and without any loss of generality, we may assume that spurious signals in fractional-N Σ-Δ frequency synthesizers exceed all the noises generated in the PLL in dividers, phase detectors, amplifiers, and so on (cf. Section 9.4.7).

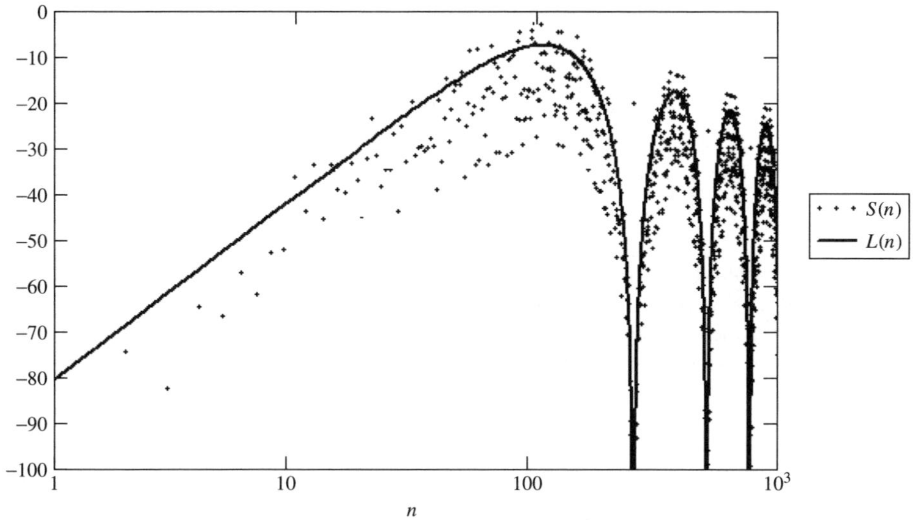

Figure 12.18 Sine wave approximation of the spurious signals generated by the Σ-Δ chain ($X = 13$, $Y = 128$). (Reproduced from V.F. Kroupa, J. Štursa, V. Čížek and H. Švandová, "Direct digital frequency synthesizers with the – arrangement in PLL systems", *2001 IEEE International Frequency Control Symposium and PDA Exhibition*. Proceedings, pp. 799–805 [14]).

Consequently, we shall consider that the PSD of the output signal, $S_{\phi,\text{out}}(f)$ will be formed only by the PSD of the reference source $S_{\phi,\text{ref}}$, the VCO $S_{\phi,\text{VCO}}$, and spurious signals generated by the Σ-Δ process. In this case we get

$$S_{\phi,\text{out}}(f) \approx \left[S_{\phi,\text{ref}}(f)N^2 + \sum_{n=1}^{\infty} L(n) \right] |H_{\text{eff}}(f)|^2 + S_{\phi,\text{VCO}}(f)|1 - H_{\text{eff}}|^2 \quad (12.85)$$

$$|H_{\text{eff}}(f)|^2 = \frac{S_{\Phi,\text{i}}(f)}{S_{\Phi,\text{out}}(f)} \quad (12.86)$$

where $|H_{\text{eff}}|^2$ is the PLL transfer function.

Example 12.5

In the following example, we shall investigate behavior of a fractional-N Σ-Δ frequency synthesizer using the type 2, third-order PLL shown schematically in Fig. 12.19. We shall assume the reference frequency to be $f_r = 6.4\,\text{MHz}$, the VCO frequency $f_o \approx 900\,\text{MHz}$, and the channel spacing $f_{ch} = 12.5\,\text{kHz}$. Evidently the overall division factor is

$$N_{\text{eff}} \cdot P = \frac{f_o}{f_{ch}} = \frac{9 \cdot 10^8}{12.5 \cdot 10^3} = 72000 \quad (12.87)$$

The division factor of the prescaler is $P = 4$. By choosing an SDQ with the accumulator capacity $Y = 512$ we get $N \approx 35$ from relation (12.75). The next step is to select the

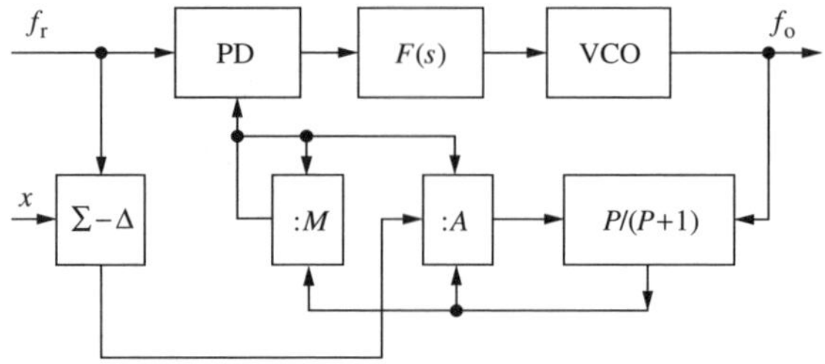

Figure 12.19 Simplified block diagram of the PLL Σ-Δ frequency synthesizer. (Reproduced from V.F. Kroupa, J. Štursa, V. Čížek and H. Švandová, "Direct digital frequency synthesizers with the – arrangement in PLL systems", *2001 IEEE International Frequency Control Symposium and PDA Exhibition*. Proceedings, pp. 799–805 [14]).

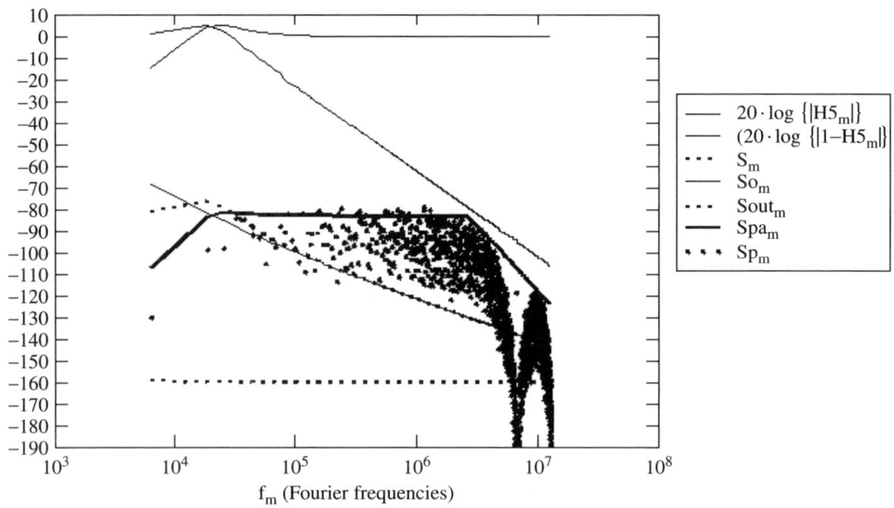

Figure 12.20 Simulation of spurious signals in a fractional-N frequency synthesizer with a three-stage Σ-Δ modulator: with PLL transfer characteristics and PSDs of the reference, VCO, envelope and sine wave spurious approximations. (Reproduced from V.F. Kroupa, J. Štursa, V. Čížek and H. Švandová, "Direct digital frequency synthesizers with the – arrangement in PLL systems", *2001 IEEE International Frequency Control Symposium and PDA Exhibition*. Proceedings, pp. 799–805 [14]).

effective (natural) loop frequency: for the first approximation we take $f_L \approx 15\,\text{kHz}$. The expected spurious signals, computed with the assistance of (12.76) and (12.77), are plotted in Fig. 12.20 together with the straight line approximation in accordance with relations (12.81), (12.82), and (12.83) for the case in which the loop divider noises are small.

The characteristics from the top indicate: PLL transfer loop characteristics $|H_{\text{eff}}(f)|^2$ and $|1 - H_{\text{eff}}(f)|^2$, phase-noise PSD of the open loop VCO and of the closed loop, the envelope approximation of the spurious signals, all expected spurious signals, and PSD of the reference oscillator.

In instances in which the channel spacing f_{ch} and the loop f_L are "design constants", the increase of the reference frequency f_r would require larger Y and in accordance with (12.85) and (12.86) the plateau of the spurious signals in Fig. 12.20 would be lower. On the other hand, a larger natural (loop) frequency f_L will shift the cut-off frequency of the loop transfer function $|H_{\text{eff}}(f)|^2$ to higher frequencies and, vice versa, shift the plateau to a higher level. In addition, the use of the simple envelope approximation makes it possible to investigate properties of the DDFS-PLL at the stage of the preliminary design in a large range of frequencies.

Example 12.6

Inspection of Fig. 12.20 reveals that spurious signals exceed the VCO noise in the range above 10 kHz. We have tried a remedy by reducing the loop frequency by one order to 1.5 kHz and by adding another RC section to the loop filter (see Fig. 12.21). Since the loop is of the fourth order, we have plotted the open loop characteristic instead of the transfer function $|H_{\text{eff}}|^2$ and the corresponding phase characteristic Ψ, to appreciate the phase margin that is found to be about 40°. Note that we have plotted the envelope in accordance with relation (12.84).

12.4 APPENDIX

In the view of our discrete solution of the spurious signals in the PLL-Σ-Δ fractional-N frequency synthesizers, we shall try to discuss earlier results [12]. The starting relation, in the z-transform, is

$$f_{\text{out}}(z) = (N + \Delta N)f_{\text{ref}} + (1 - z^{-1})^3 f_{\text{ref}} e_{q^3}(z) \tag{12.88}$$

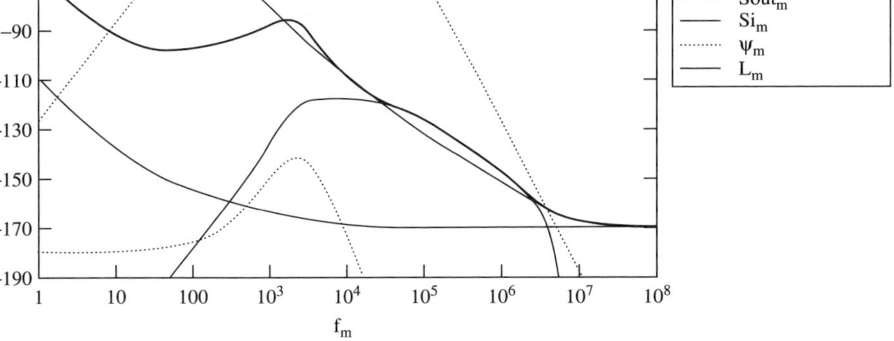

Figure 12.21 Rearrangement of Fig. 12.20 by reducing the loop frequency and by adding another filtering section RC. (Reproduced from V.F. Kroupa, J. Štursa, V. Čížek and H. Švandová, "Direct digital frequency synthesizers with the – arrangement in PLL systems", *2001 IEEE International Frequency Control Symposium and PDA Exhibition*. Proceedings, pp. 799–805 [14]).

where the second term represents the "frequency noise" due to fractional division. To arrive at the "phase noise" we have to integrate the above equation, that is, to sum individual phase contributions $\Delta\phi_i(z)$

$$\Delta\phi_i(z) = 2\pi f_{\text{out}}(z)T_{\text{ref}} = 2\pi(1 - z^{-1})^3 z^{-i} e_{q^3}(z), \quad i = 1, 2, \ldots \tag{12.89}$$

These individual phase contributions, $\Delta\phi_i(z)$, form a geometric series and after its summation, that is, after integration, we get

$$\Delta\phi(z) \approx \frac{2\pi(1 - z^{-1})^3}{(1 - z^{-1})} e_{q^3}(z) = 2\pi(1 - z^{-1)^2} e_{q^3}(z) \tag{12.90}$$

The next step is evaluation of the rms of the spurious phase noise

$$\langle(\Delta\phi(z))^2\rangle = [2\pi(1 - z^{-1})^2]^2 \langle e_{q^3}(z) \cdot e_{q^3}(z)\rangle \tag{12.91}$$

Squaring of the $e_{q3}(z)$ results in the autocorrelation

$$\langle e_{q^3}(z) \cdot e_{q^3}(z)\rangle = R_{\text{eq}}(0) \tag{12.92}$$

For discrete spurious signals, the autocorrelation $R_{\text{eq}}(0)$ is the sum of individual spectral components $e_{q,n}^2$

$$R_{\text{eq}}(0) = \sum_{n=0}^{\infty} e_{q,n}{}^2; \quad n = 0, 1, \ldots \tag{12.93}$$

Since we are not in a position to evaluate the individual spectral lines $e_{q,n}^2$, we shall replace them with a mean value taken from the repetition period $T_{\text{L}} = 2YT_{\text{r}}$. We limit their number to $2Y$.

The other information we have is knowledge of the variance (1/12) of the quantized signal e_{q^3} with maximum amplitude 1 and of its squared mean value (1/4). Evidently [15]

$$R_{\text{eq}}(0) = \frac{1}{12} + \frac{1}{4} = \frac{1}{3} \tag{12.94}$$

Then the relation for $\langle(\Delta\phi(z))^2\rangle$, is given by

$$\langle[\Delta\phi(z)]^2\rangle = [2\pi(1 - z^{-1})^2]^2 \cdot \frac{1}{3} \tag{12.95}$$

and the mean value of one spurious spectral line by

$$\langle e_{q,n}^2\rangle = \frac{1}{3 \cdot 2Y} \tag{12.96}$$

To evaluate the PSD of the output signal, we must replace the z-transform with the Laplace notation and take into account the sampled form of the signal, that is, to introduce the transfer function $H_h(s)$ (cf. eq. (10.3))

$$H_h(s) = \frac{1 - e^{-sT_{ref}}}{s} = \frac{T_{ref}}{2} \cdot e^{-sT_{ref}/2} \cdot \frac{\sinh(sT_{ref}/2)}{sT_{ref}/2} \tag{12.97}$$

By substituting, $z = e^{j\omega T_{ref}}$ we get

$$\Delta\phi(s) \approx H_h(s) \cdot 2\pi \cdot [e^{-sT_{ref/2}} \cdot 2 \cdot \sin(sT_{ref/2})]^2 \frac{1}{T_{ref}} \cdot \sum_{m=-\infty}^{\infty} e_{q^3}(s - jm\omega_{ref}) \tag{12.98}$$

And after inserting for $H_h(s)$ we arrive at

$$\phi(s) \approx 2\pi \frac{e^{-sT_{ref/2}}}{2} \frac{\sinh(sT_{ref/2})}{sT_{ref/2}} [e^{-sT_{ref/2}} \cdot 2 \cdot \sin(sT_{ref/2})]^2 \cdot \sum_{m=-\infty}^{\infty} e_{q^3}(s - jm\omega_{ref}) \tag{12.99}$$

and when neglecting all $jm\omega_{ref}$ signals for $m > 0$, the rms of the spurious phase is

$$\langle [\Delta\phi(s)]^2 \rangle = \left[\frac{1}{2} \sinh(sT_{ref/2})\right]^2 \cdot [2\pi \cdot (2\sin(sT_{ref/2}))^2]^2 \cdot \frac{1}{3} \tag{12.100}$$

Further, by taking into account the fact that relation (12.98) represents a periodic system with the period $T_L = 2YT_r$ by replacing the variable s with $j\omega$, and introducing the smallest frequency to be encountered in this system, that is, $f_{Lowest} = f_1 = f_r/2Y$, we arrive at powers of individual spurious phase components

$$S_\phi(f_n) = \pi^2 \left[\sin\left(\pi \frac{f_n}{f_r}\right)\right]^2 \sin^4\left(\pi \frac{f_n}{f_r}\right) \frac{1}{3 \cdot 2Y} \tag{12.101}$$

where we have rejected all the terms with $m > 0$ in eq. (12.99) and introduced

$$f_n = n \frac{f_r}{2Y} \tag{12.102}$$

REFERENCES

[1] T.-H. Linn and W.J. Kaiser, "A900-MHz 2.5 mA CMOS frequency synthesizer with an automatic SC tuning loop", *IEEE J. Solid-State Circuits*, **36**, 424–430, 2001.
[2] W.F. Egan, *Frequency Synthesis by Phase Lock*. 2nd ed., New York: John Wiley, 2000.
[3] B.-U.H. Klepser, M. Scholz and E. Götz, "A 10 GHz SiGe BiCMOS phase-locked-loop frequency synthesizer", *IEEE J. Solid-State Circuits*, **37**, 328–335, 2002.
[4] J. Gibbs and R. Temple, "Frequency domain yields its data to phase-locked synthesizer", *Electronics*, **27**, 107–113, 1978.

[5] D.D. Danielson and S.E. Froseth, "A synthesized signal source with function generator capabilities", *Hewlett-Packard J.*, **30**, 18–26, 1979.
[6] U.L. Rohde, *Digital PLL Frequency Synthesizers*. Englewood Cliffs: Prentice Hall, 1983.
[7] V.F. Kroupa, ed. *Direct Digital Frequency Synthesizers*. New York: IEEE Press, 1999.
[8] V.F. Kroupa, "Spectra of pulse rate frequency synthesizers", *Proc. IEEE*, **67**(12), 1680–1682, 1979 (reprinted in [6]).
[9] G. Becker, Quasiperiodic frequency synthesis, *Proc. of the 26th Ann. Frequency Control Symposium*, 1972.
[10] Marconi Instruments Ltd (inventor: U.N. Wells), European Patent No. 0-125-790-B2, Date of Filing: 11.04.1984, Priority 17, May 1983.
[11] Y. Matsuya, K. Uchimura, A. Iwata, T. Kobayashi, M. Ishikawa and T. Yoshitome, "A 16-bit oversampling A-to-D conversion technology using triple-integration noise shaping", *IEEE J. Solid-State Circuits*, **SC 22**, 921–928, 1987.
[12] B. Miller and R.J. Conley, "A multiple modulator fractional divider", *IEEE Trans.*, **IM-40**, 578–583, 1991.
[13] T.A.D. Riley,A. Copeland and T.A. Kwasniewski, "Delta-sigma modulation in fractional-N frequency synthesis", *IEEE J. Solid-State Circuits*, **28**, 553–559, 1993.
[14] V.F. Kroupa, J. Štursa, V. Čížek and H. Švandová, Direct digital frequency synthesizer with the Σ-Δ arrangement in the PLL systems, *2001 IEEE International Frequency Control Symposium and PDA Exhibition*. Proceedings, pp. 799–805.
[15] A. Papoulis, *Probability, Random Variables and Stochastic Processes*. New York: McGraw-Hill, 1965.
[16] E.J. Angelo, "Digital signal processor: a tutorial introduction to digital filtering", *Bell Syst. Tech. J.*, **60**(7), September, 1499–1546, 1981.

Appendix

List of Symbols

α	Normalized frequency $(\omega_n/\omega_{\mathrm{nf}})$ or $(\omega_n/\omega_{\mathrm{rf}})$, auxiliary exponent
$\alpha_1, \alpha_2, \alpha_3, \alpha_4, \ldots$	Coefficients of the $P(s)$ polynomial
$\alpha_n, \alpha(t)$	Normalized spurious amplitude perturbation
$\beta = \omega_{\mathrm{n}}/\mathrm{K}$	Rosenkreuz constant, friction force constant
γ	Constant in the type 3 PLL; normalized frequency of the second or the loop pass filter
$\gamma_0, \gamma_1, \gamma_2$	Normalized frequencies in modified Engel series
δ	Damping constant; normalized time delay
$\Delta\nu$	Difference frequency of $\nu_c - \nu_b$ (Greek nu)
$\Delta\Phi_{\mathrm{e}}$	Phase difference; viz. Φ_{e}
$\Delta\phi_{\mathrm{i}}, \Delta\phi_{\mathrm{o}}$	Input (output) phase step
$\Delta\Phi_{\mathrm{i}}(s)$	Input phase step
$\Delta\Psi_{\mathrm{e}}$	Phase difference; viz. Ψ_{e}
$\Delta\psi_{\mathrm{RMS}}$	rms phase difference
$\Delta\omega$	Frequency difference $(\omega_{\mathrm{i}} - \omega_{\mathrm{o}})$, spurious frequency modulation (shift)
$\Delta\omega_{\mathrm{H}}$	Hold in range
$\Delta\omega_{\mathrm{i}}$	Frequency step
$\Delta\dot{\omega}_{\mathrm{i}}$	Frequency ramp
$\Delta\omega_{\mathrm{L}}$	Lock in range
$\Delta\omega_{\mathrm{P}}$	Pull-in range
$\Delta\omega_{\mathrm{P},1}, \Delta\omega'_{\mathrm{P},2}, \Delta\omega''_{\mathrm{P},3}$	Pull-in frequencies
$\Delta\omega_{\mathrm{PO}}$	Pull-out range
$\Delta_1, \Delta_2, \ldots, \Delta_n$	Subdeterminants (minors) of the Hurwitz determinant
ΔC	Capacity fluctuation
Δf	Frequency bandwidth

Phase Lock Loops and Frequency Synthesis V.F. Kroupa
 ISBN: 0-470-84866-9

Δf_a	Frequency instability of the mixed signal
Δf_i	Frequency instability of the input (reference signal)
Δf_o	Frequency instability of the VCO
Δf_{xo}	The smallest frequency step
ΔG	Conductivity fluctuation
ΔL	Inductance fluctuation
ΔR	Resistance fluctuation
ΔT	Duration of the sampling impulse, sampling time
Δt	Time jitter
Δv	Spurious voltage pulse, voltage different
ζ, ζ_{eff}, ζ_{orig}	Damping factor; effective damping factor; original damping factor of the second order loop
η	Design factor (T_4/T_3)
η_k	Noise samples
Θ	Angle in the root locus plot; flow angles
κ	Design factor (T_3/T_2)
λ	Ratio of time constants
μ	Ratio of time constants
μ_{max}	Maximum of μ
ξ, ξ_x	Normalized frequency
σ, σ_1, σ_2	Normalized Laplace operator
$\sigma_y^2(\tau)$	Allan variance
τ, τ_0, τ_1, τ_2	Time unit, time constants, time delays,
ν, ν_b	Beat frequency
ν_c	VCO free running frequency
$\nu_{c,min}$	Pull-in frequency
ν_m	Minimum beat note frequency
ν_{VCO}	VCO frequency
$\Phi(\nu)$	Real part of the loop filter gain
$\Phi(x)$	Normalized real part of the loop filter gain
ϕ_{-1}, ϕ_{-2}	Complex factors of the ϕ_e
$\phi_1, \phi_2, \ldots, \phi_r$	Phases of the open loop gain factors; harmonics of ϕ_e
ϕ_a	Phase introduced by VCO tunning
$\Phi_{DN,n}$, $\Phi_{DQ,n}$	Divider noise factors
ϕ_{dr}, ϕ_{sp}	Phase of the spurious voltage
ϕ_e, $\phi_{e,0}$	Steady state phase difference; starting phase difference of $\phi_e(t)$
$\Phi_e(s)$, $\Phi_i(s)$, $\Phi_o(s)$	Laplace transform of the ϕ_e, ϕ_i, ϕ_o
$\phi_{e1}(t)$, $\phi_{e2}(t)$, $\phi_{e3}(t)$	Phase differences due to the transient
$\phi_{e2\,max}$	Maximum phase error in the PLL second order
$\phi_i(t)$, $\phi_o(t)$, $\phi_{out}(t)$	Input (output) phase
$\Phi_{i,n}$, $\Phi_{MI,n}$, $\Phi_{m,n}$, $\Phi_{MU,n}$, Φ_n	Noise contribution in the PLL

ϕ_o	Output phase
$\Phi_o(s)$	Laplace transform of the $\phi_o(t)$
$\Phi_{o,n}$, $\Phi'_{o,n}$	Noises in PLL
$\Phi_{osc,n}$	VCO noise
$\Phi_{o,sp}(s)$	Output spurious phase
ϕ_{pm}, Φ_{pm}	Phase margin
Ψ	Additional phase shift due to the second or the loop pass filter
ψ	Phase of the open loop gain
Ψ_e	Phase difference between input and output voltages
Ψ_{e0}, $\Psi_{e\infty}$	Phase difference between input and output voltages for $t = 0, t \to \infty$
Ψ_G	Phase of the zero open loop gain
Ψ_i, Ψ_o	Overall input (output) phase
Ψ_{pm}, $\Psi_{p,n}$, ψ_{pm}	Phase margin
Ψ_{tot}	Total phase shift
Ω	Frequency of the modulation signal
ω	(Fourier) frequency, angular velocity
ω_a	Input frequency to the mixer
ω_c	Free running frequency
$\omega_{cut,0}$	Cut off frequency or $1/T_0$
ω_{dr}	Frequency of the spurious voltage
ω_i, ω_o	Input (output) frequency
Ω_{LO}	Frequency of the local oscillator
ω_m	Fourier frequency
ω_n, $\omega_{n,eff}$	Natural (reduced natural loop) frequency
ω_{nf}	Natural frequency of the second or the loop pass filter
ω_{oi}	Lower side band frequency of the output voltage mixer
ω_{ref}	Reference frequency
ω_{rf}	Infinite attenuation frequency of the Twin-T RC filter
ω_s, ω_{samp}	Sampling frequency
ω_{sp}	Frequency of the spurious modulation signal
ω_{VCO}	VCO frequency
A	Operation amplifier gain, denominator of the $H_{5T}(\sigma)$
$A(s)$	Numerator of the $G(s)$
$a(t)$	Spurious amplitude modulation
A_0	Midband gain of the amplifier
$A_1, A_2, \ldots, A_r$	Amplitudes of the open loop gain factors; numerators in continued fraction expansion

AC	Alternating current
a_e, a_r	Flicker noise constants
A_k	Amplitude of spurious signal
$a_n, a_{n-1}, \ldots$	Elements in the Hurwitz determinant
a_0, a_1, a_2, a_3, a_4	Coefficients in the polynomial $P_n(s)$, auxiliary constants, noise constants
B	Neglected bits in DDFS accumulators
b	Friction force
$B(s)$	Denominator of the $G(s)$
$B(t)$	Collision force
$B_1, B_2, \ldots, B_r$	Denominators in continued fraction expansion
b2	Roots of the quadratic equation
b4	Roots of the fourth order equation
$b_0, b_1, b_2, b_3, \ldots$	Partial quotients
C, C_1, C_2, C_3	Capacitors
c0, c1, c2	Coefficients in the third order equation
C_H, C_{H1}, C_{H2}	Memory capacitors
CMOS	Complementary metal oxide semiconductor
C_R	Auxiliary capacitor
d	Damping factor of the second or the loop pass filter
DAC	Digital analog convertor
dB	Decibel
DC	Direct current
DDFS	Direct digital frequency synthesizers
DFD	Digital frequency divider
DPLL	Digital PLL
$e(t)$	Noise voltage
$e_c(t)$, $e_s(t)$	Slowly varying noise components
ECL	Emitter coupled logic
e_n, e_{n1}, e_{n2}	Noise voltage
$E_{n,out}$	Total noise output voltage
E_{out}	Output signal voltage
E_s	Signal voltage
$f(\zeta)$	Inverse of the pull-out frequency x_{PO}
$F(jy)$	Normalized transfer function of the second or the loop pass filter or Twin-T RC filter
$F(s)$, $F_L(s)$	Transfer function of the loop filter
f_1, f_2	Input frequencies to mixer
f_c	Clock frequency
$F_{dl}(s)$	Sampling transfer function of the time delay
$F_h(s)$	Additional sample and hold transfer function
f_i, f_{out}, f_o	Input (reference), output signal frequency
$F_L(s)$, $F_L(f)$	Transfer function of the loop filter

$F_M(j\omega)$	Transfer function of the feed back loop filter
F_n	Pass band frequency
f_n	Natural frequency
$f_{P,max}$	Upper band of the pull-in range
$f_{P,min}$	Lower band of the pull-in range
$F_{RC}(j\omega)$	Transfer function of the additional RC filter
f_{rf}	Resonant frequency
g	Base of systematic fractions
$G(\sigma)$, $G(jx)$	Normalized loop gain
$G(s)$	Open loop gain
$\hat{G}(s)$	z-transform modified open loop transfer function
$G_\tau(\sigma)$	Additional open loop gain due to sampling process
$G1_m$	Open loop gain of the first order loop
$G_2(jx)$	Open loop gain of the second order loop
$G_{2,i}(s)$	Open loop gain of the sampled PLL
$G2_m$	Open loop gain of the second order loop
$G_{2,T}(jx)$	Overall open loop gain
$G_3(\sigma)$	Open loop gain of the third order loop
$G_{3,3}(\sigma)$	Open loop gain of the third order loop type 3
$G_{3,i}(\sigma)$	Open loop gain of the third order sampled PLL
$G_{3,RC}$	Transfer function of the third order loop
$G_4(\sigma)$	Transfer function of the third order loop
$G_{4,AF}(\sigma)$	Open loop gain due to the fourth order loop with active low pass filter
$G_{4,T}(\sigma)$	Open loop gain due to the fourth order loop with Twin-T RC filter
$G_{5,AF}(\sigma)$	Open loop gain due to the fifth order loop with active low pass filter
$G_{5T}(\sigma)$	Open loop gain of the fifth order loop the Twin -T RC filter
$G_{AF}(\sigma)$	Additional open loop gain due to the active low pass filter
GAF_m	Additional gain to the second order RC filter
$G_{individual}$	Sum of individual gains
$G_{mod}(s)$	Modified open loop gain
G_{RC}	Additional open loop gain due to the RC filter
Grc_m	Gain of the additional RC filter
$G_T(jx)$	Additional open loop gain due to Twin-T RC filter
GT_m	Additional gain to the Twin-T RC filter
Gt_m	Additional gain due to the sampling
$G_{tot}(\sigma)$	Total open loop gain
h	Planck constant

$H(\sigma)$, $H(\mathrm{j}x)$	Normalize transfer function
$H(s)$	PLL transfer function
$h(t)$	Time sampling factor
$h_{-2}, h_{-1}, h_0, h_1, h_2$	Fractional frequency noise constants
$H_3(\mathrm{j}x)$	Transfer function of the third order the loop
$H_{3,3}(\sigma)$	Transfer function of the third order loop type 3
$H_{3,\mathrm{RC}}(\sigma)$	Transfer function of the third order the loop
$H_4(\sigma)$	Transfer function of the fourth order loop
$H_{4,\mathrm{AF}}(\sigma)$	Transfer function of the fourth order loop with the active low pass filter
$H_{4\mathrm{T}}(\sigma)$	Transfer function of the fourth order loop with Twin-T RC filter
$H_{5,\mathrm{AF}}(\sigma)$	Transfer function of the fifth order loop with the active low pass filter
$H_{5\mathrm{T}}(\sigma)$	Transfer function of the fifth order loop with Twin-T RC filter
$H_i(\mathrm{j}x)$	Is equal to $20\log(\|H(\mathrm{j}x)\|)$
$H_\mathrm{o}(\mathrm{j}x)$	Is equal to $20\log(\|1 - H(\mathrm{j}x)\|)$
$H_\mathrm{tot}(\sigma)$	Total transfer function
$H'(s)$	Effective PLL transfer function
i_0, i_1, i_2, i_3	Currents
IC	Integrated circuits
i_d	PD current
$I_d(s)$	Gain of the current PD
$-i_G$	Oscillator sustain current
i_n	Noise current
I_p	Peak current of the PD
K, K'	Gain (effective gain) of the PLL [2π.Hz]
k	Counting number
K_{-1}	Is equal to $-K_\mathrm{d}/2\mathrm{j}$; $K_1 = K_\mathrm{d}/2\mathrm{j}$ complex components of the PD gain K_d
$K_1, K_2, \ldots, K_r, \ldots, K_n$	Nominators of the simple fraction expansion of $R(s)/S(s)$
K_A	Additional DC gain
K_a	Acceleration or dynamic tracking error constant
K_d	Phase detector gain [V.radian^{-1}]
K_DC	DC loop gain
K_di	Phase detector gain [A.radian^{-1}]
K_fd	Gain of the frequency detector
K_m	Multiplication transfer constant [V^{-1}]
K_o	VCO Oscillator gain [2π.Hz.V^{-1}]
K_r	Reduced gain of the PLL [2π.Hz]

$K_{r,m}, \ldots, K_{r,1}$	Additional numerators of the simple fraction expansion of $R(s)/S(s)$
K_v	Velocity error constant
L	Inductance, conversion loss, logic state
M	Multiplication factor
m	Exponent, mass of the particle
m_1, m_2	First, second moment, number of store red pulses
$m_\mathrm{a}, m_\mathrm{p}, m_\mathrm{po}$	Amplitude, phase modulation index, output phase modulation index
M_p	Overshoot magnitude of the transfer function $H(s)$ and others
MSB	Most significant bit
N	Division factor; number of encirclements
n	Exponent superscript, number of charges
n	Ratio of resistance in Twin-T RC filter
NAND, NOR, OR	Logic operation
NCO	Number controlled oscillator
NF	Noise figure
OP	Operation amplifier
P	Number of poles
$p(n)$	Poisson distribution
$P(s)$, $P_\mathrm{n}(s)$	Polynomial
$p(t)$	Rectangular periodic function
$p_1, p_2, \ldots$	Partial divisors in Engel series
PD	Phase detector
PLLS	Phase locked loops
P_n	Noise power
P_r	Power dissipated in the resonant circuits
P_s	Signal power
PSD	Power spectral density
$Q, Q_\mathrm{L}, Q_\mathrm{U}$	Quality factor or the resonance circuits, loaded (unloaded)
q	Mixing ratio
$\boldsymbol{R}$	Complex constant
R	Resistance; number of roots
r	Radius of a root locus plot
$R(s)$	Numerator of the $1/[1 + G(s)]$
R-S	Flip flop
$R_\phi(\tau)$	Autocorrelation
R_1, R_2, R_3, .., R_a, R_b	Resistance
RC	Time constant
RCC, RRC	Phase lag-lead or proportional-integral filter
RF	Radio frequency

r_f, r_b	Forward, backward resistance
RG	Reference generator
rhs	Right hand side (often rhs)
rms	Root mean square
R_n	Approximation error (remainder)
R_out	Output resistance
$R_{s,\mathrm{leak}}$	Leaking resistor
R_v	Input resistance of the OP amplifier
s	Laplace operator
$\boldsymbol{s}$, $\boldsymbol{s}_{k,i}$, $\boldsymbol{s}_{p,i}$	Complex vectors
$S(s)$	Denominator of the $1/[1+G(\mathrm{s})]$
$S_{\Delta\omega}(f)$	PSD of frequency fluctuations
$S_\phi(f)$, $S'_\phi(f)$	One-sided, two-sided spectral density
$S_{\phi,\mathrm{D}}$, $S_{\phi,\mathrm{DN}}$, $S_{\phi,\mathrm{DQ}}$	Noise PSD of dividers
$S_{\phi,\mathrm{e}}$	Noise PSD of the phase detector
$S_{\phi,\mathrm{i}}$, $S_{\phi,\mathrm{o}}$, $S_{\phi,\mathrm{out}}$	Input, output PSD
$S_{\phi,\mathrm{in}}$	Noise PSD at the PLL input
$S_{\phi,\mathrm{L}}$	Noise PSD in the PLL
$S_{\phi,\mathrm{osc}}$	Noise PSD of the VCO
$s_1, s_2, \ldots, s_r, \ldots, s_n$	Roots of the polynomial $P(s)$ or $S(s)$
$S_{\mathrm{e,n}}(f)$, $S_{\mathrm{i,n}}(f)$	Voltage current noise PSD
$S_\mathrm{n}(f)$	Noise PSD
$\mathrm{SNR_L}$	Loop signal to noise ratio
$\mathrm{SSB_{dr}}$, $\mathrm{SSB_{d,sp}}$	Spurious side band signal
$S_\mathrm{v}(f)$	Voltage PSD
$S_{\mathrm{v,add}}(f)$	Additive voltage PSD
S_VF	Noise voltage PSD of the loop filter
S_VPD	Noise voltage PSD of the phase detector
$S_y(f)$	Fractional frequency PSD
T	Time constant, absolute temperature, sampling period
t_0	Integration time constant
T_1	Time constant in simple RC filter
t_1, t_2	Number of teeth
T_{13}, T_2, T_3, T_4	Addition time constant
$T_{2,\mathrm{APLL}}$	Time constant T_2 in the analog PLL
T_av	Average time to unlock
T_b	Period of the beat frequency
T_IF	Time constant due to the feed back IF filter
$T_{\mathrm{L},1}$, $T_{\mathrm{L},2}$	Lock-in time
T_M	Time delay due to the IF filter, resonance circuits
T_o	Time constant in the factor with one zero in the transfer function

T_p	Time constant in the factor with one pole in the transfer function; pull-in time
T_{ref}	Reference period
T_s	Signal period; sampling time
TTL	Transistor–transistor logic
T_{xr}, T_{xo}	Periods of the pulse train in DDFS arrange
$v(t)$, $v_i(t)$, $v_o(t)$	Time-dependent voltage: input, output
$V_2(s)$	VCO input voltage (in the Laplace notation)
$v_2(t)$	VCO input voltage (time-dependent)
v_{20}	VCO input voltage in the looked state
$v_{2,DC}(t)$	Output voltage of the integrator
$v_{2,max}$	Maximum VCO input voltage
V_{2p}	Peak input voltage to the VCO due to the sampling
$v_a(t)$	Output voltage of an additional feed back mixer
V_{cc}	Supply DC voltage
VCO	Voltage-controlled oscillator
V_d	Amplitude voltage of the phase detector
v_d	Output voltage of the phase detector
v_d	Input voltage at the integrator input
$V_{dr}(s)$	Laplace transform of the spurious voltage
$v_{dr}(t)$	Spurious voltage
$V_{d,sp}$	Amplitude of the spurious voltage
V_i, V_0	Amplitude of the input (reference), output voltage
V_i, V_1	Amplitude of the input voltage, etc.
V_m	Amplitude of the output voltage of the feed back mixer
$v_m(t)$	Output voltage of the feed back mixer
$v_{mf}(t)$	Output voltage of the IF filter
V_n, V_{sp}	Amplitude of spurious signal voltage; amplitude of the spurious modulation signal
$v_s(t)$, V_s	Signal voltage
X	Numerator of the normalized frequency
x	Normalized frequency
x_c	Normalized free running frequency VCO
$x_{d,sp}$, x_{dr},	Normalized frequency of the spurious signal
x_k	Is equal to K/ω_n
x_L	Normalized lock-in range
x_m	Normalized Fourier frequency; normalized minimum beat note frequency
$x_{m,3}$	Normalized minimum beat note frequency
$x_{m,crit}$	Critical normalized minimum beat note frequency
x_o	Normalized frequency for zero open loop gain
x_p	Normalized pull-in frequency

$x_{P,1}$, $x'_{P,2}$, $x'_{P,3}$	Approximate normalized pull-in frequency
$x_{P,4}$	Normalized pull-in range
x_{PO}	Normalized pull-out range
$<y_k>$	Fractional frequency deviation
$y(t)$	Fractional frequency fluctuations
y_1, y_2, y_3	Partial dividers in modified Engel series
Z	Number of zeros
$Z(s)$	Loop impedance
Z_1, Z_3, Z_4	Input impedance to the operation amplifier
Z_2	Feedback impedance over the operation amplifier
ε, ε_1, ε_2	Small voltage; small voltage error
$\mathscr{L}(f) \approx S'_\phi(f) \approx 1/2S_\phi(f)$	Noise power spectral density

Index

Phase Lock Loops and Frequency Synthesis V.F. Kroupa
 ISBN: 0-470-84866-9